Electrical properties of materials

SEVENTH EDITION

L. Solymar and D. Walsh

Department of Engineering Science
University of Oxford

OXFORD
UNIVERSITY PRESS

OXFORD

UNIVERSITY PRESS

Great Clarendon Street, Oxford OX2 6DP

Oxford University Press is a department of the University of Oxford.

It furthers the University's objective of excellence in research, scholarship, and education by publishing worldwide in

Oxford New York

Auckland Bangkok Buenos Aires Cape Town Chennai
Dar es Salaam Delhi Hong Kong Istanbul Karachi Kolkata
Kuala Lumpur Madrid Melbourne Mexico City Mumbai Nairobi
São Paulo Shanghai Taipei Tokyo Toronto

Oxford is a registered trade mark of Oxford University Press
in the UK and in certain other countries

Published in the United States
by Oxford University Press Inc., New York

© Oxford University Press, 1970, 1979, 1984, 1988, 1993, 1998, 2004
First edition 1970
Second edition 1979
Third edition 1984
Fourth edition 1988
Fifth edition 1993
Sixth edition 1998, reprinted 1999
Seventh edition 2004

A catalogue record for this title is available from the British Library

Library of Congress Cataloging in Publication Data

(Data available)

ISBN 0 19 9267936

Typeset by Newgen Imaging Systems (P) Ltd, Chennai, India
Printed in Great Britain by
Antony Rowe Ltd, Chippenham

Preface to the seventh edition

We have taken the opportunity of a new edition to include some subjects which reached maturity in the last 5 years, like organic semiconductors and artificial materials. Other additions are mainly in the device field. We have added microelectromechanical systems (MEMS), integrated circuits growing in the third dimension, expanded considerably nanoelectronics heralding the advent of radically new devices still at the laboratory stage like nanotube transistors, single electron transistors, and organic transistors. We added quantum wires, quantum dots, and the quantum cascade laser to the section on semiconductor lasers, introduced a new set of magnetic devices, whose operation is based on electron spin, under the heading of spintronics, added a new ferroelectric random access memory to the dielectric chapter and described a very fast new memory in the superconductivity chapter.

We can claim sustained progress on all fronts. A possible exception is the theory of superconducting mechanisms that continues to exasperate the experts.

The biggest change has probably occurred in the field of lighting with remarkable advances in light emitting diodes based on GaN and with the surprising emergence of organic molecular devices glowing brightly. It can be argued that the tungsten filament electric light, invented by Swan and Edison in 1861 and 1879, did more to improve the lives of people than most inventions (oil lamps were very tedious-Second World War experience). The basic technology was similar to thermionic valves, which came later and have long since almost vanished. So is it time for solid state electronics to take over with more efficient lighting, reduce greenhouse emissions and perhaps save the planet? We hope so.

In preparing this edition, we again benefited from comments received from students and lecturers. We are grateful to Professor Tung Hsu of the National Tsinghua University who called our attention to a mistake that has been there since the first edition. Our thanks are due to Richard Syms who not only introduced us to a subject we knew little about but also provided us with a draft version of the sections on MEMS and on its application in optical switching. We are indebted to Hisashi Masui for the anthropomorphic analogy of the vander Waals bond. We also wish to acknowledge the help we received from Harry Anderson, Kristel Fobelets, and Paul Stavrinou in the fields of organic semiconductors, semiconductor devices, and lasers, respectively. And thanks to Steffi Friedrichs for help with nanotubes.

L.S.

D.W.

Oxford

May 2003

Contents

Data on specific materials in text　　　　　　　　　　　　　　　　xi
Introduction　　　　　　　　　　　　　　　　　　　　　　　　xiii

1　The electron as a particle

1.1	Introduction	1
1.2	The effect of an electric field—conductivity and Ohm's law	2
1.3	The hydrodynamic model of electron flow	4
1.4	The Hall effect	5
1.5	Electromagnetic waves in solids	6
1.6	Waves in the presence of an applied magnetic field: cyclotron resonance	13
1.7	Plasma waves	16
1.8	Heat	18
Exercises		**20**

2　The electron as a wave

2.1	Introduction	22
2.2	The electron microscope	25
2.3	Some properties of waves	26
2.4	Applications to electrons	28
2.5	Two analogies	30
Exercises		**32**

3　The electron

3.1	Introduction	33
3.2	Schrödinger's equation	35
3.3	Solutions of Schrödinger's equation	36
3.4	The electron as a wave	37
3.5	The electron as a particle	38
3.6	The electron meeting a potential barrier	38
3.7	Two analogies	41
3.8	The electron in a potential well	42
3.9	The potential well with a rigid wall	44
3.10	The uncertainty relationship	44
3.11	Philosophical implications	45
Exercises		**47**

4   The hydrogen atom and the periodic table

4.1	The hydrogen atom	50
4.2	Quantum numbers	55
4.3	Electron spin and Pauli's exclusion principle	56
4.4	The periodic table	56
Exercises		**60**

5   Bonds

5.1	Introduction	63
5.2	General mechanical properties of bonds	64
5.3	Bond types	66
	5.3.1 Ionic bonds	66
	5.3.2 Metallic bonds	67
	5.3.3 The covalent bond	67
	5.3.4 The van der Waals bond	70
	5.3.5 Mixed bonds	71
	5.3.6 Carbon again	71
5.4	Feynman's coupled mode approach	72
5.5	Nuclear forces	77
5.6	The hydrogen molecule	77
5.7	An analogy	78
Exercises		**79**

6   The free electron theory of metals

6.1	Free electrons	80
6.2	The density of states and the Fermi–Dirac distribution	81
6.3	The specific heat of electrons	84
6.4	The work function	85
6.5	Thermionic emission	85
6.6	The Schottky effect	88
6.7	Field emission	91
6.8	The field-emission microscope	91
6.9	The photoelectric effect	92
6.10	Quartz-halogen lamps	94
6.11	The junction between two metals	94
Exercises		**95**

7   The band theory of solids

7.1	Introduction	97
7.2	The Kronig–Penney model	98
7.3	The Ziman model	101
7.4	The Feynman model	105
7.5	The effective mass	108
7.6	The effective number of free electrons	110
7.7	The number of possible states per band	111
7.8	Metals and insulators	113
7.9	Holes	113

7.10   Divalent metals   115
7.11   Finite temperatures   116
7.12   Concluding remarks   117
Exercises   **118**

8   Semiconductors

8.1   Introduction   119
8.2   Intrinsic semiconductors   119
8.3   Extrinsic semiconductors   124
8.4   Scattering   128
8.5   A relationship between
        electron and hole densities   130
8.6   III–V and II–VI compounds   132
8.7   Non-equilibrium processes   136
8.8   Real semiconductors   137
8.9   Amorphous semiconductors   138
8.10   Measurement of semiconductor properties   139
        8.10.1   Mobility   139
        8.10.2   Hall coefficient   142
        8.10.3   Effective mass   142
        8.10.4   Energy gap   143
        8.10.5   Carrier lifetime   146
8.11   Preparation of pure and controlled-impurity single-crystal semiconductors   147
        8.11.1   Crystal growth from the melt   147
        8.11.2   Zone refining   148
        8.11.3   Floating zone purification   149
        8.11.4   Epitaxial growth   149
        8.11.5   Molecular beam epitaxy   150
        8.11.6   Metal–organic chemical vapour deposition   152
        8.11.7   Hydride vapour phase epitaxy (HVPE) for nitride devices 153
Exercises   **153**

9   Principles of semiconductor devices

9.1   Introduction   156
9.2   The p–n junction in equilibrium   156
9.3   Rectification   161
9.4   Injection   163
9.5   Junction capacity   165
9.6   The transistor   165
9.7   Metal–semiconductor junctions   171
9.8   The role of surface states; real metal–semiconductor junctions   173
9.9   Metal–insulator–semiconductor junctions   175
9.10   The tunnel diode   178
9.11   The backward diode   181
9.12   The Zener diode and the avalanche diode   181
        9.12.1   Zener breakdown   182
        9.12.2   Avalanche breakdown   182
9.13   Varactor diodes   183
9.14   Field-effect transistors   184
9.15   Heterostructures   188
9.16   Charge-coupled devices   192
9.17   Silicon controlled rectifier   195

9.18	The Gunn effect	196
9.19	Strain gauges	198
9.20	Measurement of magnetic field by the Hall effect	199
9.21	Gas sensors	200
9.22	Microelectronic circuits	200
9.23	Building in the third dimension	204
9.24	Microelectro-Mechanical Systems (MEMS)	205
9.25	Nanoelectronics	206
9.26	Social implications	210
Exercises		**211**

10 Dielectric materials

10.1	Introduction	213
10.2	Macroscopic approach	213
10.3	Microscopic approach	214
10.4	Types of polarization	215
10.5	The complex dielectric constant and the refractive index	216
10.6	Frequency response	217
10.7	Polar and non-polar materials	218
10.8	The Debye equation	221
10.9	The effective field	222
10.10	Dielectric breakdown	223
	10.10.1 Intrinsic breakdown	223
	10.10.2 Thermal breakdown	223
	10.10.3 Discharge breakdown	224
10.11	Piezoelectricity	224
10.12	Ferroelectrics	229
10.13	Optical fibres	230
10.14	The xerox process	232
10.15	Liquid crystals	232
Exercises		**234**

11 Magnetic materials

11.1	Introduction	237
11.2	Macroscopic approach	238
11.3	Microscopic theory (phenomenological)	238
11.4	Domains and the hysteresis curve	242
11.5	Soft magnetic materials	246
11.6	Hard magnetic materials (permanent magnets)	248
11.7	Microscopic theory (quantum-mechanical)	252
	11.7.1 The Stern–Gerlach experiment	256
	11.7.2 Paramagnetism	256
	11.7.3 Paramagnetic solids	258
	11.7.4 Antiferromagnetism	259
	11.7.5 Ferromagnetism	259
	11.7.6 Ferrimagnetism	260
	11.7.7 Garnets	260
	11.7.8 Helimagnetism	260
11.8	Magnetic resonance	260
	11.8.1 Paramagnetic resonance	260
	11.8.2 Electron spin resonance	261

	11.8.3	Ferromagnetic, antiferromagnetic, and ferrimagnetic resonance	261
	11.8.4	Nuclear magnetic resonance	261
	11.8.5	Cyclotron resonance	262
	11.8.6	The quantum Hall effect	262
11.9		Some applications	264
	11.9.1	Magnetic bubbles	264
	11.9.2	Spintronics	266
	11.9.3	Isolators	267
	11.9.4	Sensors	268
	11.9.5	Medical imaging	268
	11.9.6	Electric motors	269
Exercises			**269**

12 Lasers

12.1		Equilibrium	271
12.2		Two-state systems	271
12.3		Lineshape function	275
12.4		Absorption and amplification	277
12.5		Resonators and conditions of oscillation	277
12.6		Some practical laser systems	278
	12.6.1	Solid state lasers	279
	12.6.2	The gaseous discharge laser	280
	12.6.3	Dye lasers	281
	12.6.4	Gas-dynamic lasers	282
	12.6.5	Excimer lasers	283
	12.6.6	Chemical lasers	283
12.7		Semiconductor lasers	283
12.8		Laser modes and control techniques	294
	12.8.1	Transverse modes	294
	12.8.2	Axial modes	295
	12.8.3	Q switching	295
	12.8.4	Cavity dumping	296
	12.8.5	Mode locking	296
12.9		Parametric oscillators	297
12.10		Optical fibre amplifiers	298
12.11		Masers	299
12.12		Noise	299
12.13		Applications	301
	12.13.1	Nonlinear optics	302
	12.13.2	Spectroscopy	302
	12.13.3	Photochemistry	302
	12.13.4	Study of rapid events	302
	12.13.5	Plasma diagnostics	302
	12.13.6	Plasma heating	302
	12.13.7	Acoustics	302
	12.13.8	Genetics	302
	12.13.9	Metrology	303
12.14		The atom laser	308
Exercises			**309**

13 Optoelectronics

13.1		Introduction	310
13.2		Light detectors	311
13.3		Light emitting diodes (LEDs)	313

13.4	Electro-optic, photorefractive, and nonlinear materials	316
13.5	Volume holography and phase conjugation	318
13.6	Acousto-optic interaction	323
13.7	Integrated optics	325
	13.7.1 Waveguides	326
	13.7.2 Phase shifter	326
	13.7.3 Directional coupler	327
	13.7.4 Filters	329
13.8	Spatial light modulators	329
13.9	Nonlinear Fabry–Perot cavities	331
13.10	Optical switching	334
13.11	Electro-absorption in quantum well structures	336
	13.11.1 Excitons	336
	13.11.2 Excitons in quantum wells	337
	13.11.3 Electro-absorption	337
	13.11.4 Applications	339
Exercises		**341**

14 Superconductivity

14.1	Introduction	343
14.2	The effect of a magnetic field	345
	14.2.1 The critical magnetic field	345
	14.2.2 The Meissner effect	346
14.3	Microscopic theory	347
14.4	Thermodynamical treatment	348
14.5	Surface energy	352
14.6	The Landau–Ginzburg theory	354
14.7	The energy gap	360
14.8	Some applications	364
	14.8.1 High-field magnets	364
	14.8.2 Switches and memory elements	365
	14.8.3 Magnetometers	365
	14.8.4 Metrology	366
	14.8.5 Suspension systems and motors	366
	14.8.6 Radiation detectors	366
	14.8.7 Heat valves	367
14.9	High-T_c superconductors	367
14.10	New superconductors	371
Exercises		**373**
Epilogue		375
Appendix I:	**Organic semiconductors**	377
Appendix II:	**Artificial materials**	383
Appendix III:	**Physical constants**	387
Appendix IV:	**Variational calculus. Derivation of Euler's equation**	389
Appendix V:	**Suggestions for further reading**	391
Answers to exercises		393
Index		396

Data on specific materials in text

Errors using inadequate data are much
less than using no data at all
Charles Babbage

Table	1.1	Threshold wavelength for alkali metals	11
Table	1.2	Electrical and thermal conductivites at 293 K	18
Table	4.1	The electronic configuration of the elements	59
Figure	4.5	The periodic table of the elements	61
Table	5.1	Mohs hardness scale (modified)	69
Table	6.1	Fermi levels of metals	83
Table	6.2	Work function of metals	87
Table	8.1	Energy levels of impurities in Si and Ge	124
Figure	8.7	Electron and hole mobilities in Ge and Si	131
Figure	8.8	Electron and hole mobilities in GaAs	133
Table	8.2	Size of atoms in tetrahedral bonds	134
Table	8.3	Covalent and ionic bonds	135
Table	8.4	Properties of semiconductors	141
Exercise	9.6	Specific doping data in Ge and Si	211
Table	10.1	Dielectric constant and refractive index of polar and non-polar materials	219
Exercise	10.5	Dielectric loss in thoria	234
Figure	11.9	Hysteresis loop for Supermalloy and Alnico	247
Table	11.1	Soft magnetic materials with typical properties	249
Table	11.2	Hard magnetic materials	251
Figure	11.12	Hysteresis curves of some rare-earth magnets	251
Exercise	11.6	Magnetic susceptibility of Ni at varying temperature	270
Figure	12.15	Energy gap and lattice spacing of III–V semiconductors	291
Table	12.1	Compounds for laser diodes	291
Table	13.1	Electro negativities of elements	314
Table	13.2	Properties of electro-optic materials	317
Table	13.3	Properties of acousto-optic materials	323
Table	14.1	The critical temperature and critical magnetic field of a number of superconducting elements	346
Figure	14.9	Temperature dependence of the specfic heat of tin near the critical temperature	352

Figure	**14.13**	Temperature variation of the energy gap as a function of T/T_c	360
Table	**14.2**	The critical temperature and magnetic field of the more important hard superconductors	364
Figure	**14.21**	The maximum critical temperature against time for traditional and oxide superconductors	367
Figure	**14.24**	Critical current densities as a function of magnetic field at 77 and 4.2 K for BSCCO, Nb–Ti and Nb_3 Sn	371
Figure	**A.1.4**	The benzene series	379
Appendix	**III**	Physical constants	387

Introduction

Till now man has been up against Nature;
from now on he will be up against his own nature.
Dennis Gabor *Inventing the future*

It is a good thing for an uneducated man to read books of quotations.
W.S. Churchill *Roving commission in my early life* (1930)

Engineering used to be a down-to-earth profession. The Roman engineers, who provided civilized Europe with bridges and roads, did a job comprehensible to all. And this is still true in most branches of engineering today. Bridge-building has become a sophisticated science, the mathematics of optimum structures is formidable; nevertheless, the basic relationships are not far removed from common sense. A heavier load is more likely to cause a bridge to collapse, and the use of steel instead of wood will improve the load-carrying capacity.

Solid-state electronic devices are in a different category. In order to understand their behaviour, you need to delve into quantum mechanics. Is quantum mechanics far removed from common sense? Yes, for the time being, it is. We live in a classical world. The phenomena we meet every day are classical phenomena. The fine details represented by quantum mechanics are averaged out; we have no first-hand experience of the laws of quantum mechanics; we can only infer the existence of certain relationships from the final outcome. Will it be always this way? Not necessarily. There are quantum phenomena known to exist on a macroscopic scale as, for example, superconductivity, and it is quite likely that certain biological processes will be found to represent macroscopic quantum phenomena. So, a ten-year-old might be able to give a summary of the laws of quantum mechanics—half a century hence. For the time being there is no easy way to quantum mechanics; no short cuts and no broad highways. We just have to struggle through. I believe it will be worth the effort. It will be your first opportunity to glance behind the scenes, to pierce the surface and find the grandiose logic of a hidden world.

Should engineers be interested at all in hidden mysteries? Isn't that the duty and privilege of the physicists? I do not think so. If you want to invent new electronic devices, you must be able to understand the operation of the existing ones. And perhaps you need to more than merely understand the physical mechanism. You need to grow familiar with the world of atoms and electrons, to feel at home among them, to appreciate their habits and characters.

We shall not be able to go very deeply into the subject. Time is short, and few of you will have the mathematical apparatus for the frontal assault. So we shall approach the subject in carefully planned steps. First, we shall try to deduce as much information as possible on the basis of the classical picture. Then, we shall talk about a number of phenomena that are clearly in contrast with classical ideas and introduce quantum mechanics, starting with Schrödinger's equation. You will become acquainted with the properties of individual atoms and what happens when they conglomerate and take the form of a solid. You will hear about conductors, insulators, semiconductors, p–n junctions, transistors, lasers, superconductors, and a number of related solid-state devices. Sometimes the statement will be purely qualitative but in most cases we shall try to give the essential quantitative relationships.

These lectures will not make you an expert in quantum mechanics nor will they enable you to design a computer the size of a matchbox. They will give you no more than a general idea.

If you elect to specialize in solid-state devices you will, no doubt, delve more deeply into the intricacies of the theory and into the details of the technology. If you should work in a related subject then, presumably, you will keep alive your interest, and you may occasionally find it useful to be able to think in quantum-mechanical terms. If your branch of engineering has nothing to do with quantum mechanics, would you be able to claim in ten years' time that you profited from this course? I hope the answer to this question is *yes*. I believe that once you have been exposed (however superficially) to quantum-mechanical reasoning, it will leave permanent marks on you. It will influence your ideas on the nature of physical laws, on the ultimate accuracy of measurements, and, in general, will sharpen your critical faculties.

The electron as a particle

And I laugh to see them whirl and flee,
Like a swarm of golden bees.

Shelley *The Cloud*

1.1 Introduction

In the popular mind the electron lives as something very small that has something to do with electricity. Studying electromagnetism does not change the picture appreciably. You learn that the electron can be regarded as a negative point charge and it duly obeys the laws of mechanics and electromagnetism. It is a particle that can be accelerated or decelerated but cannot be taken to bits.

Is this picture likely to benefit an engineer? Yes, if it helps him to produce a device. Is it a *correct* picture? Well, an engineer is not concerned with the truth; that is left to philosophers and theologians: the prime concern of an engineer is the utility of the final product. If this physical picture makes possible the birth of the vacuum tube, we must deem it useful; but if it fails to account for the properties of the transistor then we must regard its appeal as less alluring. There is no doubt, however, that we can go quite far by regarding the electron as a particle even in a solid—the subject of our study.

What does a solid look like? It consists of atoms. This idea originated a few thousand years ago in Greece, and has had some ups and downs in history, but today its truth is universally accepted. Now if matter consists of atoms, they must be somehow piled upon each other. The science that is concerned with the spatial arrangement of atoms is called crystallography. It is a science greatly revered by crystallographers; engineers are respectful, but lack enthusiasm. This is because the need to visualize structures in three dimensions adds to the hard enough task of thinking about what the electron will do next. For this chapter, let us assume that all materials crystallize in the simple cubic structure of Fig. 1.1, with the lattice ions fixed (it is a solid) and some electrons are free to wander between them. This will shortly enable us to explain Ohm's law, the Hall effect and several other important events. But if you are sceptical about over simplification, look forward to Fig. 5.3 to see how the elemental semiconductors crystallize in the diamond structure, or get a greater shock with Fig. 5.4 which shows a form of carbon that was discovered in meteorites but has only recently been fabricated in laboratories.

Let us specify our model a little more closely. If we postulate the existence of a certain number of electrons capable of conducting electricity, we must also say that a corresponding amount of positive charge exists in the solid. It must look electrically neutral to the outside world. Second, in analogy with our picture of gases, we may assume that the electrons bounce around in the

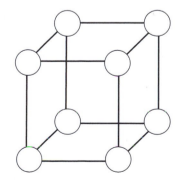

Fig. 1.1
Atoms crystallizing in a cubical lattice.

interatomic spaces, colliding occasionally with lattice atoms. We may even go further with this analogy and claim that in equilibrium the electrons follow the same statistical distribution as gas molecules (that is, the Maxwell–Boltzmann distribution) which depends strongly on the temperature of the system. The average kinetic energy of each degree of freedom is then $\frac{1}{2}kT$ where T is absolute temperature and k is Boltzmann's constant. So we may say that the mean thermal velocity of electrons is given by the formula*

$$\tfrac{1}{2}mv_{\text{th}}^2 = \tfrac{3}{2}kT \tag{1.1}$$

because because particles moving in three dimensions have three degrees of freedom.[†]

We shall now calculate some observable quantities on the basis of this simplest model and see how the results compare with experiment. The success of this simple model is somewhat surprising, but we shall see as we proceed that viewing a solid, or at least a metal, as a fixed lattice of positive ions held together by a jelly-like mass of electrons approximates well to the modern view of the electronic structure of solids. Some books discuss mechanical properties in terms of dislocations that can move and spread; the solid is then pictured as a fixed distribution of negative charge in which the lattice ions can move. These views are almost identical; only the external stimuli are different.

1.2   The effect of an electric field—conductivity and Ohm's law

Suppose a potential difference U is applied between the two ends of a solid length L. Then an electric field

$$\mathcal{E} = \frac{U}{L} \tag{1.2}$$

is present at every point in the solid, causing an acceleration

$$a = \frac{e}{m}\mathcal{E}. \tag{1.3}$$

Thus, the electrons, in addition to their random velocities, will acquire a velocity in the direction of the electric field. We may assume that this directed velocity is completely lost after each collision, because an electron is much lighter than a lattice atom. Thus, only the part of this velocity that is picked up in between collisions counts. If we write τ for the average time between two collisions, the final velocity of the electron will be $a\tau$ and the average velocity

$$v_{\text{average}} = \tfrac{1}{2}a\tau. \tag{1.4}$$

This is simple enough but not quite correct. We should not use the *average* time between collisions to calculate the average velocity but the actual times and then the average. The correct derivation is fairly lengthy, but all it gives is a factor of 2.[‡] Numerical factors like 2 or 3 or π are generally not worth worrying

---

[*] We shall see later that this is not so for metals but it is nearly true for conduction electrons in semiconductors.

v_{th} is the thermal velocity, and m is the mass of the electron.

[†] This is often called the Drude model after the man who proposed it a good hundred years ago.

[‡] See, for example, W. Shockley, *Electrons and holes in semiconductors*, D. van Nostrand, New York, 1950, pp. 191–5.

about in simple models, but just to agree with the formulae generally quoted in the literature, we shall incorporate that factor 2, and use

$$v_{\text{average}} = a\tau. \tag{1.5}$$

The average time between collisions, τ, has many other names; for example, mean free time, relaxation time, and collision time. Similarly, the average velocity is often referred to as the mean velocity or drift velocity. We shall call them 'collision time' and 'drift velocity', denoting the latter by v_{D}.

The relationship between drift velocity and electric field may be obtained from eqns (1.3) and (1.5), yielding

$$v_{\text{D}} = \left(\frac{e}{m}\tau\right)\mathcal{E}, \tag{1.6}$$

where the proportionality constant in parentheses is called the 'mobility'. This is the only name it has, and it is quite a logical one.

The higher the mobility, the more mobile the electrons.

Assuming now that all electrons drift with their drift velocity, the total number of electrons crossing a plane of unit area per second may be obtained by multiplying the drift velocity by the density of electrons, N_e. Multiplying further by the charge on the electron we obtain the electric current density

$$J = N_e e v_{\text{D}}. \tag{1.7}$$

Notice that it is only the drift velocity, created by the electric field, that comes into the expression. The random velocities do not contribute to the electric current because they average out to zero.*

We can derive similarly the relationship between current density and electric field from eqns (1.6) and (1.7) in the form

$$J = \frac{N_e e^2 \tau}{m}\mathcal{E}. \tag{1.8}$$

* They give rise, however, to *electrical noise* in a conductor. Its value is usually much smaller than the signals we are concerned with so we shall not worry about it, although some of the most interesting engineering problems arise just when signal and noise are comparable.

This is a linear relationship which you may recognize as Ohm's law

$$J = \sigma\mathcal{E}, \tag{1.9}$$

where σ is the electrical conductivity. When first learning about electricity you looked upon σ as a bulk constant; now you can see what it comprises of. We can write it in the form

$$\sigma = \left(\frac{e}{m}\tau\right)(N_e e)$$
$$= \mu_e(N_e e). \tag{1.10}$$

That is, we may regard conductivity as the product of two factors, charge density ($N_e e$) and mobility (μ_e). Thus, we may have high conductivities because there are lots of electrons around or because they can acquire high drift velocities, by having high mobilities.

Ohm's law further implies that σ is a constant, which means that τ must be independent of electric field.† From our model so far it is more reasonable

In metals, incidentally, the mobilities are quite low, about two orders of magnitude below those of semiconductors; so their high conductivity is due to the high density of electrons.

† It seems reasonable at this stage to assume that the charge and mass of the electron and the number of electrons present will be independent of the electric field.

to assume that l, the distance between collisions (usually called the mean free path) in the regularly spaced lattice, rather than τ, is independent of electric field. But l must be related to τ by the relationship,

$$l = \tau(v_{\text{th}} + v_{\text{D}}). \tag{1.11}$$

Since v_{D} varies with electric field, τ must also vary with the field unless

$$v_{\text{th}} \gg v_{\text{D}}. \tag{1.12}$$

In a typical metal $\mu_{\text{e}} = 5 \times 10^{-3}\,\text{m}^2\,\text{V}^{-1}\,\text{s}^{-1}$, which gives a drift velocity v_{D} of $5 \times 10^{-3}\,\text{m}\,\text{s}^{-1}$ for an electric field of $1\,\text{V}\,\text{m}^{-1}$.

As Ohm's law is accurately true for most metals, this inequality should hold. The thermal velocity at room temperature according to eqn (1.1) (which actually gives too low a value for metals) is

$$v_{\text{th}} = \left(\frac{3kT}{m}\right)^{1/2} \cong 10^5\,\text{m}\,\text{s}^{-1}. \tag{1.13}$$

* This is less true for semiconductors as they violate Ohm's law at high electric fields.

Thus, there will be a constant relationship between current and electric field accurate to about 1 part in 10^8.*

This important consideration can be emphasized in another way. Let us draw the graph (Fig. 1.2) of the distribution of particles in velocity space, that is with rectilinear axes representing velocities in three dimensions, v_x, v_y, v_z. With no electric field present, the distribution is spherically symmetric about the origin. The surface of a sphere of radius v_{th} represents all electrons moving in all possible directions with that r.m.s. speed. When a field is applied along the x-axis (say), the distribution is minutely perturbed (the electrons acquire some additional velocity in the direction of the x-axis) so that its centre shifts from $(0, 0, 0)$ to about $(v_{\text{th}}/10^8, 0, 0)$.

Taking copper, a field of $1\,\text{V}\,\text{m}^{-1}$ causes a current density of $10^8\,\text{A}\,\text{m}^{-2}$. It is quite remarkable that a current density of this magnitude can be achieved with an almost negligible perturbation of the electron velocity distribution.

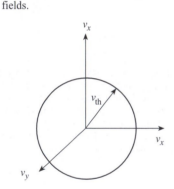

Fig. 1.2

Distributions of electrons in velocity space.

1.3   The hydrodynamic model of electron flow

By considering the flow of a charged fluid, a sophisticated model may be developed. We shall use it only in its crudest form, which does not give much of a physical picture but leads quickly to the desired result.

The equation of motion for an electron is

$$m\frac{\mathrm{d}v}{\mathrm{d}t} = e\mathscr{E}. \tag{1.14}$$

If we now assume that the electron moves in a viscous medium, then the forces trying to change the momentum will be resisted. We may account for this by adding a 'momentum-destroying' term, proportional to v. Taking the proportionality constant as ζ eqn (1.14) modifies to

$$m\left(\frac{dv}{dt} + \zeta v\right) = e\mathcal{E}.$$ (1.15)

ζ may be regarded here as a measure of the viscosity of the medium.

In the limit, when viscosity dominates, the term dv/dt becomes negligible, resulting in the equation

$$mv\zeta = e\mathcal{E},$$ (1.16)

which gives for the velocity of the electron

$$v = \frac{e}{m}\frac{1}{\zeta}\mathcal{E}.$$ (1.17)

It may be clearly seen that by taking $\zeta = 1/\tau$ eqn (1.17) agrees with eqn (1.6); hence we may regard the two models as equivalent and, in any given case, use whichever is more convenient.

1.4 The Hall effect

Let us now investigate the current flow in a rectangular piece of material, as shown in Fig. 1.3. We apply a voltage so that the right-hand side is positive. Current, by convention, flows from the positive side to the negative side, that is in the direction of the negative z-axis. But electrons, remember, flow in a direction opposite to conventional current, that is from left to right. Having sorted this out let us now apply a magnetic field in the positive y-direction. The force on an electron due to this magnetic field is

$$e(\mathbf{v} \times \mathbf{B}).$$ (1.18)

To get the resultant vector, we rotate vector $\mathbf{v}$ into vector $\mathbf{B}$. This is a clockwise rotation, giving a vector in the negative x-direction. But the charge of the electron, e, is negative; so the force will point in the positive x-direction; the electrons are deflected upwards. They cannot move farther than the top end of the slab, and they will accumulate there. But if the material was electrically neutral before, and some electrons have moved upwards, then some positive ions at the bottom will be deprived of their compensating negative charge. Hence an electric field will develop between the positive bottom layer and the

Equilibrium is established when the force due to the transverse electric field just cancels the force due to the magnetic field.

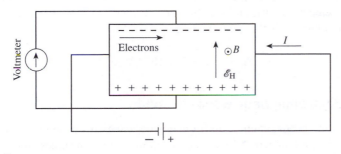
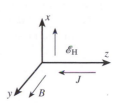

Fig. 1.3
Schematic representation of the measurement of the Hall effect.

negative top layer. Thus, after a while, the upward motion of the electrons will be prevented by this internal electric field. This happens when

$$\mathcal{E}_H = vB. \tag{1.19}$$

Expressed in terms of current density,

R_H is called the Hall coefficient.

$$\mathcal{E}_H = R_H J B, \qquad R_H = \frac{1}{N_e e}. \tag{1.20}$$

In this experiment $\mathcal{E}_H$, J, and B are measurable; thus R_H, and with it the density of electrons, may be determined.

What can we say about the direction of $\mathcal{E}_H$? Well, we have taken meticulous care to find the correct direction. Once the polarity of the applied voltage and the direction of the magnetic field are chosen, the electric field is well and truly defined. So if we put into our measuring apparatus one conductor after the other, the measured transverse voltage should always have the same polarity. Yes . . . the logic seems unassailable. Unfortunately, the experimental facts do not conform. For some conductors and semiconductors the measured transverse voltage is in the *other* direction.

How could we account for the different sign? One possible way of explaining the phenomenon is to say that in certain conductors (and semiconductors) electricity is carried by positively charged particles. Where do they come from? We shall discuss this problem in more detail some time later; for the moment just accept that mobile positive particles may exist in a solid. They bear the unpretentious name 'holes'.

To incorporate holes in our model is not at all difficult. There are now two species of charge carriers bouncing around, which you may imagine as a mixture of two gases. Take good care that the net charge density is zero, and the new model is ready. It is actually quite a good model. Whenever you come across a new phenomenon, try this model first. It might work.

Returning to the Hall effect, you may now appreciate that the experimental determination of R_H is of considerable importance. If only one type of carrier is present, the measurement will give us immediately the sign and the density of the carrier. If both carriers are simultaneously present it still gives useful information but the physics is a little more complicated (see Examples 1.7 and 1.8).

In our previous example we took a typical metal where conduction takes place by electrons only, and we got a drift velocity of $5 \times 10^{-3}\,\text{m s}^{-1}$. For a magnetic field of 1 T the transverse electric field is

The corresponding electric field in a semiconductor is considerably higher because of the higher mobilities.

$$\mathcal{E}_H = Bv = 5 \times 10^{-3}\,\text{V m}^{-1}. \tag{1.21}$$

1.5   Electromagnetic waves in solids

So far as the propagation of electromagnetic waves is concerned, our model works very well indeed. All we need to assume is that our holes and electrons obey the equations of motion, and when they move, they give rise to fields in accordance with Maxwell's theory of electrodynamics.

It is perfectly simple to take holes into account, but the equations, with holes included, would be considerably longer, so we shall confine our attention to electrons.

We could start immediately with the equation of motion for electrons, but let us first review what you already know about wave propagation in a medium characterized by the constants permeability, μ, dielectric constant, ϵ, and conductivity, σ (it will not be a waste of time).

First of all we shall need Maxwell's equations:

$$\frac{1}{\mu}\mathbf{\nabla} \times \mathbf{B} = \mathbf{J} + \epsilon\frac{\partial \mathcal{E}}{\partial t}, \tag{1.22}$$

$$\mathbf{\nabla} \times \mathcal{E} = -\frac{\partial \mathbf{B}}{\partial t}. \tag{1.23}$$

Second, we shall express the current density in terms of the electric field as

$$\mathbf{J} = \sigma\mathcal{E}. \tag{1.24}$$

It would now be a little more elegant to perform all the calculations in vector form, but then you would need to know a few vector identities, and tensors (quite simple ones, actually) would also appear. If we use coordinates instead, it will make the treatment a little lengthier, but not too clumsy if we consider only the one-dimensional case, when

$$\frac{\partial}{\partial x} = 0, \quad \frac{\partial}{\partial y} = 0. \tag{1.25}$$

Assuming that the electric field has only a component in the x-direction (see the coordinate system in Fig. 1.3), then

$$\mathbf{\nabla} \times \mathcal{E} = \begin{vmatrix} \mathbf{e}_x & \mathbf{e}_y & \mathbf{e}_z \\ 0 & 0 & \dfrac{\partial}{\partial z} \\ \mathcal{E}_x & 0 & 0 \end{vmatrix} = \frac{\partial \mathcal{E}_x}{\partial z}\mathbf{e}_y, \tag{1.26}$$

where $\mathbf{e}_x$, $\mathbf{e}_y$, $\mathbf{e}_z$ are the unit vectors. It may be seen from this equation that the magnetic field can have only a y-component. Thus, eqn (1.23) takes the simple form

$$\frac{\partial \mathcal{E}_x}{\partial z} = -\frac{\partial B_y}{\partial t}. \tag{1.27}$$

We need further

$$\mathbf{\nabla} \times \mathbf{B} = \begin{vmatrix} \mathbf{e}_x & \mathbf{e}_y & \mathbf{e}_z \\ 0 & 0 & \dfrac{\partial}{\partial z} \\ 0 & B_y & 0 \end{vmatrix} = \frac{\partial B_y}{\partial z}\mathbf{e}_x, \tag{1.28}$$

which, combined with eqn (1.24), brings eqn (1.22) to the scalar form

$$-\frac{\partial B_y}{\partial z} = \mu\sigma\mathcal{E}_x + \mu\epsilon\frac{\partial \mathcal{E}_x}{\partial t}. \tag{1.29}$$

Thus, we have two fairly simple differential equations to solve. We shall attempt the solution in the form*

$$\mathcal{E}_x = \mathcal{E}_{x_0} \exp\{-i(\omega t - kz)\} \tag{1.30}$$

and

$$B_y = B_{y_0} \exp\{-i(\omega t - kz)\}. \tag{1.31}$$

Then,

$$\frac{\partial}{\partial z} \equiv ik, \quad \frac{\partial}{\partial t} \equiv -i\omega, \tag{1.32}$$

which reduces our differential equations to the algebraic equations

$$ik\mathcal{E}_x = i\omega B_y \tag{1.33}$$

and

$$-ikB_y = (\mu\sigma - i\omega\mu\epsilon)\mathcal{E}_x. \tag{1.34}$$

This is a homogeneous equation system. By the rules of algebra, there is a solution, apart from the trivial $\mathcal{E}_x = B_y = 0$, only if the determinant of the coefficients vanishes, that is

$$\begin{vmatrix} -ik & i\omega \\ \mu\sigma - i\omega\mu\epsilon & ik \end{vmatrix} = 0. \tag{1.35}$$

Expanding the determinant we get

$$k^2 - i\omega(\mu\sigma - i\omega\mu\epsilon) = 0. \tag{1.36}$$

Essentially, the equation gives a relationship between the frequency, ω, and the wavenumber, k, which is related to phase velocity by $v_P = \omega/k$. Thus, unless ω and k are linearly related, the various frequencies propagate with different velocities and at the boundary of two media are refracted at different angles. Hence the name dispersion.

A medium for which $\sigma = 0$ and μ and ϵ are independent of frequency is nondispersive. The relationship between k and ω is simply

$$k = \omega\sqrt{\mu\epsilon} = \frac{\omega}{c_m}. \tag{1.37}$$

Solving eqn (1.36) formally, we get

$$k = (\omega^2\mu\epsilon + i\omega\mu\sigma)^{1/2}. \tag{1.38}$$

Thus, whenever $\sigma \neq 0$, the wavenumber is complex. What is meant by a complex wavenumber? We can find this out easily by looking at the exponent of eqn (1.30). The spatially varying part is

$$\exp(ikz) = \exp\{i(k_{\text{real}} + ik_{\text{imag}})z\}$$
$$= \exp(ik_{\text{real}}z)\exp(-k_{\text{imag}}z). \tag{1.39}$$

Hence, if the imaginary part of k is positive, the amplitude of the electromagnetic wave declines exponentially.[†]

ω represents frequency, and k is the wavenumber.

* We have here come face to face with a dispute that has raged between physicists and engineers for ages. For some odd reason the physicists (aided and abetted by mathematicians) use the symbol i for $\sqrt{-1}$ and the exponent $-i(\omega t - kz)$ to describe a wave travelling in the z-direction. The engineers' notation is j for $\sqrt{-1}$ and $j(\omega t - kz)$ for the exponent. In this course we have, rather reluctantly, accepted the physicists' notations so as not to confuse you further when reading books on quantum mechanics.

Different people call this equation by different names. Characteristic, determinantal, and dispersion equation are among the names more frequently used. We shall call it the *dispersion equation* because that name describes best what is happening physically.

$c_m \leqslant c$ is the velocity of the electromagnetic wave in the medium.

† The negative sign is also permissible though it does not give rise to an exponentially increasing wave as would follow from eqn (1.39). It would be very nice to make an amplifier by putting a piece of lossy material in the way of the electromagnetic wave. Unfortunately, it violates the principle of conservation of energy. Without some source of energy at its disposal no wave can grow. So the wave which seems to be exponentially growing is in effect a decaying wave which travels in the direction of the negative z-axis.

If the conductivity is large enough, the second term is the dominant one in eqn (1.38) and we may write

$$k \cong (i\omega\mu\sigma)^{1/2} = \frac{\pm(i+1)}{\sqrt{2}}(\omega\mu\sigma)^{1/2}. \qquad (1.40)$$

So if we wish to know how rapidly an electromagnetic wave decays in a good conductor, we may find out from this expression. Since

$$k_{\text{imag}} = \left(\frac{\omega\mu\sigma}{2}\right)^{1/2} \qquad (1.41)$$

the amplitude of the electric field varies as

$$|\mathcal{E}_x| = \mathcal{E}_{x_0} \exp\left\{-\left(\frac{\omega\mu\sigma}{2}\right)^{1/2} z\right\}. \qquad (1.42)$$

The distance δ at which the amplitude decays to $1/e$ of its value at the surface is called the *skin depth* and may be obtained from the equation

$$1 = \left(\frac{\omega\mu\sigma}{2}\right)^{1/2} \delta, \qquad (1.43)$$

yielding

$$\delta = \left(\frac{2}{\omega\mu\sigma}\right)^{1/2}. \qquad (1.44)$$

You have seen this formula before. You need it often to work out the resistance of wires at high frequencies. I derived it solely to emphasize the major steps that are common to all these calculations.

We can now go further, and instead of taking the constant σ, we shall look a little more critically at the mechanism of conduction. We express the current density in terms of velocity by the equation

$$\mathbf{J} = N_e e \mathbf{v}. \qquad (1.45)$$

The symbol $\mathbf{v}$ still means the average velocity of electrons, but now it may be a function of space and time, whereas the notation v_D is generally restricted to d.c. phenomena.

This is really the same thing as eqn (1.7). The velocity of the electron is related to the electric and magnetic fields by the equation of motion

$$m\left(\frac{d\mathbf{v}}{dt} + \frac{\mathbf{v}}{\tau}\right) = e(\mathcal{E} + \mathbf{v} \times \mathbf{B}). \qquad (1.46)$$

$1/\tau$ is introduced again as a 'viscous' or 'damping' term

We are looking for linearized solutions leading to waves. In that approximation the quadratic term $\mathbf{v} \times \mathbf{B}$ can be clearly neglected and the total derivative can be replaced by the partial derivative to yield

$$m\left(\frac{\partial\mathbf{v}}{\partial t} + \frac{\mathbf{v}}{\tau}\right) = e\mathcal{E}. \qquad (1.47)$$

Assuming again that the electric field is in the x-direction, eqn (1.47) tells us that the electron velocity must be in the same direction. Using the rules set out in eqn (1.32) we get the following algebraic equation

$$mv_x \left(-i\omega + \frac{1}{\tau} \right) = e\mathcal{E}_x. \tag{1.48}$$

The current density is then also in the x-direction:

$$\begin{aligned} J_x &= N_e e v_x \\ &= \frac{N_e e^2 \tau}{m} \frac{1}{1 - i\omega\tau} \mathcal{E}_x \\ &= \frac{\sigma}{1 - i\omega\tau} \mathcal{E}_x, \end{aligned} \tag{1.49}$$

where σ is defined as before. You may notice now that the only difference from our previous $(J - \mathcal{E})$ relationship is a factor $(1 - i\omega\tau)$ in the denominator. Accordingly, the whole derivation leading to the expression of k in eqn (1.38) remains valid if σ is replaced by $\sigma/(1 - i\omega\tau)$. We get

$$\begin{aligned} k &= \left(\omega^2 \mu\epsilon + i\omega\mu \frac{\sigma}{1 - i\omega\tau} \right)^{1/2} \\ &= \omega(\mu\epsilon)^{1/2} \left(1 + \frac{i\sigma}{\omega\epsilon(1 - i\omega\tau)} \right)^{1/2}. \end{aligned} \tag{1.50}$$

If $\omega\tau \ll 1$, we are back where we started from, but what happens when $\omega\tau \gg 1$? Could that happen at all? Yes, it can happen if the signal frequency is high enough or the collision time is long enough. Then, unity is negligible in comparison with $i\omega\tau$ in eqn (1.50), leading to

$$k = \omega(\mu\epsilon)^{1/2} \left(1 - \frac{\sigma}{\omega^2 \epsilon\tau} \right)^{1/2}. \tag{1.51}$$

Introducing the new notation

$$\omega_p^2 \equiv \frac{N_e e^2}{m\epsilon} = \frac{(N_e e^2/m)\tau}{\epsilon\tau} = \frac{\sigma}{\epsilon\tau} \tag{1.52}$$

we get

$$k = \omega(\mu\epsilon)^{1/2} \left(1 - \frac{\omega_p^2}{\omega^2} \right)^{1/2}. \tag{1.53}$$

Hence, as long as $\omega > \omega_p$, the wavenumber is real. If it is real, it has (by the rules of the game) no imaginary component; so the wave is not attenuated. This is quite interesting. By introducing a slight modification into our model, we may come to radically different conclusions. Assuming previously $J = \sigma\mathcal{E}$, we worked out that if any electrons are present at all, the wave is bound to decay. Now we are saying that for sufficiently large $\omega\tau$ an electromagnetic wave may travel across our conductor without attenuation. Is this possible? It seems to contradict the empirical fact that radio waves cannot penetrate metals. True; but that is because radio waves have not got high enough frequencies; let us try light waves. Can they penetrate a metal? No, they can not. It is another empirical fact that metals are not transparent. So we should try even higher frequencies. How high? Well, there is no need to go on guessing, we can work out the threshold frequency from eqn (1.52). Taking the electron density in a typical metal as 6×10^{28} per m^3, we then get

$$
\begin{aligned}
f_p &= \frac{1}{\pi} \left(\frac{N_e e^2}{m\epsilon_0} \right)^{1/2} \\
&= \frac{1}{2\pi} \left\{ \frac{6 \times 10^{28}(1.6 \times 10^{-19})^2}{9.11 \times 10^{-31} \times 8.85 \times 10^{-12}} \right\}^{1/2} \\
&= 2.2 \times 10^{15} \, \text{Hz.}
\end{aligned}
\tag{1.54}
$$

At this frequency range you are probably more familiar with the wavelengths of electromagnetic waves. Converting the above calculated frequency into wavelength, we get

$$
\lambda = \frac{c}{f_p} = \frac{3 \times 10^8}{2.2 \times 10^{15}} = 136 \, \text{nm.}
\tag{1.55}
$$

Thus, the threshold wavelength is well below the edge of the visible region (400 nm). It is gratifying to note that our theory is an agreement with our everyday experience; metals are not transparent.

There is one more thing we need to check. Is the condition $\omega\tau \gg 1$ satisfied? For a typical metal at room temperature, the value of τ is usually above 10^{-14} s, making $\omega\tau$ of the order of hundreds at the threshold frequency.

By making transmission experiments through a thin sheet of metal, the critical wavelength can be determined. The measured and calculated values are compared in Table 1.1. The agreement is not too bad, considering how simple our model is.

Before going further I would like to say a little about the relationship of transmission, reflection, and absorption to each other. The concepts are simple and one can always invoke the principle of conservation of energy if in trouble.

Let us take the case when $\omega\tau \gg 1$; k is given by eqn (1.54), and our conductor fills half the space, as shown in Fig. 1.4. What happens when an electromagnetic wave is incident from the left?

1. $\omega > \omega_p$. The electromagnetic wave propagates in the conducxtor. There is also some reflection, depending on the amount of mismatch. Energy

Table 1.1 *Threshold wavelengths for alkali metals*

Metal	Observed wavelength (nm)	Calculated wavelength (nm)
Cs	440	360
Rb	360	320
K	315	290
Na	210	210
Li	205	150

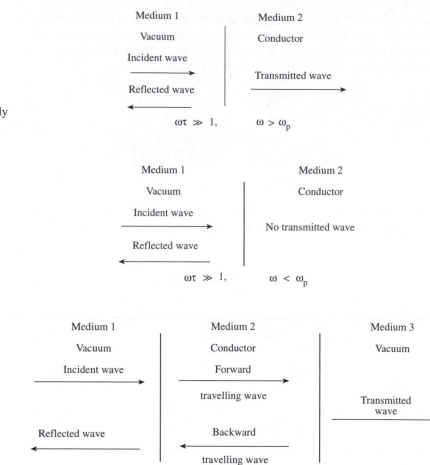

Fig. 1.4
Incident electromagnetic wave partly
reflected and partly transmitted.

Fig. 1.5
Incident electromagnetic wave
reflected by the conductor.

Fig. 1.6
Incident electromagnetic wave
transmitted to medium 3. The
amplitude of the wave decays in
medium 2 but without any energy
absorption taking place.

conservation says

$$\text{energy in the incident wave} = \text{energy in the transmitted wave}$$
$$+ \text{ energy in the reflected wave.}$$

Is there any absorption? No, because $\omega\tau \gg 1$.

2. $\omega < \omega_p$. In this case k is purely imaginary; the electromagnetic wave decays exponentially. Is there any absorption? No. Can the electromagnetic wave decay then? Yes, it can. Is this not in contradiction with something or other? The correct answer may be obtained by writing out the energy balance. Since the wave decays and the conductor is infinitely long, no energy goes out at the right-hand side. So everything must go back. The electromagnetic wave is reflected, as shown in Fig. 1.5. The energy balance is energy in the incident wave = energy in the reflected wave.

3. Let us take now the case shown in Fig. 1.6 when our conductor is of finite dimension in the z-direction. What happens now if $\omega < \omega_p$? The wave

now has a chance to get out at the other side, so there is a flow of energy, forwards and backwards, in the conductor. The wider the slab, the smaller is the amplitude of the wave that appears at the other side because the amplitude decays exponentially in the conductor. There is decay, but no absorption. The amplitudes of the reflected and transmitted waves rearrange themselves in such a way as to conserve energy.

If we choose a frequency such that $\omega\tau \ll 1$, then, of course, dissipative processes do occur and some of the energy of the electromagnetic wave is converted into heat. The energy balance in the most general case is

energy in the wave = energy in the transmitted wave

+ energy in the reflected wave

+ energy absorbed.

If there is a smaller amplitude transmitted, there will be a larger amplitude reflected.

A good example of the phenomena enumerated above is the reflection of radio waves from the ionosphere. The ionosphere is a layer which, as the name suggests, contains ions. There are free electrons and positively charged atoms, so our model should work. In a metal, atoms, and electrons are closely packed; in the ionosphere, the density is much smaller, so that the critical frequency ω_p is also smaller. Its value is a few hundred megahertz. Thus, radio waves below this frequency are reflected by the ionosphere (this is why short radio waves can be used for long-distance communication) and those above this frequency are transmitted into space (and so can be used for space or satellite communication). The width of the ionosphere also comes into consideration, but at the wavelengths used (it is the width in wavelengths that counts) it can well be regarded as infinitely wide.

1.6 Waves in the presence of an applied magnetic field: cyclotron resonance

In the presence of a constant magnetic field, the characteristics of electromagnetic waves will be modified, but the solution can be obtained by exactly the same technique as before. The electromagnetic eqns (1.22) and (1.23) are still valid for the a.c. quantities; the equation of motion should, however, contain the constant magnetic field, which we shall take in the positive z-direction. The applied magnetic field, $\mathbf{B_0}$, may be large, hence $\mathbf{v} \times \mathbf{B_0}$ is not negligible; it is a first-order quantity. Thus, the linearized equation of motion for this case is

$$m\left(\frac{\partial \mathbf{v}}{\partial t} + \frac{\mathbf{v}}{\tau}\right) = e(\mathcal{E} + \mathbf{v} \times \mathbf{B_0}). \tag{1.56}$$

In order to satisfy this vector equation, we need both the v_x and v_y components. That means that the current density, and through that the electric and magnetic fields, will also have both x and y components.

Writing down all the equations is a little lengthy, but the solution is not more difficult in principle. It may again be attempted in the exponential form, and $\partial/\partial z$ and $\partial/\partial t$ may again be replaced by ik and $-i\omega$, respectively. All the differential equations are then converted into algebraic equations, and by making the determinant of the coefficients zero we get the dispersion equation. I shall not go through the detailed derivation here because it would take up a great deal of time, and the resulting dispersion equation is hardly more complicated than eqn (1.50). All that happens is that ω in the $\omega\tau$ term is replaced by $\omega \pm$

ω_c. Thus, the dispersion equation for transverse electromagnetic waves in the presence of a longitudinal d.c. magnetic field is

$$k = \omega(\mu\epsilon)^{1/2}\left(1 + \frac{i\sigma}{\omega\epsilon\{1 - i(\omega \pm \omega_c)\tau\}}\right)^{1/2}, \qquad (1.57)$$

where

$$\omega_c = \frac{e}{m}B_0. \qquad (1.58)$$

The plus and minus signs give circularly polarized electromagnetic waves rotating in opposite directions. To see more clearly what happens, let us split the expression under the square root into its real and imaginary parts. We get

$$k = \omega\sqrt{(\mu\epsilon)}\left(1 - \frac{\omega_p^2\tau^2(1 - \omega_c/\omega)}{1 + (\omega - \omega_c)^2\tau^2} + i\frac{\omega_p^2\tau}{\omega}\frac{1}{1 + (\omega - \omega_c)^2\tau^2}\right)^{1/2}. \qquad (1.59)$$

This looks a bit complicated. In order to get a simple analytical expression, let us confine our attention to semiconductors where ω_p is not too large and the applied magnetic field may be large enough to satisfy the conditions,

$$\omega_c \gg \omega_p \quad \text{and} \quad \omega_c\tau \gg 1. \qquad (1.60)$$

We intend to investigate now what happens when ω_c is close to ω. The second and third terms in eqn (1.59) are then small in comparison with unity; so the square root may be expanded to give

$$k = \omega\sqrt{\mu\epsilon}\left(1 + \frac{i}{2}\frac{\omega_p^2\tau}{\omega}\frac{1}{1 + (\omega - \omega_c)^2\tau^2}\right). \qquad (1.61)$$

* After an accelerating device, the cyclotron, which works by accelerating particles in increasing radii in a fixed magnetic field.

The role of $\omega_c\tau$ is really analogous to that of Q in a resonant circuit. For good resonance we need a high value of $\omega_c\tau$.

The attenuation of the electromagnetic wave is given by the imaginary part of k. It may be seen that it has a maximum when $\omega_c = \omega$. Since ω_c is called the cyclotron* frequency this resonant absorption of electromagnetic waves is known as *cyclotron resonance*. The sharpness of the resonance depends strongly on the value of $\omega_c\tau$, as shown in Fig. 1.7, where Im k, normalized to its value at $\omega/\omega_c = 1$, is plotted against ω/ω_c. It may be seen that the resonance is hardly noticeable at $\omega_c\tau = 1$.

The curves have been plotted using the approximate eqn (1.61); nevertheless the conclusions are roughly valid for any value of ω_p. If you want more accurate resonance curves, use eqn (1.59).

Why is there such a thing as cyclotron resonance? The calculation from the dispersion equation provides the figures, but if we want the reasons, we should look at the following physical picture.

Suppose that at a certain point in space the a.c. electric field is at right angles to the constant magnetic field, B_0. The electron that happens to be at that point will experience a force at right angles to B_0 and will move along the arc of a

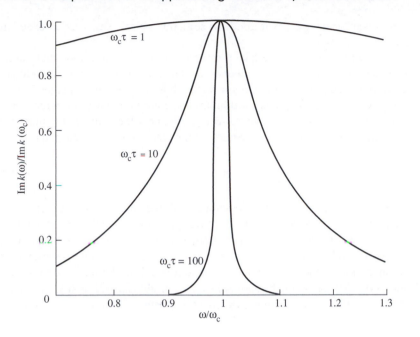

Fig. 1.7
Cyclotron resonance curves computed from eqn (1.61). There is maximum absorption when the frequency of the electromagnetic wave agrees with the cyclotron frequency.

circle. We can write a force equation. When the direction of motion is along the direction of $\mathcal{E}$ the magnetic and centrifugal forces are both at right angles to it, thus

$$B_0 e v = \frac{m v^2}{r}. \tag{1.62}$$

r is the instantaneous radius of curvature of the electron's path.

Consequently, the electron will move with an angular velocity

$$\omega_c = \frac{v}{r} = \frac{e}{m} B_0. \tag{1.63}$$

The orbits will *not* be circles, for superimposed on this motion is an acceleration varying with time in the direction of the electric field. Now if the frequency of the electric field, ω, and the cyclotron frequency, ω_c, are equal, the amplitude of the oscillation builds up. An electron that is accelerated north in one half-cycle will be ready to go south when the electric field reverses, and thus its speed will increase again. Under resonance conditions, the electron will take up energy from the electric field; and that is what causes the attenuation of the wave. Why is the $\omega_c \tau > 1$ condition necessary? Well, τ is the collision time; $\tau = 1/\omega_c$ means that the electron collides with a lattice atom after going round one radian. Clearly, if the electron is exposed to the electric field for a considerably shorter time than a cycle, not much absorption can take place. The limit might be expected when the electron covers one radian of a circle between collisions. This gives $\omega_c \tau = 1$.

Notice that any increase in *speed* must come from the electric field; the acceleration produced by a magnetic field changes direction, not speed, since the force is always at right angles to the direction of motion.

Now we may again ask the question: what is cyclotron resonance good for? There have been suggestions for making amplifiers and oscillators with the aid of cyclotron resonance, where by clever means the sign of attenuation is reversed, turning it into gain. As far as I know none of these devices reached the ultimate glory of commercial exploitation. If cyclotron resonance is no good for devices, is it good for something else? Yes, it is an excellent measurement tool.

It is used as follows: we take a sample, put it in a waveguide and launch an electromagnetic wave of frequency, ω. Then we apply a magnetic field and measure the output electromagnetic wave while the strength of magnetic field is varied. When the output is a minimum, the condition of cyclotron resonance is satisfied. We know ω so we know ω_c; we know the value of the magnetic field, B_0 so we can work out the mass of the electron from the formula

$$m = \frac{e B_0}{\omega}. \tag{1.64}$$

But, you would say, what is the point in working out the mass of the electron? That's a fundamental constant, isn't it? Well, it is, but not in the present context. When we put our electron in a crystal lattice, its mass will appear to be different. The actual* value can be measured directly with the aid of cyclotron resonance. So once more, under the pressure of experimental results we have to modify our model. The bouncing billiard balls have variable mass. Luckily, the charge of the electron does remain a fundamental constant. We must be grateful for small favours.

* The actual value is called, quite reasonably, the effective mass.

The charge of the electron is a fundamental constant; the mass of an electron is not.

1.7 Plasma waves

Electromagnetic waves are not the only type of waves that can propagate in a solid. There are sound waves and plasma waves as well. We know about sound waves; but what are plasma waves? In their simplest form they are density waves of charged particles in an electrically neutral medium. So they exist in a solid that has some mobile carriers. The main difference between this case and the previously considered electromagnetic case is that now we permit the accumulation of space charge. At a certain point in space, the local density of electrons may exceed the local density of positive carriers. Then an electric field arises, owing to the repulsive forces between these 'unneutralized' electrons. The electric field tries to restore the equilibrium of positive and negative charges. It drives the electrons away from the regions where they accumulated. The result is, of course, that the electrons overshoot the mark, and some time later, there will be a deficiency of electrons in the same region. An opposite electric field is then created which tries to draw back the electrons, etc. This is the usual case of harmonic oscillation. Thus, as far as an individual electron is concerned, it performs simple harmonic motion.

If we consider a one-dimensional model again, where everything is the same in the transverse plane, then the resulting electric field has a longitudinal component only. A glance at eqn (1.26), where $\nabla \times \mathcal{E}$ is worked out, will convince you that if the electric field has a z-component, only then $\nabla \times \mathcal{E} = 0$, that is $\mathbf{B} = 0$. There is no magnetic field present; the interplay is solely between the charges and the electric field. For this reason these density waves are often referred to as electrostatic waves.

If $\mathbf{B} = 0$, then eqn (1.22) takes the simple form

$$\mathbf{J} + \epsilon \frac{\partial \mathcal{E}}{\partial t} = 0. \tag{1.65}$$

We need the equation of motion, which for longitudinal motion will have exactly the same form as for transverse motion, namely

$$m\frac{\partial \mathbf{v}}{\partial t} = e\mathcal{E},$$ (1.66)

where we have neglected the damping term $m\mathbf{v}/\tau$.*

Current density and velocity are related again by

$$J = N_{e_0}ev.$$ (1.67)

We have changed over to scalar quantities.

Finally, there is the equation giving the relationship between the local density of electrons and the electric field, Poisson's equation,

$$\frac{\partial \mathcal{E}}{\partial z} = \frac{1}{\epsilon}N_e e.$$ (1.68)

Substituting $\mathcal{E}$ from eqn (1.66) and J from eqn (1.67) into (1.65), we obtain

$$N_{e_0}ev + \frac{\epsilon m}{e}\frac{\partial^2 v}{\partial t^2} = 0.$$ (1.69)

Following again our favourite method of replacing $\partial/\partial t$ by $-i\omega$, eqn (1.69) reduces to

$$v\left\{N_{e_0}e + \frac{\epsilon m}{e}(-\omega^2)\right\} = 0.$$ (1.70)

Since v must be finite, this means

$$N_{e_0}e - \frac{\epsilon m}{e}\omega^2 = 0,$$ (1.71)

or, rearranging,

$$\omega^2 = \frac{N_{e_0}e^2}{m\epsilon}.$$ (1.72)

This is our dispersion equation. It is a rather odd one because k does not appear in it. A relationship between k and ω gives the allowed values of k for a given ω. If k does not appear in the dispersion equation, *all* values of k are allowed. On the other hand, there is only a single value of ω allowed. Looking at it more carefully, we may recognize that it is nothing else but ω_p, the frequency we met previously as the critical frequency of transparency for electromagnetic waves. Historically, it was first discovered in plasma oscillations (in gas discharges by Langmuir); so it is more usual to call it the 'plasma frequency', and that is where the subscript p comes from.

Summarizing, a lossless plasma wave may have any wavelength and may propagate at any velocity, but only at one single frequency, the plasma frequency.

What are plasmas good for? This is a very difficult question to answer. We have touched only the simplest case. Plasma physics covers a vast field with immense potentialities (e.g. fusion), but the number of devices based on some sort of charge imbalance is very limited.[†] This is because plasma waves are generally harmful. It is more the concern of physicists and engineers to *get rid* of plasma waves than to set them up.

* Ignoring losses will considerably restrict the applicability of the formulae derived, but our aim here is to show no more than the simplest possible case.

N_{e_0} is the equilibrium density of electrons.

† Microwave tubes belong to a special category. They do employ density waves but the electrons' space charge is not compensated.

1.8 Heat

When the aim is to unravel the electrical properties of materials, should we take a detour and discuss heat? In general, no, we should not do that but when the two subjects overlap a little digression is permissible. I want to talk here first about the relationship between the electrical conductivity and heat conductivity, and then point out some discrepancies suggesting that something is seriously wrong with our model.

We have already discussed electrical conductivity. Heat conductivity is the same kind of thing but involves heat. An easy but rather unpleasant way of learning about it is to touch a piece of metal in freezing weather. The heat from your finger is immediately conducted away and you may get frostbite. Now back to that relationship. Denoting heat conductivity by K, it was claimed around the middle of the nineteenth century that for metals

$$\frac{K}{\sigma} = C_{WF} T \tag{1.73}$$

where C_{WF}, the so-called Wiedemann–Franz constant, was empirically derived. It was taken as

$$C_{WF} = 2.31 \times 10^{-8} \, \text{W} \, \text{S}^{-1} \text{K}^{-2}. \tag{1.74}$$

How well is the Wiedemann–Franz law satisfied? Very well, as Table 1.2 shows. Can it be derived from our model in which our electrons bounce about in the solid? Yes, that is what Drude did in about 1900. Let us follow what he did.

At equilibrium, the average energy of an electron (eqn 1.1) is $E = \frac{3}{2}kT$. The specific heat C_V is defined as the change in the average energy per unit volume with temperature

$$C_V = N_e \frac{dE}{dT} = N_e \left(\frac{3}{2}\right) k. \tag{1.75}$$

Let us now consider heat flow, assuming that all the heat is carried by the electrons. We shall take a one-dimensional model in which the electrons move only in the x direction. If there is a heat flow the average energy may change slightly from point to point. Taking an interval from $x - \ell$ to $x + \ell$ (remember ℓ

Table 1.2 *Electrical and thermal conductivities measured at 293 K*

Metal	σ $(10^7 \, \Omega^{-1} \text{m}^{-1})$	K $(\text{W} \, \text{m}^{-1} \text{K}^{-1})$	C_{WF} $(10^{-8} \, \text{W} \, \Omega \, \text{K}^{-2})$
Silver	6.15	423	2.45
Copper	5.82	387	2.37
Aluminium	3.55	210	2.02
Sodium	2.10	135	2.18
Cadmium	1.30	102	2.64
Iron	1.00	67	2.31

is the mean free path) the average at the two boundaries will be $E - (\mathrm{d}E/\mathrm{d}x)\ell$ and $E + (\mathrm{d}E/\mathrm{d}x)\ell$ respectively. Referring now to a result from the kinetic theory of gases that the number of particles flowing in a given direction per unit surface per unit time is $\frac{1}{6}N_e v_{\text{th}}$, the net flow across the plane at x is

$$\text{net energy flow} = (1/3)N_e v_{\text{th}} \left(\frac{\mathrm{d}E}{\mathrm{d}x} \right) \ell. \tag{1.76}$$

According to the simple theory of heat, the flow of heat energy is proportional to the gradient of temperature where the proportionality constant is the heat conductivity, K, yielding

$$\text{net heat energy flow} = K \left(\frac{\mathrm{d}T}{\mathrm{d}x} \right). \tag{1.77}$$

Equating now eqn (1.76) with (1.77) we obtain

$$K = \frac{1}{2} N_e v_{\text{th}} \ell k. \tag{1.78}$$

We may now relate the heat conductivity to the electrical conductivity as follows

$$\frac{K}{\sigma} = \frac{\frac{1}{2} N_e V_{\text{th}} \ell k}{N_e (e^2 \tau)/m} = \frac{3}{2} \left(\frac{k}{e} \right)^2 T \tag{1.79}$$

where the relation $\ell = v_{\text{th}} \tau$ (neglect v_D in eqn (1.11)) has been used. The functional relationship is exactly the same: the ratio of the two conductivities is indeed proportional to T as was stipulated by the empirical formula. But what is the value of the constant? Inserting the values of e and k into eqn (7.79), we obtain for the Wiedemann–Franz constant a value of $1.22 \times 10^{-8} \, \text{W S}^{-1} \, \text{K}^{-2}$, about a factor 2 smaller than the experimental value. This was regarded at the time as extremely good and as a justification of the electron as a particle model. Well, the factor of 2 was bothersome and the separate variation of the two conductivities with temperature did not fit very well either. All that was perhaps acceptable, but a closer look at the specific heat of metals versus insulators revealed that something was seriously wrong.

Up to now we have talked only about the electronic contribution to the specific heat and quoted it as being $\frac{3}{2}N_e k$, but classically the lattice will also contribute a term* $3Nk$ where N is the density of atoms. Thus, we should expect an alkali metal (in which $N_e = N$) to have a 50% greater specific heat than an insulator having the same number of lattice atoms because of the electronic contribution. These expectations are, however, wrong. It turns out that metals and insulators have about the same specific heat. Our model fails again to explain the experimentally observed value. What shall we do? Modify our model. But how? Up to now the modifications have been fairly obvious. The 'wrong sign' of the Hall voltage could be explained by introducing positive carriers, and when cyclotron resonance measurements showed that the mass of an electron in a solid was different from the 'free' electron mass, we simply said: 'all right, the electron's mass is not a constant. How should we modify our model now?' There seems to be no simple way of doing so. An entirely new start is needed.

* Valid at room temperature but fails at low temperatures.

Metals behave as if the free electrons make practically no contribution to the specific heat.

There is no quick fix for this real dilemma. We have to go quite deeply into wave theory and quantum mechanics. Finally, all is revealed in Chapter 6.3 when we find that electrons do make a quantifiable contribution to specific heat, which turns out to be very small.

Exercises

1.1. A 10 mm cube of germanium passes a current of 6.4 mA when 10 mV is applied between two of its parallel faces. Assuming that the charge carriers are electrons that have a mobility of $0.39 \, \text{m}^2\text{V}^{-1}\text{s}^{-1}$, calculate the density of carriers. What is their collision time if the electron's effective mass in germanium is $0.12 \, m_0$ where m_0 is the free electron mass?

1.2. An electromagnetic wave of free space wavelength 0.5 mm propagates through a piece of indium antimonide that is placed in an axial magnetic field. There is resonant absorption of the electromagnetic wave at a magnetic field, $B = 0.323 \, \text{wb m}^{-2}$.

(i) What is the effective mass of the particle in question?
(ii) Assume that the collision time is 15 times longer (true for electrons around liquid nitrogen temperatures) than in germanium in the previous example. Calculate the mobility.
(iii) Is the resonance sharp? What is your criterion?

1.3. If both electrons and holes are present the conductivities, add. This is because under the effect of an applied electric field the holes and electrons flow in opposite directions, and a negative charge moving in the (say) $+z$-direction is equivalent to a positive charge moving in the $-z$-direction.

Assume that in a certain semiconductor the ratio of electronic mobility, μ_e, to hole mobility, μ_h, is equal to 10, the density of holes is $N_h = 10^{20} \, \text{m}^{-3}$, and the density of electrons is $N_e = 10^{19} \, \text{m}^{-3}$. The measured conductivity is $0.455 \, \text{ohm}^{-1}\text{m}^{-1}$. Calculate the mobilities.

1.4. Measurements on sodium have provided the following data: resistivity $4.7 \times 10^{-8} \, \text{ohm m}$, Hall coefficient $-2.5 \times 10^{-10} \, \text{m}^3 \, \text{C}^{-1}$, critical wavelength of transparency 210 nm, and density 971 kg m^{-3}.

Calculate (i) the density of electrons, (ii) the mobility, (iii) the effective mass, (iv) the collision time, (v) the number of electrons per atom available for conduction.

Electric conduction in sodium is caused by electrons. The number of atoms in a kg mole is 6.02×10^{26} and the atomic weight of sodium is 23.

1.5. For an electromagnetic wave propagating in sodium plot the real and imaginary part of the wave number k as a function of frequency (use a logarithmic scale) from 10^6 to 10^{16} Hz.

Determine the penetration depth for 10^6, 10^{15}, and 2×10^{15} Hz.

Use the conductivity and the collision time as obtained from example 1.4.

1.6. A cuboid of Ge has contacts over all of its 2 mm $\times$ 1 mm ends and point contacts approximately half way along its 5 mm length, at the centre of the 5 mm $\times$ 1 mm faces. A magnetic field can be applied parallel to this face. A current of 5 mA is passed between the end contacts when a voltage of 310 mV is applied. This generates a voltage across the point contacts of 3.2 mV with no magnetic field and 8.0 mV when a field of 0.16 T is applied.

(i) Suggest why an apparent Hall voltage is observed with no magnetic field.
(ii) Using the corrected Hall voltage find the carrier density in the Ge sample.
(iii) Estimate the conductivity of the Ge.
(iv) What is the mobility of the carriers?
(v) Is it a p or n type semiconductor?

1.7. The Hall effect (see Fig. 1.3) is measured in a semiconductor sample in which both electrons and holes are present. Under the effect of the magnetic field both carriers are deflected in the same transverse direction. Obviously, no electric field can stop simultaneously both the electrons and the holes, hence whatever the Hall voltage there will always be carrier motion in the transverse direction. Does this mean that there will be an indefinite accumulation of electrons and holes on the surface of one of the boundaries? If not, why not?

1.8. Derive an expression for the Hall coefficient R_H [still defined by eqn (1.20)] when both electrons and holes are present.

The experimentally determined Hall coefficient is found to be negative. Can you conclude that electrons are the dominant charge carriers?

(Hint: Write down the equation of motion (neglect inertia) for both holes and electrons in vectorial form. Resolve the equations in the longitudinal (z-axis in Fig. 1.3) and in the transverse (x-axis in Fig. 1.3) directions. Neglect the product of transverse velocity with the magnetic field. Find the transverse velocities for electrons and holes. Find the transverse current, and finally find the transverse field from the condition that the transverse current is zero.)

1.9. An electromagnetic wave is incident from Medium 1 upon Medium 2 as shown in Figs 1.4 and 1.5. Derive expressions for the reflected and transmitted power. Show that the transmitted electromagnetic power is finite when $\omega > \omega_p$ and zero when $\omega < \omega_p$.

[Hint: Solve Maxwell's equations separately in both media. Determine the constants by matching the electric and magnetic fields at the boundary. The power in the wave (per unit surface) is given by the Poynting vector.]

1.10. An electromagnetic wave is incident upon a medium of width d, as shown in Fig. 1.6. Derive expressions for the reflected and transmitted power. Calculate the transmitted power for the cases $d = 0.25\,\mu\text{m}$ and $d = 2.5\,\mu\text{m}$ when $\omega = 6.28 \times 10^{15}\,\text{rad s}^{-1}$, $\omega_p = 9 \times 10^{15}\,\text{rad s}^{-1}$ (take $\epsilon = \epsilon_0$ and $\mu = \mu_0$).

1.11. In a medium containing free charges the total current density may be written as $\mathbf{J}_{\text{total}} = \mathbf{J} - i\omega\epsilon\mathcal{E}$, where $\mathbf{J}$ is the particle current density, $\mathcal{E}$ electric field, ϵ dielectric constant, and ω frequency of excitation. For convenience, the above expression is often written in the form $\mathbf{J}_{\text{total}} = -i\omega\bar{\bar{\epsilon}}_{\text{eqv}}\mathcal{E}$, defining thereby an equivalent dielectric tensor $\bar{\bar{\epsilon}}_{\text{eqv}}$. Determine $\bar{\bar{\epsilon}}_{\text{eqv}}$ for a fully ionized electron-ion plasma to which a constant magnetic field B_0 is applied in the z-direction.

2 The electron as a wave

The old order changeth, yielding place to new.
Tennyson *The Idylls of the King*

Vezess uj utakra, Lucifer.
Madách *Az Ember Tragédiája*

2.1 Introduction

We have considered the electron as a particle and managed to explain successfully a number of interesting phenomena. Can we explain the rest of the electronics by gentle modifications of this model? Unfortunately (for students if not lecturers), the answer is no. The experimental results on specific heat have already warned us that something is wrong with our particles, but the situation is, in fact, a lot worse. We find that the electron has wavelike properties too. The chief immigrant in this particular woodpile, the experiment that could not possibly be explained by a particle model, was the electron diffraction experiment of Davisson and Germer in 1927. The electrons behaved as waves.

We shall return to the experiment a little later; let us see first what the basic difference is between particle and wave behaviour. The difference can best be illustrated by the following 'thought' experiment. Suppose we were to fire bullets at a bullet-proof screen with two slits in it (Fig. 2.1). We will suppose that the gun barrel is old and worn, so that the bullets bespatter the screen around the slits uniformly after a fairly large number of shots. If at first, slit B is closed by a bullet-proof cover, the bullets going through A make a probability pattern on the target screen, shown graphically in Fig. 2.1 as a plot of probability against

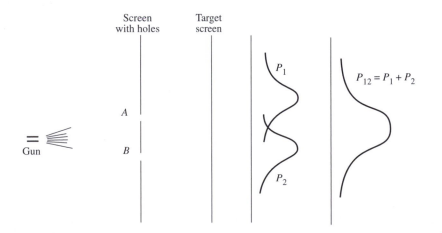

Fig. 2.1
An experiment with bullets.

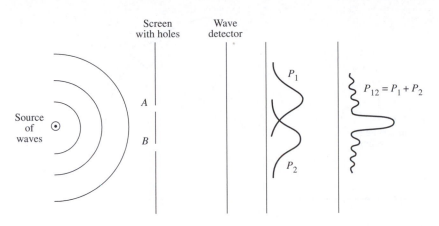

Fig. 2.2
An experiment with waves.

distance from the gun nozzle–slit A axis. Calling this pattern P_1, we expect (and get) a similar but displaced pattern P_2 if slit B is opened and slit A is closed. Now if both slits are open, the combined pattern, P_{12}, is simply

$$P_{12} = P_1 + P_2. \tag{2.1}$$

We will now think of a less dangerous and more familiar experiment, with waves in a ripple tank (Fig. 2.2). The gun is replaced by a vibrator or ripple-generator, the slits are the same, and in the target plane there is a device to measure the ripple intensity, that is a quantity proportional to the square of the height of the waves produced. Then, with one slit open we find

$$P_1 = |h_1|^2, \tag{2.2}$$

or with the other open

$$P_2 = |h_2|^2, \tag{2.3}$$

where we have taken h_1 and h_2 as complex vectors and the constant of proportionality is unity. The probability functions are similar to those obtained with one slit and bullets. So far, waves and bullets show remarkable similarity. But with both slits open we find

$$P_{12} \neq P_1 + P_2. \tag{2.4}$$

Instead, as we might intuitively suppose, the instantaneous values of the wave heights from each slit add; and as the wavelength of each set of ripples is the same, they add up in the familiar way

$$P_{12} = |h_1 + h_2|^2 = |h_1|^2 + |h_2|^2 + 2|h_1||h_2|\cos\delta. \tag{2.5}$$

δ is the phase difference between the two interfering waves.

Thus, the crucial difference between waves and particles is that waves interfere, but particles do not.

The big question in the 1920s was: are electrons like bullets or do they follow a theoretical prediction by L. de Broglie (in 1924)? According to de Broglie, the electrons should have wavelike properties with a wavelength inversely

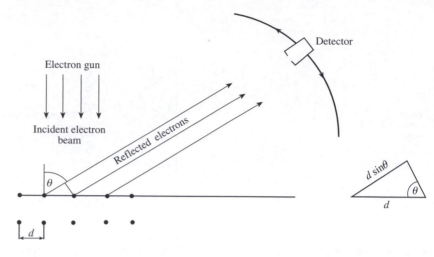

Fig. 2.3
Schematic representation of Davisson and Germer's experiment with low energy electrons. The electrons are effectively reflected by the surface layer of the crystal. The detector shows maximum intensity when the individual reflections add in phase.

h is Planck's constant* (not the height of the waves in the ripple tank) with a rather small numerical value, namely $6.6 \times 10^{-34}\,\mathrm{J\,s}$, and m and v are the mass and speed of the electron.

proportional to particle momentum, namely

$$\lambda = \frac{h}{mv}. \tag{2.6}$$

To test de Broglie's hypothesis, Davisson and Germer fired a narrow beam of electrons at the surface of a single crystal of nickel (Fig. 2.3). The wave-like nature of the electron was conclusively demonstrated. The reflected beam displayed an interference pattern.

The arrangement is analogous to a reflection grating in optics; the grating is replaced by the regular array of atoms and the light waves are replaced by electron waves. Maximum response is obtained when the reflections add in phase, that is when the condition

n is an integral number, d is the lattice spacing, and λ is the wavelength to be determined as a function of electron-gun accelerating voltage.

$$n\lambda = d \sin\theta \tag{2.7}$$

is satisfied.

From eqn (2.7) the difference in angle between two successive maxima is of the order of λ/d. Thus, if the wavelength of the radiation is too small, the maxima lie too close to each other to be resolved. Hence, for good resolution, the wavelength should be about equal to the lattice spacing, which is typically a fraction of a nanometre. The electron velocity corresponding to a wavelength of 0.1 nm is

*Planck introduced this quantity in 1901 in a theory to account for discrepancies encountered in the classical picture of radiation from hot bodies. He considered a radiator as an assembly of oscillators whose energy could not change continuously, but must always increase or decrease by a quantum of energy, hf. This was the beginning of the twentieth century for science and science has not been the same since. The confidence and assurance of nineteenth-century physicists disappeared, probably forever. The most we can hope nowadays is that our latest models and theories go one step further in describing Nature.

$$v = \frac{h}{m\lambda} = \frac{6.6 \times 10^{-34}}{9.1 \times 10^{-31} \times 10^{-10}}\,\mathrm{J\,s\,kg^{-1}\,m^{-1}} = 7.25 \times 10^6\,\mathrm{ms^{-1}}. \tag{2.8}$$

The accelerating voltage may be obtained from the condition of energy conservation

$$\tfrac{1}{2}mv^2 = eV,$$

whence

$$V = \frac{mv^2}{2e} = \frac{9.1 \times 10^{-31}(7.25 \times 10^6)^2}{2 \times 1.6 \times 10^{-19}}\,\mathrm{kg\,m^2\,s^{-2}\,C^{-1}} = 150\,\mathrm{V}. \tag{2.9}$$

The voltages used by Davisson and Germer were of this order.

So electrons are waves. Are protons waves? Yes, they are; it can be shown experimentally. Are neutrons waves? Yes, they are; it can be shown experimentally. Are bullets waves? Well, they should be, but there are some experimental difficulties in proving it. Take a bullet which has a mass of 10^{-3} kg and travels at a velocity 10^3 m s^{-1}. Then the bullet's wavelength is 6.6×10^{-34} m. Thus, our reflecting agents or slits should be about 10^{-34} m apart to observe the diffraction of bullets, and that would not be easily realizable. Our bullets are obviously too fast. Perhaps with slower bullets we will get a diffraction pattern with slits a reasonable distance apart. Taking 10 mm for the distance between the slits and requiring the same wavelength for the bullets, their velocity comes to 10^{-28} m s^{-1}; that is, the bullet would travel 1 m in about 10^{21} years. Best modern estimates give the age of the universe as 10^{10} years so this way of doing the experiment runs again into practical difficulties.

The conclusion from this rather eccentric aside is of some importance. It seems to suggest that everything, absolutely everything, that we used to regard as particles may behave like waves if the right conditions are ensured. The essential difference between electrons and particles encountered in some other branches of engineering is merely one of size. Admittedly, the factors involved are rather large. The bullet in our chosen example has a mass 10^{27} times the electron mass, so it is not entirely unreasonable that they behave differently.

2.2   The electron microscope

Particles are waves, waves are particles. This outcome of a few simple experiments mystifies the layman, delights the physicist, and provides the philosopher with material for a couple of treatises. What about the engineer? The engineer is supposed to ask the consequential (though grammatically slightly incorrect) question; what is this good for?

Well, one well-known practical effect of the wave nature of light is that the resolving power of a microscope is fundamentally limited by the wavelength of the light. If we want greater resolution, we need a shorter wavelength. Let me use X-rays then. Yes, but they can not be easily focused. Use electrons then; they have short enough wavelengths. An electron accelerated to a voltage of 150 V has a wavelength of 0.1 nm. This is already four thousand times shorter than the wavelength of violet light, and using higher voltages we can get even shorter wavelengths. Good, but can electrons be focused? Yes, they can. Very conveniently, just about the same time that Davisson and Germer proved the wave properties of electrons, Busch discovered that electric and magnetic fields of the right configuration can bring a diverging electron beam to a focus. So all we need is a fluorescent screen to make the incident electrons visible, and the electron microscope is ready.

You know, of course, about the electron microscope, that it has a resolving power so great that it is possible to see large molecules with it, and using the latest techniques even individual atoms can be made visible. Our aim is mainly to emphasize the mental processes that lead from scientific discoveries to practical applications. But besides, there is one more interesting aspect of the electron microscope. It provides perhaps the best example for what is known as the 'duality of the electron'. To explain the operation of the electron microscope, both the 'wave' and the 'particle' aspects of the electron are needed.

The focusing is possible because the electron is a charged *particle*, and the great resolution is possible because it is a wave of extremely short wavelength.

In conclusion, it must be admitted that the resolving power of the electron microscope is not as large as would follow from the available wavelengths of the electrons. The limitation in practice is caused by lens aberrations.

2.3 Some properties of waves

You are by now familiar with all sorts of waves, and you know that a wave of frequency, ω, and wave number, k, may be described by the formula

$$u = a \exp i\varphi; \quad \varphi = -(\omega t - kz), \tag{2.10}$$

where the positive z-axis is chosen as the direction of propagation.

The phase velocity may be defined as

$$v_{\mathrm{p}} = \left.\frac{\partial z}{\partial t}\right|_{\varphi=\text{constant}} = \frac{\omega}{k} = f\lambda. \tag{2.11}$$

This is the velocity with which any part of the wave moves along. For a single frequency wave this is fairly obvious. One can easily imagine how the crest moves. But what happens when several waves are superimposed? The resultant wave is given by

$$u = \sum_n a_n \exp\{-\mathrm{i}(\omega_n t - k_n z)\}, \tag{2.12}$$

where to each value of k_n belongs an a_n and an ω_n. Going over to the continuum case, when the number of components within an interval Δk tends to infinity, we get

$$u = \int_{-\infty}^{\infty} a(k) \exp\{-\mathrm{i}(\omega t - kz)\}\, \mathrm{d}k, \tag{2.13}$$

where $a(k)$ and ω are functions of k.

We shall return to the general case later; let us take for the time being, $t = 0$, then

$$u(z) = \int_{-\infty}^{\infty} a(k) \exp(\mathrm{i}kz)\, \mathrm{d}k \tag{2.14}$$

and investigate the relationship between $a(k)$ and $u(z)$. We shall be interested in the case when the wave number and frequency of the waves do not spread out too far, that is $a(k)$ is zero everywhere with the exception of a narrow interval Δk. The simplest possible case is shown in Fig. 2.4 where

$$a(k) = 1 \quad \text{for } k_0 - \frac{\Delta k}{2} < k < k_0 + \frac{\Delta k}{2} \tag{2.15}$$

and

$$a(k) = 0$$

outside this interval. The integral (2.14) reduces then to

$$u(z) = \int_{k_0-\Delta k/2}^{k_0+\Delta k/2} \exp(\mathrm{i}kz)\, \mathrm{d}k, \tag{2.16}$$

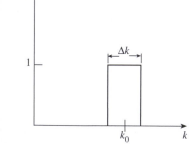

Fig. 2.4
The amplitude of the waves as a function of wave number, described by eqn (2.15).

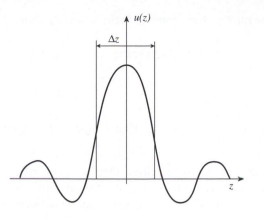

Fig. 2.5
The spatial variation of the amplitude
of the wave packet of Fig. 2.4.

which can be easily integrated to give

$$u(z) = \Delta k \, \exp(ik_0 z) \frac{\sin \frac{1}{2}(\Delta k z)}{\frac{1}{2}(\Delta k z)}. \tag{2.17}$$

We have here a wave whose *envelope* is given by the function

$$\frac{\sin \frac{1}{2}(\Delta k z)}{\frac{1}{2}(\Delta k z)}, \tag{2.18}$$

plotted in Fig. 2.5. It may be seen that the function is rapidly decreasing outside a certain interval Δz. We may say that the wave is essentially contained in this 'packet', and in future we shall refer to it as a wave packet. We chose the width of the packet, rather arbitrarily, to be determined by the points where the amplitude drops to 0.63 of its maximum value, that is where

$$\frac{\Delta k z}{2} = \pm \frac{\pi}{2}. \tag{2.19}$$

Hence, the relationship between the spread in wave number Δk, and the spread in space Δz, is as follows

$$\Delta k \, \Delta z = 2\pi. \tag{2.20}$$

An obvious consequence of this relationship is that by making Δk large, Δz must be small, and vice versa.

Let us return now to the time-varying case, still maintaining that $a(k)$ is essentially zero beyond the interval Δk. Equation (2.13) then takes the form

For having a narrow wave packet in space, we need a larger spread in wave number.

$$u = \int_{k_0 - \Delta k/2}^{k_0 + \Delta k/2} a(k) \exp\{-i(\omega t - kz)\}dk. \tag{2.21}$$

Let us now rewrite the above formula in the following form

ω_0 is the frequency at $k = k_0$.

$$u(z,t) = A(z,t) \exp\{-\mathrm{i}(\omega_0 t - k_0 z)\}, \tag{2.22}$$

where

$$A(z,t) = \int_{k_0 - \Delta k/2}^{k_0 + \Delta k/2} a(k) \exp[-\mathrm{i}\{(\omega - \omega_0)t - (k - k_0)z\}]\mathrm{d}k. \tag{2.23}$$

We may now define two velocities. One is ω_0/k_0, which corresponds to the previously defined phase velocity, and is the velocity with which the central components propagate. The other velocity may be defined by looking at the expression for A. Since A represents the envelope of the wave, we may say that the envelope has the same shape whenever

$$(\omega - \omega_0)t - (k - k_0)z = \text{constant.} \tag{2.24}$$

Hence, we may define a velocity,

v_g is called the *group velocity* because it gives the velocity of the wave packet.

$$v_g = \frac{\partial z}{\partial t} = \frac{\omega - \omega_0}{k - k_0}, \tag{2.25}$$

which, for sufficiently small δk, reduces to

$$v_g = \left(\frac{\partial \omega}{\partial k}\right)_{k=k_0}. \tag{2.26}$$

2.4 Applications to electrons

We have discussed some properties of waves. It has been an exercise in mathematics. Now we take a deep plunge and will try to apply these properties to the particular case of the electrons. The first step is to identify the wave packet with an electron in your mental picture. This is not unreasonable. We are saying, in fact, that where the ripples are, *there* must be the electron. If the ripples are uniformly distributed in space, as is the case for a single frequency wave, the electron can be anywhere. If the ripples are concentrated in space in the form of a wave packet, the presence of an electron is indicated. Having identified the wave packet with an electron, we may identify the velocity of the wave packet with the electron velocity.

What can we say about the energy of the electron? We know that a photon of frequency ω has an energy

$$E = hf = \hbar\omega, \tag{2.27}$$

where f is the frequency of the electromagnetic wave and $\hbar = h/2\pi$. Analogously, it may be suggested that the energy of an electron in a wave packet centred at the frequency ω is given by the same formula. Hence, we may write down the energy of the electron, taking the potential energy as zero, in the form

$$\hbar\omega = \tfrac{1}{2}mv_g^2. \tag{2.28}$$

We can differentiate this partially with respect to k to get

$$\hbar\frac{\partial\omega}{\partial k} = mv_g\frac{\partial v_g}{\partial k} \tag{2.29}$$

which, with the aid of eqn (2.26), reduces to

$$\hbar = m\frac{\partial v_g}{\partial k}. \tag{2.30}$$

Integrating, and taking the integration constant as zero, we get

$$\hbar k = mv_g, \tag{2.31}$$

which can be expressed in terms of wavelength as

$$\lambda = \frac{h}{mv_g}, \tag{2.32}$$

and this is nothing else but de Broglie's relationship. Thus, if we assume the validity of the wave picture, identify the group velocity of a wave packet with the velocity of an electron, and assume that the centre frequency of the wave packet is related to the energy of the electron by Planck's constant, de Broglie's relationship automatically drops out.

This proves, of course, nothing. There are too many assumptions, too many identifications, representations, and interpretations; but, undeniably, the different pieces of the jigsaw puzzle do show some tendency to fit together. We have now established some connection between the wave and particle aspects, which seemed to be entirely distinct not long ago.

What can we say about the electron's position? Well, we identified the position of the electron with the position of the wave packet. So, wherever the wave packet is, there is the electron. But remember, the wave packet is not infinitely narrow; it has a width Δz, and there will thus be some uncertainty about the position of the electron.

Let us look again at eqn (2.20). Taking note of the relationship expressed in eqn (2.31) between wave number and momentum, eqn (2.20) may be written as

$$\Delta p\,\Delta z = h. \tag{2.33}$$

If we know the position of the electron with great precision, that is if Δz is very small, then the uncertainty in the velocity of the electron must be large.

This is the celebrated uncertainty relationship. It means that the uncertainty in the position of the electron is related to the uncertainty in the momentum of the electron. Let us put in a few figures to see the orders of magnitude involved. If we know the position of the electron with an accuracy of 10^{-9} m then the uncertainty in momentum is

$$\Delta p = 6.6 \times 10^{-25}\ \text{kg}\,\text{m}\,\text{s}^{-1}, \tag{2.34}$$

corresponding to

$$\Delta v \cong 7 \times 10^5 \mathrm{\ m\,s^{-1}}, \qquad (2.35)$$

that is, the uncertainty in velocity is quite appreciable.

Taking macroscopic dimensions, say 10^{-3} m for the uncertainty in position, and a bullet with a mass 10^{-3} kg, the uncertainty in velocity decreases to

$$\Delta v = 6.6 \times 10^{-28} \mathrm{\ m\,s^{-1}}, \qquad (2.36)$$

which is something we can easily put up with in practice. Thus, whenever we come to very small distances and very light particles, the uncertainty in velocity becomes appreciable, but with macroscopic objects and macroscopic distances the uncertainty in velocity is negligible. You can see that everything here depends on the value of h, which happens to be rather small in our universe. If it were larger by a factor of, say 10^{40}, the police would have considerable difficulty in enforcing the speed limit.

The uncertainty relationship has some fundamental importance. It did away (probably for ever) with the notion that distance and velocity can be simultaneously measured with arbitrary accuracy. It is applicable not only to position and velocity, but to a number of other related pairs of physical quantities.* It may also help to explain qualitatively some complicated phenomena. We may, for example, ask the question why there is such a thing as a hydrogen atom consisting of a negatively charged electron and a positively charged proton. Why doesn't the electron eventually fall into the proton? Armed with our knowledge of the uncertainty relationship, we can now say that this event is energetically unfavourable. If the electron is too near to the proton then the uncertainty in its velocity is high; so it may have quite a high velocity, which means high kinetic energy. Thus the electron's search for low potential energy (by moving near to the proton) is frustrated by the uncertainty principle, which assigns a large kinetic energy to it. The electron must compromise and stay at a certain distance from the proton (see example 4.4).

* You might find it interesting to learn that electric and magnetic intensities are also subject to this law. They cannot be simultaneously measured to arbitrary accuracy.

2.5   Two analogies

The uncertainty relationship is characteristic of quantum physics. We would search in vain for anything similar in classical physics. The derivation is, however, based on certain mathematical formulae that also appear in some other problems. Thus, even if the phenomena are entirely different, the common mathematical formulation permits us to draw analogies.

Analogies may or may not be helpful. It depends to a certain extent on the person's imagination or lack of imagination and, of course, on familiarity or lack of familiarity with the analogue.

We believe in the use of analogies. We think they can help, both in memorizing a certain train of thought, and in arriving at new conclusions and new combinations. Even such a high-powered mathematician as Archimedes resorted to mechanical analogies when he wanted to convince himself of the truth of certain mathematical theorems. So this is quite a respectable

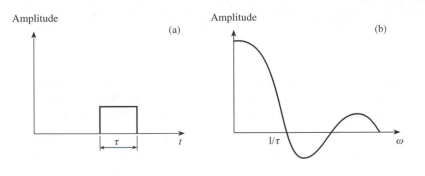

Fig. 2.6
A rectangular pulse and its frequency spectrum.

method, and as we happen to know two closely related analogies, we shall describe them.

Notice first of all that $u(z)$ and $a(k)$ are related to each other by a Fourier integral in eqn (2.14). In deriving eqn (2.20), we made the sweeping assumption that $a(k)$ was constant within a certain interval, but this is not necessary. We would get the same sort of final formula, with slightly different numerical constants, for any reasonable $a(k)$. The uncertainty relationship, as derived from the wave concept, is a consequence of the Fourier transform connection between $a(k)$ and $u(z)$. Thus, whenever two functions are related in the same way, they can readily serve as analogues.

Do such functions appear in engineering practice? They do. The time variation of a signal and its frequency spectrum are connected by Fourier transform. A pulse of the length τ has a spectrum (Fig. 2.6) exactly like the envelope we encountered before. The width of the frequency spectrum, referred to as bandwidth in common language, is related to the length of the pulse. All communication engineers know that the shorter the pulse the larger is the bandwidth to be transmitted. For television, for example, we need to transmit lots of pulses (the light intensity for some several hundred thousand spots twenty-five times per second), so the pulses must be short and the bandwidth large. This is why television works at much higher frequencies than radio broadcasting.

In the mathematical formulation, k and z of eqn (2.20) are to be replaced by the frequency ω and time t. Hence, the relationship for communication engineering takes the form,

$$\Delta\omega\,\Delta t = 2\pi. \tag{2.37}$$

The analogy is close indeed.

In the second analogue the size of an aerial and the sharpness of the radiation pattern are related. It is the same story. In order to obtain a sharp beam one needs a big aerial. So if you have ever wondered why radio astronomers use such giant aerials, here is the answer. They need narrow beams to be able to distinguish between the various radio stars, and they must pay for them by erecting (or excavating) big antennas.*

The mathematical relationship comes out as follows:

$$\Delta\theta\,\Delta z = \lambda. \tag{2.38}$$

* Incidentally there is another reason why radio telescopes must be bigger than, say, radar aerials. They do a lot of work at a wavelength of 210 mm, which is seven times longer than the wavelength used by most radars. Hence for the same resolution an aerial seven times bigger is needed.

$\Delta\theta$ is the beamwidth, Δz is the linear dimension of the aerial and λ is the wavelength of the electromagnetic radiation (transmitted or received).

Exercises

2.1. Find the de Broglie wavelength of the following particles:

(i) an electron in a semiconductor having average thermal velocity at $T = 300\,\text{K}$ and an effective mass of $m_e^* = a m_0$,

(ii) a helium atom having thermal energy at $T = 300\,\text{K}$,

(iii) an α-particle (He^4 nucleus) of kinetic energy $10\,\text{MeV}$.

2.2. A finite wave train moves with constant velocity v. Its profile as a function of space and time is given as

$$f(a) = \begin{array}{ll} \exp iu & \\ 0 & \end{array} \quad \text{if} \quad \begin{array}{l} |u| < u_0 \\ |u| > u_0 \end{array}$$

where $u = \Omega t - kx$ and $k = \Omega/v$.

At $t = 0$ the wave packet extends from $x = -u_0 v\,\Omega^{-1}$ to $x = u_0 v\,\Omega^{-1}$, that is, it has a length of $\Delta l = 2u_0 v\,\Omega^{-1}$.

Find the spectral composition of the wave train at $x = 0$ and find from that the spectral width $\Delta\omega$ (range within which the amplitude drops to 63% of its maximum value).

Prove the uncertainty relationship

$$\Delta p\,\Delta l = h.$$

2.3. A typical operating voltage of an electron microscope is $50\,\text{kV}$.

(i) What is the smallest distance that it could possibly resolve?

(ii) What energy of neutrons could achieve the same resolution?

(iii) What are the main factors determining the actual resolution of an electron microscope?

2.4. Electrons accelerated by a potential of $70\,\text{V}$ are incident perpendicularly on the surface of a single crystal metal.

The crystal planes are parallel to the metal surface and have a (cubic) lattice spacing of $0.352\,\text{nm}$. Sketch how the intensity of the scattered electron beam would vary with angle.

2.5. A beam of electrons of $10\,\text{keV}$ energy passes perpendicularly through a very thin (of the order of a few nanometres) foil of our previous single crystal metal. Determine the diffraction pattern obtained on a photographic plate placed $0.1\,\text{m}$ behind the specimen. How will the diffraction pattern be modified for a polycrystalline specimen? (Hint: Treat the lattice as a two-dimensional array.)

2.6. Consider again an electron beam incident upon a thin metal foil but look upon the electrons as particles having a certain kinetic energy. In experiments with aluminium foils (J. Geiger and K. Wittmaack, *Zeitschrift für Physik*, **195**, 44, 1966) it was found that a certain fraction of the electrons passing through the metal had a loss of energy of $14.97\,\text{eV}$. We could explain this loss as being the creation of a particle of that much energy. But what particle? It cannot be a photon (a transverse electromagnetic wave in the wave picture) because an electron in motion sets up no transverse waves. It must be a particle that responds to a longitudinal electric field. So it might be a plasma wave of frequency ω_p which we could call a 'plasmon' in the particle picture. The energy of this particle would be $\hbar\omega_p$.

Calculate the value of $\hbar\omega_p$ for aluminium assuming three free electrons per atom. Compare it with the characteristic energy loss found.

The density of aluminium is $2700\,\text{kg m}^{-3}$ and its atomic weight is 27.

The electron

That's how it is, says Pooh.
A.A. Milne *Now we are six*

3.1 Introduction

We have seen that some experimental results can be explained if we regard the electron as a particle, whereas the explanation of some other experiments is possible only if we look upon our electron as a wave. Now which is it? Is it a particle or is it a wave? It is neither, it is an electron.

An electron is an electron; this seems a somewhat tautological definition. What does it mean? I want to say by this that we don't have to regard the electron as something else, something we are already familiar with. It helps, of course, to know that the electron sometimes behaves as a particle because we have some intuitive idea of what particles are supposed to do. It is helpful to know that the electron may behave as a wave because we know a lot about waves. But we do not have to look at the electron as something else. It is sufficient to say that an electron is an electron as long as we have some means of predicting its properties.

How can we predict what an electron will do? Well, how can we predict any physical phenomenon? We need some mathematical relationship between the variables. Prediction and mathematics are intimately connected in science—or are they? Can we make predictions without any mathematics at all? We can. Seeing, for example, dark heavy clouds gathering in the sky we may say that 'it is going to rain' and on a large number of occasions we will be right. But this is not really a very profound and accurate prediction. We are unable to specify *how* dark the clouds should be for a certain amount of rain, and we would find it hard to guess the temporal variation of the positions of the clouds. So, as you know very well, meteorology is not yet an exact science.

In physics fairly good predictions are needed because otherwise it is difficult to get further money for research. In engineering the importance of predictions can hardly be overestimated. If the designer of a bridge or of a telephone exchange makes some wrong predictions, this mistake may bring upon him the full legal apparatus of the state or the frequent curses of the subscribers. Thus, for engineers, prediction is not a trifling matter.

Now what about the electron? Can we predict its properties? Yes, we can because we have an equation which describes the behaviour of the electron in mathematical terms. It is called Schrödinger's equation. Now I suppose you would like to know where Schrödinger's equation came from? It came from nowhere; or more correctly it came straight from Schrödinger's head, not unlike Pallas Athene who is reputed to have sprung out of Zeus' head (and in full armour too!). Schrödinger's equation is a product of Schrödinger's imagination; it cannot be derived from any set of physical assumptions. Schrödinger's equation is, of course, not unique in this respect. You have met similar cases before.

In the sixth form you learned Newton's equation. At the time you had just gained your first glances into the hidden mysteries of physics. You would not have dared to question your schoolmaster about the origin of Newton's equation. You were probably more reverent at the time, more willing to accept the word of authorities, and besides, Newton's equation looks so simple that one's credulity is not seriously tested. Force equals mass times acceleration; anyone is prepared to believe that much. And it seems to work in practice.

At the university you are naturally more inquisitive than in your schooldays; so you may have been a bit more reluctant to accept Maxwell's equations when you first met them. It must have been very disturbing to be asked to accept the equation,

$$\nabla \times \mathcal{E} = -\frac{\partial \mathbf{B}}{\partial t}, \tag{3.1}$$

as the truth and nothing but the truth. But then you were shown that this equation is really identical with the familiar induction law,

$$V = -\frac{\partial \phi}{\partial t}, \tag{3.2}$$

and the latter merely expresses the result of a simple experiment. Similarly

$$\nabla \times \mathbf{H} = \mathbf{J} \tag{3.3}$$

is only a rewriting of Ampère's law. So all is well again or rather all would be well if there was not another term on the right-hand side, the displacement current $\partial \mathbf{D}/\partial t$. Now what is this term? Not many lecturers admit that it came into existence as a pure artifice. Maxwell felt there should be one more term there, and that was it. True, Maxwell himself made an attempt to justify the introduction of displacement current by referring to the a.c. current in a capacitor, but very probably that was just a concession to the audience he had to communicate with. He must have been more concerned with refuting the theory of instantaneous action at a distance, and with deriving a velocity with which disturbances can travel.

The extra term had no experimental basis, whatsoever. It was a brilliant hypothesis which enabled Maxwell to predict the existence of electromagnetic waves. When some years later Hertz managed to find these waves, the hypothesis became a law. It was a momentous time in history, though most history books keep silent about the event.*

I am telling you all this just to show that an equation which comes from nowhere in particular may represent physical reality. Of course, Schrödinger had good reasons for setting up his equation. He had immediate success in several directions. Whilst Maxwell's displacement current term explained no experimental observation, Schrödinger's equation could immediately account for the atomic spectrum of hydrogen, for the energy levels of the Planck oscillator, for the non-radiation of electronic currents in atoms, and for the shift

* According to most historians' definition, an event is important if it affects a large number of people to a considerable extent for a long time. If historians were faithful to this definition they should write a lot about Maxwell and Hertz because by predicting and proving the existence of electromagnetic waves, Maxwell and Hertz had more influence on the life of ordinary people nowadays than any nineteenth-century general, statesman, or philosopher.

of energy levels in strong transverse fields. He produced four papers in quick succession and noted at the end with quiet optimism:

'I hope and believe that the above attempt will turn out to be useful for explaining the magnetic properties of atoms and molecules, and also the electric current in the solid state.'

Schrödinger was right.* His equation turned out to be useful indeed. He was not *exactly* right, though. In order to explain all the properties of the solid state (including magnetism) two further requisites are needed: Pauli's principle and 'spin'. Fortunately, both of them can be stated in simple terms, so if we make ourselves familiar with Schrödinger's equation, the rest is relatively easy.

3.2 Schrödinger's equation

After such a lengthy introduction, let us have now the celebrated equation itself. In the usual notation,

$$-\frac{\hbar^2}{2m}\nabla^2\Psi + V\Psi = i\hbar\frac{\partial\Psi}{\partial t}. \tag{3.4}$$

We have a partial differential equation in Ψ. But what is Ψ? It is called the wave function, and

$$|\Psi(x, y, z; t)|^2 \, dx \, dy \, dz \tag{3.5}$$

gives the probability that the electron can be found at time, t, in the volume element, $dx \, dy \, dz$, in the immediate vicinity of the point, x, y, z. To show the significance of this function better $|\Psi|^2$ is plotted in Fig. 3.1 for a hypothetical case where $|\Psi|^2$ is independent of time and varies only in one dimension. If we make many measurements on this system, we shall find that the electron is always between z_0 and z_4 (the probability of being outside this region is zero), that it is most likely to be found in the interval dz around z_3, and it is three times as probable to find the electron at z_2 than at z_1. Since the electron must be somewhere, the probability of finding it between z_0 and z_4 must be unity, that is,

$$\int_{z_0}^{z_4} |\Psi(z)|^2 \, dz = 1. \tag{3.6}$$

The above example does not claim to represent any physical situation. It is shown only to illustrate the meaning of $|\Psi|^2$.

* In the above discussion the role of Schrödinger in setting up modern quantum physics was very much exaggerated. There were a number of others (Heisenberg, Born, Dirac, Pauli to name a few) who made comparable contributions, but since this is not a course in the history of science, and the Schrödinger formulation is adequate for our purpose, we shall not discuss these contributions.

m is the mass of the electron, and V is the potential in which the electron moves.

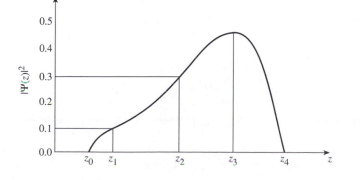

Fig. 3.1

Introducing the concept of the wave function. $|\psi(z)|^2 dz$ proportional to the probability that the electron may be found in the interval dz at the point z.

The physical content of eqn (3.4) will be clearer when we shall treat more practical problems, but there is one thing we can say immediately. Schrödinger's equation does not tell us the *position* of the electron, only the *probability* that it will be found in the vicinity of a certain point.

The description of the electron's behaviour is statistical, but there is nothing particularly new in this. After all, you have met statistical descriptions before, in gas dynamics for example, and there was considerably less fuss about it.

The main difference is that in classical mechanics, we use statistical methods in order to simplify the calculations. We are too lazy to write up 10^{27} differential equations to describe the motion of all the gas molecules in a vessel, so we rely instead on a few macroscopic quantities like pressure, temperature, average velocity, etc. We use statistical methods because we elect to do so. It is merely a question of convenience. This is not so in quantum mechanics. The statistical description of the electron is inherent in quantum theory. That is the best we can do. We cannot say much about an electron at a given time. We can only say what happens on the average when we make many observations on one system, or we can predict the statistical outcome of simultaneous measurements on identical systems. It may be sufficient to make one single measurement (specific heat or electrical conductivity) when the phenomenon is caused by the collective interaction of a large number of electrons.

We cannot even say how an electron moves as a function of time. We cannot say this because the position and the momentum of an electron cannot be simultaneously determined. The limiting accuracy is given by the uncertainty relationship, eqn (2.33).

3.3   Solutions of Schrödinger's equation

Let us separate the variables and attempt a solution in the following form

$$\Psi(\mathbf{r}, t) = \psi(\mathbf{r})w(t). \tag{3.7}$$

Now **r** represents all the spatial variables.

Substituting eqn (3.7) into eqn (3.4), and dividing by ψw we get

$$-\frac{\hbar^2}{2m}\frac{\nabla^2\psi}{\psi} + V = i\hbar\frac{1}{w}\frac{\partial w}{\partial t}. \tag{3.8}$$

Since the left-hand side is a function of **r** and the right-hand side is a function of t, they can be equal only if they are both separately equal to a constant which we shall call E, that is we obtain two differential equations as follows:

$$i\hbar\frac{\partial w}{\partial t} = Ew \tag{3.9}$$

and

$$-\frac{\hbar^2}{2m}\nabla^2\psi + V\psi = E\psi. \tag{3.10}$$

The solution of eqn (3.9) is simple enough. We can immediately integrate and get

$$w = \exp\left(-i\frac{E}{\hbar}t\right), \tag{3.11}$$

and this is nothing else but our good old wave solution, at least as a function of time, if we equate

$$E = \hbar\omega. \tag{3.12}$$

This is actually something we have suggested before [eqn (2.27)] by recourse to Planck's formula. So we may call E the energy of the electron. However, before making such an important decision let us investigate eqn (3.10) which also contains E. We could rewrite eqn (3.10) in the form:

$$\left(-\frac{\hbar^2}{2m}\nabla^2 + V\right)\psi = E\psi. \tag{3.13}$$

The second term in the bracket is potential energy, so we are at least in good company. The first term contains ∇^2, the differential operator you will have met many times in electrodynamics. Writing it symbolically in the form:

$$-\frac{\hbar^2}{2m}\nabla^2 = \frac{1}{2m}(-i\hbar\nabla)^2, \tag{3.14}$$

we can immediately see that by introducing the new notation,

$$\mathbf{p} = -i\hbar\nabla, \tag{3.15}$$

and calling it the 'momentum operator' we may arrive at an old familiar relationship:

$$\frac{\mathbf{p}^2}{2m} = \text{kinetic energy.} \tag{3.16}$$

Thus, on the left-hand side of eqn (3.13) we have the sum of kinetic and potential energies in operator form and on the right-hand side we have a constant E having the dimensions of energy. Hence, we may, with good conscience, interpret E as the total energy of the electron.

You might be a little bewildered by these definitions and interpretations, but you must be patient. You cannot expect to unravel the mysteries of quantum mechanics at the first attempt. The fundamental difficulty is that first steps in quantum mechanics are not guided by intuition. You cannot have any intuitive feelings because the laws of quantum mechanics are not directly experienced in everyday life. The most satisfactory way, at least for the few who are mathematically inclined, is to plunge into the full mathematical treatment and leave the physical interpretation to a later stage. Unfortunately, this method is lengthy and far too abstract for an engineer. So the best we can do is to digest alternately a little physics and a little mathematics and hope that the two will meet.

3.4   The electron as a wave

Let us look at the simplest case when $V = 0$ and the electron can move only in one dimension. Then eqn (3.13), which is often called the time independent

Schrödinger equation, reduces to

$$\frac{\hbar^2}{2m}\frac{\partial^2 \psi}{\partial z^2} + E\psi = 0. \tag{3.17}$$

The solution of this differential equation is a wave in space. Hence the general solution of Schrödinger's equation for the present problem is

$$\Psi = \exp\left(-i\frac{E}{\hbar}t\right)\{A\exp(ikz) + B\exp(-ikz)\}. \tag{3.18}$$

A and B are constants representing the amplitudes of the forward and backward travelling waves, and k is related to E by

$$E = \frac{\hbar^2 k^2}{2m}. \tag{3.19}$$

In this example we have chosen the potential energy of the electron as zero, thus eqn (3.19) must represent the kinetic energy. Hence, we may conclude that $\hbar k$ must be equal to the momentum of the electron. We have come to this conclusion before, heuristically, on the basis of the wave picture, but now we have the full authority of Schrödinger's equation behind us.

You may notice too that $p = \hbar k$ is an alternative expression of de Broglie's relationship, thus we have obtained from Schrödinger's equation both the wave behaviour and the correct wavelength.

What can we say about the position of the electron? Take $B = 0$ for simplicity, then we have a forward travelling wave with a definite value for k. The probability of finding the electron at any particular point is given by $|\psi(z)|^2$ which according to eqn (3.18) is unity, independently of z. This means physically that there is an equal probability of the electron being at any point on the z-axis. The electron can be anywhere; that is, the uncertainty of the electron's position is infinite. This is only to be expected. If the value of k is given then the momentum is known, so the uncertainty in the momentum of the electron is zero; hence the uncertainty in position must be infinitely great.

3.5   The electron as a particle

Equation (3.17) is a linear differential equation, hence the sum of the solutions is still a solution. We are therefore permitted to add up as many waves as we like; that is, a wave packet (as constructed in Chapter 2) is also a solution of Schrödinger's equation.

We can now be a little more rigorous than before. A wave packet represents an electron because $|\Psi(z)|^2$ is appreciably different from zero only within the packet. With the choice $a(k) = 1$ in the interval Δk, it follows from eqns (2.16) and (2.17) that the probability of finding the electron is given by*

$$|\psi(z)|^2 = K\left\{\frac{\sin\frac{1}{2}(\Delta kz)}{\frac{1}{2}(\Delta kz)}\right\}^2. \tag{3.20}$$

*The constant K may be determined by the normalization condition

$$\int_{-\infty}^{\infty}|\psi(z)|^2\,\mathrm{d}z = 1.$$

3.6   The electron meeting a potential barrier

Consider again a problem where the motion of the electron is constrained in one dimension, and the potential energy is assumed to take the form shown in Fig. 3.2.

You are familiar with the classical problem, where the electron starts somewhere on the negative z-axis (say at $-z_0$) in the positive direction with a definite

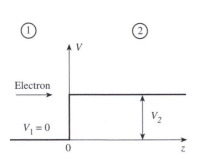

Fig. 3.2
An electron incident upon a potential barrier.

velocity. The solution may be obtained purely from energetic considerations. If the kinetic energy of the electron E is smaller than V_2, the electron is turned back by the potential barrier at $z = 0$. If $E > V_2$, the electron slows down but carries on regardless.

How should we formulate the equivalent quantum mechanical problem? We should represent our electron by a wave packet centred in space on $-z_0$ and should describe its momentum with an uncertainty Δp. We should use the wave function obtained as initial condition at $t = 0$ and should solve the time dependent Schrödinger equation. This would be a very illuminating exercise, alas much too difficult mathematically.

We have to be satisfied by solving a related problem. We shall give our electron a definite energy, that is a definite momentum, and we shall put up with the concomitant uncertainty in position. We shall not be able to say anything about the electron's progress towards the potential barrier, but we shall have a statistical solution which will give the probability of finding the electron on either side of this potential barrier.

Specifying the momentum and not caring about the position of the electron is not so unphysical as you might think. The conditions stated may be approximated in practice by shooting a sufficiently sparse* electron beam towards the potential barrier with a well defined velocity. We are not concerned then with the positions of individual electrons, only with their spatial distribution on the *average*, which we call the macroscopic charge density. Hence we may identify $e|\psi|^2$ with the charge density.

* So that the interaction between the electrons can be neglected.

This is not true in general. What is always true is that $|\psi|^2$ gives the *probability* of an electron being found at z. Be careful, $|\psi(z)|^2$ does *not* give the fraction of the electron's charge residing at z. If, however, a large number of electrons behave identically, then $|\psi|^2$ may be justifiably regarded to be proportional to the charge density.

The charge of the electron is *not* smoothed out. *When the electron is found, the whole electron is there.*

Let us proceed now to the mathematical solution. In region 1 where $V_1 = 0$ the solution is already available in eqns (3.18) and (3.19),

$$\Psi_1 = \exp\left(-i\frac{E}{\hbar}t\right)\{A\exp(ik_1z) + B\exp(-ik_1z)\}, \qquad (3.21)$$

and

$$k_1^2 = \frac{2mE}{\hbar^2}. \qquad (3.22)$$

In region 2 the equation to be solved is as follows (the time-dependent part of the solution remains the same because E is specified)

$$\frac{\hbar^2}{2m}\frac{\partial^2\psi}{\partial z^2} + (E - V_2)\psi = 0, \qquad (3.23)$$

with the general solution

$$\Psi_2 = \exp\left(-i\frac{E}{\hbar}t\right)\{C\exp(ik_2z) + D\exp(-ik_2z)\}, \qquad (3.24)$$

where

$$k_2^2 = \frac{2m}{\hbar^2}(E - V_2). \qquad (3.25)$$

Now we shall ask the question, depending on the relative magnitudes of E and V_2, what is the probability that the electron can be found in regions 1 and 2, respectively.

It is actually easier to speak about this problem in wave language because then the form of the solution is automatically suggested. Whenever a wave is incident on some sort of discontinuity, there is a reflected wave, and there is a transmitted wave. Since no wave is incident from region 2, we can immediately decide that D must be zero.

In order to determine the remaining constants, we have to match the two solutions at $z = 0$, requiring that both ψ and $\partial\psi/\partial z$ should be continuous. From eqns (3.21) and (3.24), the above conditions lead to the algebraic equations,

$$A + B = C \tag{3.26}$$

and

$$ik_1(A - B) = ik_2C, \tag{3.27}$$

whence

$$\frac{B}{A} = \frac{k_1 - k_2}{k_1 + k_2}, \quad \frac{C}{A} = \frac{2k_1}{k_1 + k_2}. \tag{3.28}$$

Let us distinguish now two cases: (i) $E > V_2$. In this case $k_2^2 > 0$, k_2 is real, which means an oscillatory solution in region 2. The values of k_2 and k_1 are, however, different. Thus, B/A is finite, that is, there is a finite amount of reflection. In contrast to the classical solution, there is some probability that the electron is turned back by the potential discontinuity. (ii) $E < V_2$. In this case $k_2^2 < 0$, k_2 is imaginary; that is, the solution declines exponentially in region 2.* Since $|C/A| > 0$, there is a finite, though declining, probability of finding the electron at $z > 0$. Classically, an electron has no chance of getting inside region 2. Under the laws of quantum mechanics the electron may penetrate the potential barrier.

A third case of interest is when the potential profile is as shown in Fig. 3.3, and $E < V_2$. Then k_2 is imaginary and k_3 is real. Hence one may expect that $|\psi|^2$ declines in region 2 and is constant in region 3. The interesting thing is that $|\psi|^2$ in region 3 is not zero. Thus, there is a finite probability that the electron crosses the potential barrier and appears at the other side with energy unchanged. Since there is an exponential decline in region 2, it is necessary that that region should be narrow to obtain any appreciable probability in region 3. If we are thinking in terms of the incident electron beam, we may say that a certain fraction of the electrons will get across the potential barrier. This tendency for

* The exponentially increasing solution cannot be present for physical reasons.

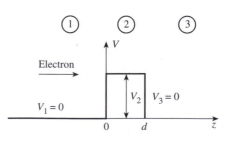

Fig. 3.3
An electron incident upon a narrow potential barrier.

electrons to escape across the potential barrier is called the tunnel effect, or simply tunnelling.

As you will see later, it is an important effect which we shall often invoke to explain phenomena as different as the bonding of the hydrogen molecule or the operation of the tunnel diode.*

3.7 Two analogies

Without the help of Schrödinger's equation, we could not have guessed how electrons behaved when meeting a potential barrier. But having found the solutions in the form of propagating and exponentially decaying waves, a physical picture, I hope, is emerging. There is always a physical picture if you are willing to think in terms of waves. Then it is quite natural that discontinuities cause reflections and that the waves might be attenuated.

The concepts are not appreciably more difficult than those needed to describe the motion of classical electrons, but you need time to make yourself familiar with them. 'Familiarity breeds contempt' may very well apply to arts subjects, but in most branches of science the saying should be reformulated as 'lack of familiarity breeds bewilderment'.

Assuming that you have already developed some familiarity with waves, it may help to stress the analogy further. If we went a little more deeply into the mathematical relationships, we would find that the problem of an electron meeting a potential barrier is entirely analogous to an electromagnetic wave meeting a new medium. Recalling the situations depicted in Figs 1.4 to 1.6, the analogies are as follows.

1. There are two semi-infinite media (Fig. 1.4); electromagnetic waves propagate in both of them. Because of the discontinuity, a certain part of the wave is reflected. This is analogous to the electron meeting a potential barrier (Fig. 3.2) with an energy $E > V_2$. Some electrons are reflected because of the presence of a discontinuity in potential energy.

2. There are two semi-infinite media (Fig. 1.5); electromagnetic waves may propagate in the first one but not in the second one. The field intensities are, however, finite in medium 2 because the electromagnetic wave penetrates to a certain extent. This is analogous to the electron meeting a potential barrier (Fig. 3.2) with an energy $E < V_2$. In spite of not having sufficient energy, some electrons may penetrate into region 2.

3. There are two semi-infinite media separated from each other by a third medium (Fig. 1.6); electromagnetic waves may propagate in media 1 and 3 but not in the middle one. The wave incident from medium 1 declines in medium 2 but a finite amount arrives and can propagate in medium 3. This is analogous to the electron meeting a potential barrier shown in Fig. 3.3, with an energy $E < V_2$. In spite of not having sufficient energy some electrons may cross region 2 and may appear and continue their journey in region 3.

Instead of taking plane waves propagating in infinite media, one might make the analogy physically more realizable though mathematically less perfect, by employing waveguides. Discontinuities can then be represented by joining two waveguides of different cross-sections, and the exponentially decaying wave may be obtained by using a cut-off waveguide (of dimension smaller

* A more mundane example of tunnelling occurs every time we switch on an electric light. The contacts are always covered with an oxide film, that in bulk would be an insulator. But it is rubbed down to a few molecules thickness by the mechanical action of the switch, and the tunnelling is so efficient that we do not notice it.

than half free-space wavelength). Then all the above phenomena can be easily demonstrated in the laboratory.

3.8   The electron in a potential well

In our previous examples, the electron was free to roam in the one-dimensional space. Now we shall make an attempt to trap it by presenting it with a region of low potential energy which is commonly called a potential well. The potential profile assumed is shown in Fig. 3.4. If $E > V_1$, the solutions are very similar to those discussed before, but when $E < V_1$, a new situation arises.

We have by now sufficient experience in solving Schrödinger's equation for a constant potential, so we shall write down the solutions without further discussions.

In region 3 there is only an exponentially decaying solution

$$\psi_3 = C \exp(-\gamma z), \tag{3.29}$$

where

$$\gamma^2 = \frac{2m}{\hbar^2}(V_1 - E). \tag{3.30}$$

In region 2 the potential is zero. The solution is either symmetric or antisymmetric.* Accordingly,

$$\psi_{2s} = A \cos kz \tag{3.31}$$

or

$$\psi_{2a} = A \sin kz \tag{3.32}$$

where

$$k^2 = \frac{2m}{\hbar^2}E. \tag{3.33}$$

In region 1 the solution must decay again, this time towards negative infinity. If we wish to satisfy the symmetry requirement as well, the wave function must look like

$$\psi_1 = \pm C \exp \gamma z. \tag{3.34}$$

Let us investigate the symmetric solution first. We have then eqns (3.29) and (3.31), and eqn (3.34) with the positive sign. The conditions to be satisfied are the continuity of ψ and $\partial\psi/\partial z$ at $L/2$ and $-L/2$, but owing to the symmetry it

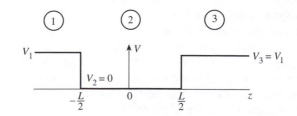

Fig. 3.4
An electron in a potential well.

is sufficient to do the matching at (say) $L/2$. From the continuity of the wave function

$$A \cos k \frac{L}{2} - C \exp\left(-\gamma \frac{L}{2}\right) = 0. \tag{3.35}$$

From the continuity of the derivative of the wave function

$$A k \sin k \frac{L}{2} - C\gamma \exp\left(-\gamma \frac{L}{2}\right) = 0. \tag{3.36}$$

We have now two linear homogeneous equations in A and C which are soluble only if the determinant vanishes, that is

$$\begin{vmatrix} \cos k \dfrac{L}{2} & -\exp\left(-\gamma \dfrac{L}{2}\right) \\ k \sin k \dfrac{L}{2} & -\gamma \exp\left(-\gamma \dfrac{L}{2}\right) \end{vmatrix} = 0, \tag{3.37}$$

leading to

$$k \tan\left(k \frac{L}{2}\right) = \gamma. \tag{3.38}$$

Thus, k and γ are related by eqn (3.38). Substituting their values from eqns (3.30) and (3.33) respectively, we get

$$E^{1/2} \tan\left(\frac{2m}{\hbar^2} E \frac{L^2}{4}\right)^{1/2} = (V_1 - E)^{1/2}, \tag{3.39}$$

which is an algebraic equation to be solved for E. Nowadays one feeds this sort of equation into a computer and has the results printed in a few seconds. But let us be old-fashioned and solve the equation graphically by plotting the left-hand side and the right-hand side separately. Putting in the numerical values, we know

$$m = 9.1 \times 10^{-31} \text{ kg}, \quad \hbar = 1.05 \times 10^{-34} \text{ J s},$$

and we shall take

$$\frac{L}{2} = 5 \times 10^{-10} \text{ m}, \quad V_1 = 1.6 \times 10^{-18} \text{ J}.$$

As may be seen in Fig. 3.5, the curves intersect each other in three points; so there are three solutions and that is the lot.

To be correct, there are three energy levels for the symmetric solution and a few more for the antisymmetric solution.

We have at last arrived at the solution of the first quantum-mechanical problem which deserves literally the name quantum mechanical. Energy is no longer continuous, it cannot take arbitrary values. Only certain discrete energy levels are permitted. In the usual jargon of quantum mechanics, it is said: energy is quantized.

We may generalize further from the above example. The discrete energy levels obtained are not a coincidence. It is true in general that whenever we try to confine the electron, the solution consists of a discrete set of wave functions and energy levels.

If $E > V_1$ the electron can have any energy it likes, but if $E < V_1$ there are only three possible energy levels.

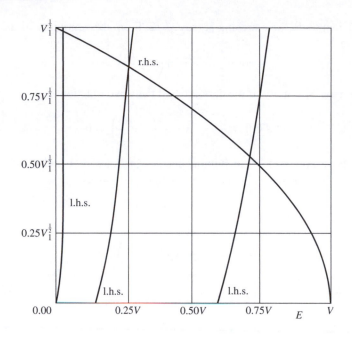

Fig. 3.5
A plot of the two sides of eqn (3.39)
against E for $L/2 = 5 \times 10^{-10}$ m and
$V_1 = 1.6 \times 10^{-18}$ J.

3.9 The potential well with a rigid wall

We call the potential wall *rigid* when the electron cannot escape from the well, not even quantum-mechanically. This happens when $V_1 = \infty$. We shall briefly investigate this case because we shall need the solution later. Equations (3.31) and (3.32) are still valid in the zero potential region, but now (since the electron cannot penetrate the potential barrier) the continuity condition is

$$\psi\left(\pm\frac{L}{2}\right) = 0. \tag{3.40}$$

The solutions are given again by eqns (3.31) and (3.32) for the symmetric and antisymmetric cases, respectively. The boundary conditions given by eqn (3.40) will be satisfied when

$$\frac{kL}{2} = (2r+1)\frac{\pi}{2} \quad \text{and} \quad \frac{kL}{2} = s\pi, \tag{3.41}$$

for the symmetric and antisymmetric wave functions, respectively, where r and s are integers. This is equivalent to saying that kL is an integral multiple of π. Hence, the expression for the energy is

$$E = \frac{\hbar^2 k^2}{2m} = \frac{h^2 n^2}{8mL^2}, \quad n = 1, 2, 3, \ldots \tag{3.42}$$

3.10 The uncertainty relationship

The uncertainty relationship may be looked upon in a number of ways. We have introduced it on the basis of the wave picture where electrons were identified with wave packets.

Can we now make a more precise statement about the uncertainty relationship? We could, if we introduced a few more concepts. If you are interested in the details, you can consult any textbook on quantum mechanics. Here, I shall merely outline one of the possible ways of deriving the uncertainty relationship.

First, the average value of a physically measurable quantity, called an observable, is defined in quantum mechanics as

$$\langle A \rangle = \frac{\int_V \psi^* A \psi \; \mathrm{d(volume)}}{\int_V |\psi|^2 \; \mathrm{d(volume)}}, \tag{3.43}$$

where the integration is over the volume of interest, wherever ψ is defined. A is in general an operator; it is $-i\hbar\Delta$ for the momentum and simply $\mathbf{r}$, the radius vector, for the position. Assuming that Schrödinger's equation is solved for a particular case, we know the wave function ψ, and hence, with the aid of eqn (3.43), we can work out the average and r.m.s. values of both the electron's position and of its momentum. Identifying Δz and Δp with

$$\{\langle (z - \langle z \rangle)^2 \rangle\}^{1/2} \quad \text{and} \quad \{\langle (p - \langle p \rangle)^2 \rangle\}^{1/2}$$

respectively, we get

$$\Delta z \, \Delta p \geqslant h. \tag{3.44}$$

There is actually another often-used form of the uncertainty relationship

$$\Delta E \, \Delta t \geqslant h, \tag{3.45}$$

which may be derived from relativistic quantum theory (where time is on equal footing with the spatial coordinates) and interpreted in the following way. Assume that an electron sits in a higher energy state of a system, for example in a potential well. It may fall to the lowest energy state by emitting a photon of energy $\hbar\omega$. So if we know the energy of the lowest state, we could work out the energy of that particular higher state by measuring the frequency of the emitted photon. But if the electron spends only a time Δt in the higher state, then the energy of the state can be determined with an accuracy not greater than $\Delta E = h/\Delta t$. This is borne out by the measurements.

The emitted radiation is not monochromatic; it covers a finite range of frequencies.

3.11 Philosophical implications

The advent of quantum mechanics brought problems to the physicist which previously belonged to the sacred domain of philosophy. The engineer can still afford to ignore the philosophical implications but by a narrow margin only. In another decade or two philosophical considerations might be relevant in the discussion of devices, so I will try to give you a foretaste of the things which might come.

To illustrate the sort of questions philosophers are *asking*, take the following one: We see a tree in the quad, so the tree must be there. We have the evidence of our senses (the eye in this particular case) that the tree exists. But what happens when we don't look at the tree, when no one looks at the tree at all; does the tree still exist? It is a good question.

Had philosophers been content *asking* this and similar questions, the history of philosophy would be an easier subject to study. Unfortunately, driven by usual human passions (curiosity, vivid imagination, vanity, ambition, the desire to be cleverer than the next man, craving for fame, etc.), philosophers did try to answer the questions. To the modern scientist, most of their answers and debates don't seem to be terribly edifying. I just want to mention Berkeley who maintained that matter would cease to exist if unobserved, but luckily there is God who perceives everything, so matter may exist after all. This view was attacked by Ronald Knox in the following limerick:

> There was a young man who said, 'God
> Must think it exceedingly odd
>   If he finds that this tree
>   Continues to be
> When there's no one about in the Quad.'

Berkeley replied in kind:

> Dear Sir:
>   Your astonishment's odd:
> *I* am always about in the Quad.
>   And that's why the tree
>   Will continue to be,
> Since observed by
>   Yours faithfully,
>   God.

You do, I hope, realize that only a minority of philosophical arguments were ever conducted in the form of limericks, and the above examples are not typical. I mention them partly for entertainment and partly to emphasize the problem of the tree in the quad a little more.

In the light of quantum mechanics we should look at the problem from a slightly different angle: The question is not so much what happens while the tree is unobserved, but rather what happens while the tree is unobservable. The tree can leave the quad because for a brief enough time it can have a high enough energy at its disposal, and no experimenter has any means of knowing about it. We are prevented by the uncertainty relationship ($\Delta E \Delta t \cong h$) from ever learning whether the tree did leave the quad or not.

You may say that this is against common sense. It is, but the essential point is whether or not it violates the Laws of Nature, as we know them today. Apparently it does not. You may maintain that for that critical Δt interval the tree stays where it always has stood. Yes, it is a possible view. You may also maintain that the tree went over for a friendly visit to the quad of another college and came back. Yes, that's another possible view.

Is there any advantage in imagining that the tree did make that brief excursion? I cannot see any, so I would opt for regarding the tree as being in the quad at all times.

But the problem remains, and becomes of more practical interest when considering particles of small size. A free electron travelling with a velocity 10^6 m s^{-1} has an energy of 2.84 eV. Assume that it wants to 'borrow' the same amount of energy again. It may borrow that much energy for an interval

$\Delta t = h/2.84\,\text{eV} \cong 1.5 \times 10^{-15}$ s. Now, you may ask, what can an electron do in a time interval as short as 1.5×10^{-15} s? Quite a lot; it can get comfortably from one atom to the next. And remember all this on borrowed energy. So if there was a barrier of (say) five electron volts, our electron could easily move *over* it, and having scaled the barrier it could return the borrowed energy, and no one would be able to find out how the electron made the journey. Thus, from a purely philosophical argument we could make up an alternative picture of tunnelling. We may say that tunnelling across potential barriers comes about because the electron can borrow energy for a limited time. Is this a correct description of what happens? I do not know, but it is a *possible* description. Is it useful? I suppose it is always useful to have various ways of describing the same event; that always improves understanding. But the crucial test is whether this way of thinking will help in arriving at new conclusions which can be experimentally tested. For an engineer the criterion is even clearer; if an engineer can think up a new device, using these sorts of arguments (e.g. violating energy conservation for a limited time) and the device works (or even better it can be sold for ready money) then the method is vindicated. The end justifies the means, as Machiavelli said.

Theoretical physicists, I believe, do use these methods. In a purely particle description of Nature, for example, the Coulomb force between two electrons is attributed to the following cause. One of the electrons borrows some energy to create a photon that goes dutifully to the other electron, where it is absorbed, returning thereby the energy borrowed. The farther the two electrons are from each other, the lesser the energy that can be borrowed, and therefore lower frequency photons are emitted and absorbed. Carrying on these arguments (if you are interested in more details ask a theoretical physicist) they do manage to get correctly the forces between electrons. So there are already some people who find it useful to play around with these concepts.

This is about as much as I want to say about the philosophical role of the electron. There are, incidentally, a number of other points where philosophy and quantum mechanics meet (e.g. the assertion of quantum mechanics that no event can be predicted with certainty, merely with a certain probability), but I think we may have already gone beyond what is absolutely necessary for the education of an engineer.

Exercises

3.1. An electron, confined by a rigid one-dimensional potential well (Fig. 3.4 with $V_1 = \infty$) may be anywhere within the interval $2a$. So the uncertainty in its position is $\Delta x = 2a$. There must be a corresponding uncertainty in the momentum of the electron and hence it must have a certain kinetic energy. Calculate this energy from the uncertainty relationship and compare it with the value obtained from eqn (3.42) for the ground state.

3.2. The wavefunction for a rigid potential well is given by eqns (3.31) and (3.32) and the permissible values of k by eqn (3.42). Calculate the average values of

$$z, \quad (z - \langle z \rangle)^2, \quad p, \quad (p - \langle p \rangle)^2.$$

[Hint: Use eqn (3.43). The momentum operator in this one-dimensional case is $-i\hbar\partial/\partial z$.]

3.3. The classical equivalent of the potential well is a particle bouncing between two perfectly elastic walls with uniform velocity.

(i) Calculate the classical average values of the quantities enumerated in the previous example.

(ii) Show that, for high-enough energies, the quantum mechanical solution tends to the classical solution.

3.4. In electromagnetic theory the conservation of charge is represented by the continuity equation

$$\nabla \cdot \mathbf{J} = -e\frac{\partial N}{\partial t}$$

where $\mathbf{J}$ = current density and N = density of electrons.

Assume that $\Psi(x, t)$ is a solution of Schrödinger's equation in a one-dimensional problem. Show that, by defining the current density as

$$J(x) = -\frac{i\hbar e}{2m}\left[\Psi^*\frac{\partial \Psi}{\partial x} - \Psi\frac{\partial \Psi^*}{\partial x}\right],$$

the continuity equation is satisfied.

3.5. The time-independent Schrödinger equation for the one-dimensional potential shown in Fig. 3.2 is solved in Section 3.6. Using the definition of current density given in the example above, derive expressions for the reflected and transmitted currents. Show that the transmitted current is finite when $E > V_2$ and zero when $E < V_2$. Comment on the analogy with example 1.9.

3.6. Solve the time-independent Schrödinger equation for the one-dimensional potential shown in Fig. 3.6;

$$V(z) = 0, \quad z < 0,$$

$$V(z) = V_2, \quad 0 < z < d,$$

$$V(z) = 0, \quad z > d.$$

Assume that an electron beam is incident from the $z < 0$ region with an energy E. Derive expressions for the reflected and transmitted current. Calculate the transmitted current when $V_2 = 2.5\,\text{eV}$, $E = 0.5\,\text{eV}$, $d = 2\,\text{Å}$ and $d = 20\,\text{Å}$.

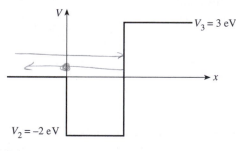

Fig. 3.6

A one-dimensional potential variation.

3.7. Exercise 3.6 may be solved approximately by assuming that the wave function in region 2 is of the form

$$\psi_2 = Ce^{-k_{2i}x}$$

(where $k_{2i} = \text{Im}\,k_2$) and then coming to the conclusion (note that the potential in region 3 is the same as that in region 1 and

therefore $k_3 = k_1$) that

$$\frac{J_3}{J_1} = e^{-2k_{2i}d}.$$

How accurate is this approximation?

If the wavefunction in region 2 is real, as it is assumed in this exercise, then the quantum mechanical current, as defined in exercise 3.4, yields zero.

How is it possible that the equation given above still gives good approximation for the ratio of the currents whereas the direct use of the proper formula leads to no current at all?

3.8. An electron beam is incident from the $x < 0$ region upon the one-dimensional potential shown in Fig. 3.6 with an energy $E = 1\,\text{eV}$.

Without detailed calculations what can you say about the reflection coefficient at $x = 0$?

3.9. Solve the time-independent Schrödinger equation for a two-dimensional rigid potential well having dimensions L_x and L_y in the x and y directions respectively.

Determine the energy of the 5 lowest lying states when $L_y = (3/2)L_x$.

3.10. The antisymmetric solution of the time-independent Schrödinger equation for the potential well of Fig. 3.4 is given by eqns (3.32) and (3.34).

Estimate the energy (a rough estimate will do) of the lowest antisymmetric state for $L = 10^{-9}$ m and $V_1 = 1.6 \times 10^{-18}$ J.

3.11. Solve the time-independent Schrödinger equation for the one-dimensional potential well shown in Fig. 3.7, restricting the analysis to even functions of x only. The solution may be expressed in determinant form. Without expanding the determinant explain what its roots represent.

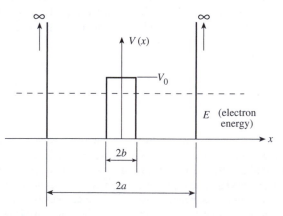

Fig. 3.7

A one-dimensional potential well.

3.12. Show that the differential equation for the electric field of a plane electromagnetic wave, assuming $\exp(-i\omega t)$

time dependence has the same form as the time-independent Schrödinger equation for constant potential. Show further that the expression for the Poynting vector of the electromagnetic wave is of the same functional form as that for the quantum mechanical current.

3.13. The potential energy of a classical harmonic oscillator is given as

$$V(x) = \tfrac{1}{2}m\omega_0^2 x^2.$$

We get the 'quantum' harmonic oscillator by putting the above potential function into Schrödinger's equation.

The solutions for the four lowest states are as follows:

$$\psi(\zeta) = H_n(\zeta) \exp(-\tfrac{1}{2}\zeta^2),$$

where

$$\zeta = \alpha x, \quad \alpha^2 = \frac{m\omega_0}{\hbar},$$

$$H_0 = 1, \quad H_1 = 2\zeta, \quad H_2 = 4\zeta^2 - 2, \quad \text{and}$$

$$H_3 = 8\zeta^3 - 12\zeta.$$

Find the corresponding energies. Compare them with the energies Planck postulated for photons.

4 The hydrogen atom and the periodic table

I see the atoms, free and fine,
That bubble like a sparkling wine;
I hear the songs electrons sing,
Jumping from ring to outer ring;
Lister The Physicist

4.1 The hydrogen atom

Up to now we have been concerned with rather artificial problems. We said: let us assume that the potential energy of our electron varies as a function of distance this way or that way without specifying the actual physical mechanism responsible for it. It was not a waste of time. It gave an opportunity of becoming acquainted with Schrödinger's equation, and of developing the first traces of a physical picture based, perhaps paradoxically, on the mathematical solution.

It would, however, be nice to try our newly acquired technique on a more physical situation where the potential is caused by the presence of some other physical 'object'. The simplest 'object' would be a proton, which, as we know, becomes a hydrogen atom if joined by an electron.

We are going to ask the following questions: (i) What is the probability that the electron is found at a distance r from the proton? (ii) What are the allowed energy levels?

The answers are again provided by Schrödinger's equation. All we have to do is to put in the potential energy due to the presence of a proton and solve the equation.

The wave function is a function of time, and one might want to solve problems, where the conditions are given at $t = 0$, and one is interested in the temporal variation of the system. These problems are complicated and of little general interest. What we should like to know is how a hydrogen atom behaves on the average, and for that purpose the solution given in eqn (3.7) is adequate. We may then forget about the temporal variation, because

$$|w(t)|^2 = 1, \tag{4.1}$$

and solve eqn (3.13), the time-independent Schrödinger equation.

The proton, we know, is much heavier than the electron; so let us regard it as infinitely heavy (that is immobile) and place it at the origin of our coordinate system.

The potential energy of the electron at a distance, r, from the proton is known from electrostatics:

$$V(r) = -\frac{e^2}{4\pi\epsilon_0 r}. \tag{4.2}$$

Thus, the differential equation to be solved is

$$\frac{\hbar^2}{2m}\nabla^2\psi + \left(\frac{e^2}{4\pi\epsilon_0 r} + E\right)\psi = 0. \tag{4.3}$$

It would be hard to imagine a physical configuration much simpler than that of a proton and an electron, and yet it is difficult to solve the corresponding differential equation. It is difficult because the $1/r$ term does not lend itself readily to analytical solutions. Thanks to the arduous efforts of nineteenth-century mathematicians, the general solution is known, but it would probably mean very little to you. Unless you have a certain familiarity with the properties of associated Legendre functions, it will not make you much happier if you learn that associated Legendre functions happen to be involved. So I will not quote the general solution because that would be meaningless, nor shall I derive it because that would be boring. But just to give an idea of the mathematical operations needed, I shall show the derivation for the simplest possible case, when the solution is spherically symmetric, and even then only for the lowest energy.

The potential energy of the electron depends only on the distance, r; it therefore seems advantageous to solve eqn (4.3) in the spherical coordinates r, θ, ϕ (Fig. 4.1). If we restrict our attention to the spherically symmetrical case, when ψ depends neither on ϕ nor on θ but only on r, then we can transform eqn (4.3) without too much trouble. We shall need the following partial derivatives

$$\frac{\partial\psi}{\partial x} = \frac{\partial r}{\partial x}\frac{\partial\psi}{\partial r} \tag{4.4}$$

and

$$\frac{\partial^2\psi}{\partial x^2} = \frac{\partial}{\partial x}\left(\frac{\partial\psi}{\partial r}\frac{\partial r}{\partial x}\right). \tag{4.5}$$

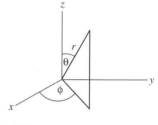

Fig. 4.1
Coordinate system used to transform eqn (4.3) to spherical coordinates.

When we differentiate eqn (4.5), we have to remember that $\partial\psi/\partial r$ is a function of r and $\partial r/\partial x$ is still a function of x; therefore,

$$\frac{\partial}{\partial x}\left(\frac{\partial\psi}{\partial r}\frac{\partial r}{\partial x}\right) = \frac{\partial r}{\partial x}\frac{\partial}{\partial r}\left(\frac{\partial\psi}{\partial r}\right)\frac{\partial r}{\partial x} + \frac{\partial\psi}{\partial r}\frac{\partial^2 r}{\partial x^2}$$

$$= \frac{\partial^2\psi}{\partial r^2}\left(\frac{\partial r}{\partial x}\right)^2 + \frac{\partial\psi}{\partial r}\frac{\partial^2 r}{\partial x^2}. \tag{4.6}$$

Obtaining the derivatives in respect with y and z in an analogous manner, we finally get

$$\nabla^2\psi = \frac{\partial^2\psi}{\partial x^2} + \frac{\partial^2\psi}{\partial y^2} + \frac{\partial^2\psi}{\partial z^2}$$

$$= \frac{\partial^2\psi}{\partial r^2}\left\{\left(\frac{\partial r}{\partial x}\right)^2 + \left(\frac{\partial r}{\partial y}\right)^2 + \left(\frac{\partial r}{\partial z}\right)^2\right\}$$

$$+ \frac{\partial\psi}{\partial r}\left(\frac{\partial^2 r}{\partial x^2} + \frac{\partial^2 r}{\partial y^2} + \frac{\partial^2 r}{\partial z^2}\right). \tag{4.7}$$

We now have to work out the partial derivatives of r. Since

$$r = (x^2 + y^2 + z^2)^{1/2}, \tag{4.8}$$

we get

$$\frac{\partial r}{\partial x} = \frac{x}{(x^2 + y^2 + z^2)^{1/2}} \tag{4.9}$$

and

$$\frac{\partial^2 r}{\partial x^2} = \frac{1}{(x^2 + y^2 + z^2)^{1/2}} - \frac{x^2}{(x^2 + y^2 + z^2)^{3/2}}, \tag{4.10}$$

and similar results for the derivatives by y and z. Substituting all of them in eqn (4.7), we get

$$\nabla^2 \psi = \frac{\partial^2 \psi}{\partial r^2} \left(\frac{x^2}{x^2 + y^2 + z^2} + \frac{y^2}{x^2 + y^2 + z^2} + \frac{z^2}{x^2 + y^2 + z^2} \right)$$

$$+ \frac{\partial \psi}{\partial r} \left\{ \frac{3}{(x^2 + y^2 + z^2)^{1/2}} - \frac{x^2}{(x^2 + y^2 + z^2)^{3/2}} \right.$$

$$\left. - \frac{y^2}{(x^2 + y^2 + z^2)^{3/2}} - \frac{z^2}{(x^2 + y^2 + z^2)^{3/2}} \right\} = \frac{\partial^2 \psi}{\partial r^2} + \frac{2}{r} \frac{\partial \psi}{\partial r}. \tag{4.11}$$

Thus, for the spherically symmetrical case of the hydrogen atom the Schrödinger equation takes the form,

$$\frac{\hbar^2}{2m} \left(\frac{\partial^2 \psi}{\partial r^2} + \frac{2}{r} \frac{\partial \psi}{\partial r} \right) + \left(E + \frac{e^2}{4\pi \epsilon_0 r} \right) \psi = 0. \tag{4.12}$$

It may be seen by inspection that a solution of this differential equation is

$$\psi = e^{-c_0 r}. \tag{4.13}$$

The constant, c_0, can be determined by substituting eqn (4.13) in eqn (4.12)

$$\frac{\hbar^2}{2m} \left\{ c_0^2 e^{-c_0 r} + \frac{2}{r} (-c_0 e^{-c_0 r}) \right\} + \left(E + \frac{e^2}{4\pi \epsilon_0 r} \right) e^{-c_0 r} = 0. \tag{4.14}$$

The above equation must be valid for every value of r, that is the coefficient of $\exp(-c_0 r)$ and that of $(1/r) \exp(-c_0 r)$ must vanish. This condition is satisfied if

$$E = -\frac{\hbar^2 c_0^2}{2m} \tag{4.15}$$

and

$$\frac{\hbar^2 c_0}{m} = \frac{e^2}{4\pi \epsilon_0}. \tag{4.16}$$

From eqn (4.16)

$$c_0 = \frac{e^2 m}{4\pi \hbar^2 \epsilon_0}, \tag{4.17}$$

which substituted in eqn (4.15) gives

$$E = -\frac{me^4}{8\epsilon_0^2 h^2}. \tag{4.18}$$

Thus, the wave function assumed in eqn (4.13) is a solution of the differential equation (4.12), provided that c_0 takes the value prescribed by eqn (4.17). Once we have obtained the value of c_0, the energy is determined as well. It can take only one single value satisfying eqn (4.18).

Let us work out now the energy obtained above numerically. Putting in the constants, we get

$$E = -\frac{(9.1 \times 10^{-31})(1.6 \times 10^{-19})^4}{8(8.85 \times 10^{-12})^2(6.63 \times 10^{-34})^2} \frac{\text{kg C}^4}{\text{F}^2 \text{ m}^{-2} \text{ J}^2 \text{ s}^2}$$

$$= -2.18 \times 10^{-18} \text{ J}. \tag{4.19}$$

Expressed in joules, this number is rather small. Since in most of the subsequent investigations this is the order of energy we shall be concerned with, and since there is a strong human temptation to use numbers only between 0.01 and 100, we abandon with regret the SI unit of energy and use instead the electron volt, which is the energy of an electron when accelerated to 1 volt. Since

$$1 \text{ eV} = 1.6 \times 10^{-19} \text{ J}, \tag{4.20}$$

the above energy in the new unit comes to the more reasonable-looking numerical value

$$E = -13.6 \text{ eV}. \tag{4.21}$$

What can we say about the electron's position? As we have discussed many times before, the probability that an electron can be found in an elementary volume (at the point r, θ, ϕ) is proportional to $|\psi|^2$—in the present case it is proportional to $\exp(-2c_0 r)$. The highest probability is at the origin, and it decreases exponentially to zero as r tends to infinity. We could, however, ask a slightly different question: what is the probability that the electron can be found in the spherical shell between r and $r + dr$? Then, the probability distribution is proportional to

$$r^2|\psi|^2 = r^2 e^{-2c_0 r}, \tag{4.22}$$

which has now a maximum, as can be seen in Fig. 4.2. The numerical value of the maximum can be determined by differentiating eqn (4.22)

$$\frac{\text{d}}{\text{d}r}(r^2 e^{-2c_0 r}) = 0 = e^{-2c_0 r}(2r - 2c_0 r^2), \tag{4.23}$$

whence

$$r = \frac{1}{c_0} = \frac{4\pi \hbar^2 \epsilon_0}{e^2 m} = 0.0528 \text{ nm}. \tag{4.24}$$

This radius was again known in pre-quantum-mechanical times and was called the radius of the first Bohr orbit, where electrons can orbit without radiating. Thus, in quantum theory, the Bohr orbit appears as the most probable position of the electron.

The negative sign of the energy means only that the energy of this state is below our chosen zero point. [By writing the Coulomb potential in the form of eqn (4.2) we tacitly took the potential energy as zero when the electron is at infinity.]

From experimental studies of the spectrum of hydrogen it was known well before the discovery of quantum mechanics that the lowest energy level of hydrogen must be -13.6 eV, and it was a great success of Schrödinger's theory that the same figure could be deduced from a respectable-looking differential equation.

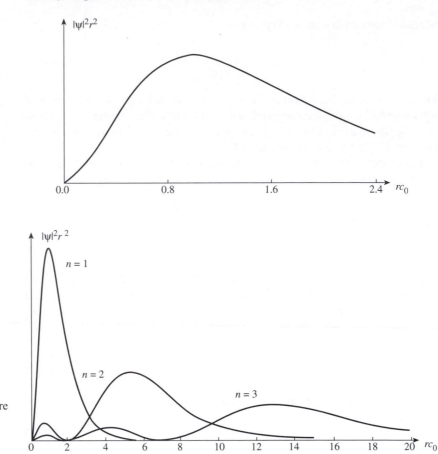

Fig. 4.2

Plot of eqn (4.22) showing the probability that an electron (occupying the lowest energy state) may be found in the spherical shell between r and $r + dr$.

Fig. 4.3

Plots of $\psi_n^2 r^2$ for the three lowest energy ($n = 1, 2, 3$) spherically symmetrical solutions. The curves are normalized so that the total probabilities (the area under curves) are equal.

We have squeezed out about as much information from our one meagre solution as is possible; we should look now at the other solutions which I shall give without any proof. Sticking for the moment to the spherically symmetrical case, the wavefunction is

$$\psi_n(r) = e^{-c_n r} L_n(r), \tag{4.25}$$

where L_n is a polynomial, and the corresponding energies are (in electron volts)

$$E_n = -13.6 \frac{1}{n^2}, \quad n = 1, 2, 3. \ldots \tag{4.26}$$

The solution we obtained before was for $n = 1$. It gives the lowest energy, and it is therefore usually referred to as the *ground state*.

If we have a large number of hydrogen atoms, most of them are in their ground state but some of them will be in excited states, which are given by $n > 1$. The probability distributions for the higher excited states have maxima farther from the origin as shown in Fig. 4.3 for $n = 1, 2, 3$. This is fair enough; for $n > 1$ the energy of the electron is nearer to zero, which is the energy of a free electron; so it is less strongly bound to the proton. If it is less strongly bound, it can wander farther away; so the radius corresponding to maximum probability increases.

4.2 Quantum numbers

So much about spherically symmetrical solutions. The general solution includes, of course, our previously obtained solutions, denoted by $R(r)$ from here on but shows variations in θ and ϕ as well. It can be written as

$$\psi_{n,l,m_l}(r,\theta,\phi) = R_{nl}(r)Y_l^{m_l}(\theta,\phi). \tag{4.27}$$

It may be shown (alas, not by simple means) from the original differential equation (eqn (4.12)) that the quantum numbers must satisfy the following relationships

$$n = 1, 2, 3 \ldots$$
$$l = 0, 1, 2, \ldots n - 1 \tag{4.28}$$
$$m_l = 0, \pm 1, \pm 2 \ldots \pm l.$$

We have met n before; l and m_l represent two more discrete sets of constants which ensure that the solutions have physical meaning. These discrete sets of constants always appear in the solutions of partial differential equations; you may remember them from the problems of the vibrating string or of the vibrating membrane. They are generally called *eigenvalues*; in quantum mechanics they are referred to as quantum numbers.

For $n = 1$ there is only one possibility: $l = 0$ and $m_l = 0$, and the corresponding wave function is the one we guessed in eqn (4.13). For the spherically symmetrical case the wave functions have already been plotted for $n=1, 2, 3$; now let us see a wave function which is dependent on direction. Choosing $n = 2$, $l = 1$, $m_l = 0$, the corresponding wave function is

$$\psi_{210} = R_{21}(r)Y_1^0(\theta,\phi)$$
$$= re^{-c_0 r/2}\cos\theta. \tag{4.29}$$

This equation tells us how the probability of finding the electron varies as a function of r and θ. Thus, the equal-probability surfaces may be determined. The spherical symmetry has gone, but there is still cylindrical symmetry (no dependence on ϕ). It is therefore sufficient to plot the curves in, say, the xz-plane. This is done in Fig. 4.4, where the unit of distance is taken as one Bohr radius, and the maximum probability (at $x = 0$ and $z = \pm 2/c_0$) is normalized to unity. It can be clearly seen that the $\theta = 0$ and $\theta = \pi$ directions are preferred (which of course follows from eqn (4.29) directly); there is a higher probability of finding the electron in those directions.

With $n = 2$ and $l = 1$ there are two more states, but they give nothing new. The preferential directions in those cases are in the direction of the $\pm x$ and $\pm y$-axes, respectively.

For higher values of n and l the equal-probability curves look more and more complicated. Since at this level of treatment they will not add much to our picture of the hydrogen atom, we can safely omit them.

I have to add a few words about notations. However convenient the parameters n and l might appear, they are never (or at least very rarely) used in that form. The usual notation is a number, equal to the value of n, followed by a

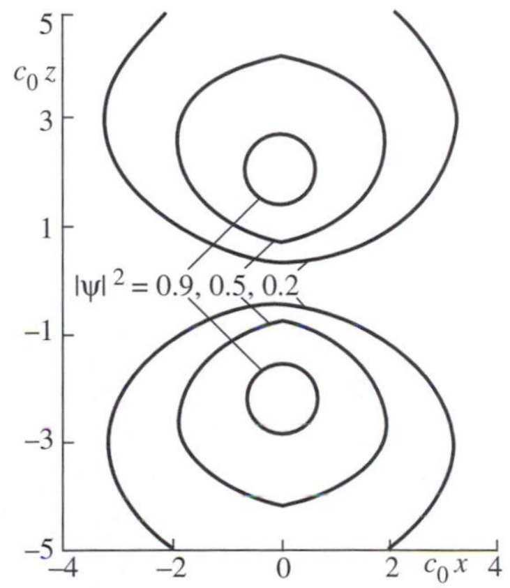

Fig. 4.4
Plots of constant $|\psi_{210}|^2$ in the xz-plane.

letter, which is related to l by the following rule

$$l = 0 \quad 1 \quad 2 \quad 3 \quad 4 \quad 5 \quad 6 \quad 7$$
$$ \text{s} \quad \text{p} \quad \text{d} \quad \text{f} \quad \text{g} \quad \text{h} \quad \text{i} \quad \text{k}$$

Thus, if you wish to refer to the states with $n = 3$ and $l = 1$, you call them the 3p states or the 3p configuration. The reason for this rather illogical notation is of course historical. In the old days when only spectroscopic information was available about these energy levels they were called s for sharp, p for principal, d for diffuse, and f for fundamental. When more energy levels were found, it was decided to introduce some semblance of order and denote them by subsequent letters of the alphabet. That is how the next levels came to bear the letters g, h, i, k, etc.

4.3   Electron spin and Pauli's exclusion principle

The quantum numbers n, l, and m_l have been obtained from the solution of Schrödinger's equation. Unfortunately, as I mentioned before, they do not represent the whole truth; there is one more quantum number to be taken into account. It is called the spin quantum number, denoted by s, and it takes the values $\pm\frac{1}{2}$.

Historically, spin had to be introduced to account for certain spectroscopic measurements, where two closely spaced energy levels were observed when only one was expected. These were explained in 1925 by Uhlenbeck and Goudsmit by assuming that the electron can spin about its own axis. This classical description is very much out of fashion nowadays, but the name spin stuck and has been universally used ever since. Today the spin is looked upon just as another quantum number obtainable from a more complicated theory which includes relativistic effects as well.

So we have now four quantum numbers: n, l, m_l, and s. Any permissible combination of these quantum numbers [eqn (4.28) shows what is permissible] gives a state; the wavefunction is determined, the electron's energy is determined, everything is determined. But what happens when we have more than one electron? How many of them can occupy the same state? One, said Pauli. *There can be no more than one electron in any given state.* This is Pauli's exclusion principle. We shall use it as a separate assumption, though it can be derived from a relativistic quantum theory.

Although both the spin and the exclusion principle are products of rather involved theories, both of them can be explained in simple terms. So even if you do not learn where they come from, you can easily remember them.

4.4   The periodic table

We have so far tackled the simplest configuration when there are only two particles: one electron and one proton. How should we attempt the solution for a more complicated case; for helium, for example, which has two protons and two electrons? (Helium has two neutrons as well, but since they are neutral they have no effect on the electrons; thus when discussing the energy levels of electrons neutrons can be disregarded.)

The answer is still contained in Schrödinger's equation, but the form of the equation is more complicated. The differential operator ∇ operated on the coordinate of the electron. If we have two electrons, we need two differential operators. Thus,

$$\nabla^2 \psi$$

is replaced by

$$\nabla_1^2 \psi + \nabla_2^2 \psi,$$

where the indices 1 and 2 refer to electrons 1 and 2, respectively. We may take the protons* as if they are infinitely heavy again and put them at the origin of the coordinate system. Thus, the potential energy of electron 1 at a distance r_1 from the protons is $-2e^2/4\pi\epsilon_0 r_1$, and similarly for electron 2. There is, however, one more term in the expression for potential energy: the potential energy due to the two electrons. If the distance between them is r_{12}, then this potential energy is $e^2/4\pi\epsilon_0 r_{12}$. It is of positive sign because the two electrons repel each other. We can now write down Schrödinger's equation for two protons and two electrons:

$$-\frac{\hbar^2}{2m}(\nabla_1^2 \psi + \nabla_2^2 \psi) + \frac{1}{4\pi\epsilon_0}\left(-\frac{2e^2}{r_1} - \frac{2e^2}{r_2} + \frac{e^2}{r_{12}}\right)\psi = E\psi. \tag{4.30}$$

Can this differential equation be solved? The answer, unfortunately, is no. No analytical solutions have been found. So we are up against mathematical difficulties even with helium. Imagine then the trouble we should have with tin. A tin atom has 50 protons and 50 electrons; the corresponding differential equation has 150 independent variables and 1275 terms in the expression for potential energy. This is annoying. We have the correct equation, but we cannot solve it because our mathematical apparatus is inadequate. What shall we do? Well, if we can't get exact solutions, we can try to find approximate solutions. This is fortunately possible. Several techniques have been developed for solving the problem of individual atoms by successive approximations. The mathematical techniques are not particularly interesting, and so I shall mention only the simplest physical model that leads to the simplest mathematical solution.

In this model we assume that there are Z positively charged protons in the nucleus, and the Z electrons floating around the nucleus are unaware of each other. If the electrons are independent of each other, then the solution for each of them is the same as for the hydrogen atom provided that the charge at the centre is taken as Ze. This means putting Ze^2 instead of e^2 into eqn (4.2) and $Z^2 e^4$ instead of e^4 into eqn (4.18). Thus, we can rewrite all the formulae used for the hydrogen atom, and in particular the formula for energy, which now stands as

$$E_n = -13.6\frac{Z^2}{n^2}. \tag{4.31}$$

That is, the energy of the electrons decreases with increasing Z. In other words, the energy is below zero by a large amount; that is, more energy is needed to liberate an electron. This is fairly easy to understand; a large positive charge in the nucleus will bind the electron more strongly.

The model of entirely independent electrons is rather crude, but it can go a long way towards a qualitative explanation of the chemical properties of the

* It is a separate story how positively charged protons and neutrons can peacefully coexist in the nucleus, and the answer is still only partly known.

Note that the wave function ψ now depends on six variables, namely on the three spatial coordinates of each electron.

There are two important points to realize:

1. We have our set of quantum numbers n, l, m_l, and s, each one specifying a state with a definite energy. The energy depends on n only, but several states exist for every value of n.

2. Pauli's exclusion principle must be obeyed. Each state can be occupied by one electron only.

elements. We shall see how Mendeleev's periodic table can be built up with the aid of the quantum-mechanical solution of the hydrogen atom.

We shall start by taking the lowest energy level, count the number of states, fill them up one by one with electrons, and then proceed to the next energy level; and so on.

According to eqn (4.31) the lowest energy level is obtained with $n = 1$. Then $l = 0$, $m_l = 0$, and there are two possible states of spin $s = \pm\frac{1}{2}$. Thus, the lowest energy level may be occupied by two electrons. Putting in one electron we get hydrogen, putting in two electrons we get helium, putting in three electrons ... No, we cannot do that; if we want an element with three electrons, then the third electron must go into a higher energy level.

With helium the $n = 1$ 'shell' is closed, and this fact determines the chemical properties of helium. If the helium atom happens to meet other electrons (in events officially termed *collisions*), it can offer only high energy states. Since all electrons look for low energy states, they generally decline the invitation. They manifest no desire to become attached to a helium atom.

If the probability of attracting an electron is small, can the helium atom give away one of its electrons? This is not very likely. It can offer to its own electrons comfortable low-energy states. The electrons are quite satisfied and stay. Thus, the helium atom neither takes up nor gives away electrons. Helium is chemically inert.

We now have to start the next energy shell with $n = 2$. The first element there is lithium, containing two electrons with $n = 1$, $l = 0$ and one electron with $n = 2$, $l = 0$. Adopting the usual notations, we may say that lithium has two 1s electrons and one 2s electron. Since the 2s electron has higher energy, it can easily be tempted away. Lithium is chemically active.

The next element is beryllium with two 1s and two 2s electrons; then comes boron with two 1s, two 2s, and one 2p electrons, which, incidentally, can be denoted in an even more condensed manner as $1s^2, 2s^2, 2p^1$. Employing this new notation, the six electrons of carbon appear as $1s^2, 2s^2, 2p^2$, the seven electrons of nitrogen as $1s^2, 2s^2, 2p^3$, the eight electrons of oxygen as $1s^2, 2s^2, 2p^4$, and the nine electrons of fluorine as $1s^2, 2s^2, 2p^5$.

Let us pause here for a moment. Recall that a 2p state means $n = 2$ and $l = 1$, which according to eqn (4.28) can have three states ($m_l = 0$ and $m_l = \pm 1$) or, taking account of spin as well, six states altogether. In the case of fluorine five of them are occupied, leaving one empty low-energy state to be offered to outside electrons. The offer is often taken up, and so fluorine is chemically active.

Lithium and fluorine are at the opposite ends, the former having one *extra* electron, the latter *needing* one more electron to complete the shell. So it seems quite reasonable that when they are together, the extra electron of lithium will occupy the empty state of the fluorine atom, making up the compound LiF. A chemical bond is born, a chemist would say.

We shall discuss bonds later in more detail. Let us return meanwhile to the rather protracted list of the elements. After fluorine comes neon. The $n = 2$ shell is completed: no propensity to take up or give away electrons. Neon is chemically inert like helium.

The $n = 3$ shell starts with sodium, which has just one 3s electron and should therefore behave chemically like lithium. A second electron fills the 3s shell in magnesium. Then come aluminium, silicon, phosphorus, sulphur, and chlorine with one, two, three, four and five 3p electrons, respectively. Chlorine is again short of one electron to fill the 3p shell, and so behaves like fluorine. The 3p shell is completed in argon, which is again inert.

So far everything has gone regularly, and by the rules of the game the next electron should go into the 3d shell. It does not. Why? Well, why should it? The electrons in potassium are under no obligation to follow the energy hierarchy of the hydrogen atom like sheep. They arrange themselves in such a way as to have the lowest energy. If there were *no* interaction between the electrons, the energy levels of the element would differ only by the factor Z^2, conforming otherwise to that of the hydrogen atom. If the interaction between the electrons mattered a lot, we should completely abandon the classification based on the energy levels of the hydrogen atom. As it happens, the electron interactions are responsible for small quantitative† changes that cause qualitative change in potassium—and in the next few elements, called the *transition elements*. First the 4s shell is filled, and only after that are the 3d states occupied. The balance between the two shells remains, however, delicate. After vanadium (with three 3d and two 4s electrons) one electron is withdrawn from the 4s shell; hence chromium has five 3d electrons but only one 4s electron. The same thing happens later with copper, but apart from that everything goes smoothly up to krypton, where the 4p shell is finally completed.

The regularity is somewhat marred after krypton. There are numerous deviations from the hydrogen-like structure but nothing very dramatic. It might be worthwhile mentioning the rare earth elements in which the 4f shell is being filled while eleven electrons occupy levels in the outer shells. Since chemical

† In the hydrogen-type solutions the energy depends only on n, whereas taking account of electron interactions the energy increases with increasing values of l. It just happens that in potassium the energy of the 3d level ($n = 3, l = 2$) is higher than that of the 4s level ($n = 4, l = 0$).

1	2
H	He

IA	IIA	IIIA	IVA	VA	VIA	VIIA	VIII			IB	IIB	IIIB	IVB	VB	VIB	VIIB	
3 Li	4 Be											5 B	6 C	7 N	8 O	9 F	10 Ne
11 Na	12 Mg											13 Al	14 Si	15 P	16 S	17 Cl	18 Ar
19 K	20 Ca	21 Sc	22 Ti	23 V	24 Cr	25 Mn	26 Fe	27 Co	28 Ni	29 Cu	30 Zn	31 Ga	32 Ge	33 As	34 Se	35 Br	36 Kr
37 Rb	38 Sr	39 Y	40 Zr	41 Nb	42 Mo	43 Tc	44 Ru	45 Rh	46 Pd	47 Ag	48 Cd	49 In	50 Sn	51 Sb	52 Te	53 I	54 Xe
55 Cs	56 Ba	57 La	72 Hf	73 Ta	74 W	75 Re	76 Os	77 Ir	78 Pt	79 Au	80 Hg	81 Tl	82 Pb	83 Bi	84 Po	85 At	86 Rn
87 Fr	88 Ra	89 Ti															

58 Ce	59 Pr	60 Nd	61 Pm	62 Sm	63 Eu	64 Gd	65 Tb	66 Dy	67 Ho	68 Er	69 Tm	70 Yb	71 Lu	IIIA
90 Th	91 Pa	92 U	93 Np	94 Pu	95 Am	96 Cm	97 Bk	98 Cf	99 Es	100 Fm	101 Md	102 No	103	IIIA

Fig. 4.5
The periodic table of the elements.

properties are mainly determined by the outer shells, all these elements are hardly distinguishable chemically.

A list of all these elements with their electron configurations is given in Table 4.1. The periodic table (in one of its more modern forms) is given in Fig. 4.5. You may now look at the periodic table with more knowing eyes. If you were asked, for example, why the alkali elements lithium, sodium, potassium, rubidium, caesium, and francium have a valency of one, you could answer in the following way.

The properties of electrons are determined by Schrödinger's equation. The solution of this equation for one electron and one proton tells us that the electron may be in one of a set of discrete states, each having a definite energy level. When there are many electrons and many protons, the order in which these states follow each other remains roughly unchanged. We may then derive the various elements by filling up the available states one by one with electrons. We cannot put more than one electron in a state because the exclusion principle forbids this.

The energy of the states varies in steps. Within a 'shell' there is a slow variation in energy but a larger energy difference between shells.

Whenever a new shell is initiated, there is one electron with considerably higher energy than the rest. Since all electrons strive for lower energy, this electron can easily be lost to another element.

All the alkali elements start new shells. Therefore each of them may lose an electron; each of them may contribute one unit to a new chemical configuration; and each of them has a valency of one.

We may pause here for a moment. You have had the first taste of the power of Schrödinger's equation. You can see now that the solution of all the basic problems that have haunted the chemists for centuries is provided by a modest-looking differential equation. The chaos prevailing before has been cleared, and a sturdy monument has been erected in its stead. If you look at it carefully, you will find that it possesses all the requisites of artistic creation. It is like a Greek temple. You can see in the background the stern regularity of the columns, but the statues placed between them are all different.

Exercises

4.1. Calculate the wavelength of electromagnetic waves needed to excite a hydrogen atom from the 1s into the 2s state.

4.2. Electromagnetic radiation of wavelength 20 nm is incident on atomic hydrogen. Assuming that an electron in its ground state is ionized, what is the maximum velocity at which it may be emitted?

4.3. An excited argon ion in a gas discharge radiates a spectral line of wavelength 450 nm. The transition from the excited to the ground state that produces this radiation takes an average time of 10^{-8} s. What is the inherent width of the spectral line?

4.4. Determine the most probable orbiting radius of the electron in a hydrogen atom from the following very crude considerations. The electron tries to move as near as possible to the nucleus in order to lower its potential energy. But

if the electron is somewhere within the region 0 to r_m (i.e. we know its position with an uncertainty, r_m), the uncertainty in its momentum must be $\Delta p \cong \hbar/r_m$. So the kinetic energy of the electron is roughly $\hbar^2/2mr_m^2$.

Determine r_m from the condition of minimum energy. Compare the radius obtained with that of the first Bohr orbit.

4.5. Determine the average radius of an electron in the ground state of the hydrogen atom.

4.6. The spherically symmetric solution for the 2s electron ($n = 2$) of the hydrogen atom may be written in the form

$$\psi(r) = A(1 + c_1 r)\exp(-rc_0/2) \tag{4.32}$$

where c_0 is the reciprocal of the Bohr radius [see eqn (4.24)].

Table 4.1 *The electronic configurations of the elements*

Atomic number	Element symbol	Number of electrons									
		1s	2s	2p	3s	3p	3d	4s	4p	4d	4f
1	H	1									
2	He	2									
3	Li	2	1								
4	Be	2	2								
5	B	2	2	1							
6	C	2	2	2							
7	N	2	2	3							
8	O	2	2	4							
9	F	2	2	5							
10	Ne	2	2	6							
11	Na	2	2	6	1						
12	Mg	2	2	6	2						
13	Al	2	2	6	2	1					
14	Si	2	2	6	2	2					
15	P	2	2	6	2	3					
16	S	2	2	6	2	4					
17	Cl	2	2	6	2	5					
18	A	2	2	6	2	6					
19	K	2	2	6	2	6	—	1			
20	Ca	2	2	6	2	6	—	2			
21	Sc	2	2	6	2	6	1	2			
22	Ti	2	2	6	2	6	2	2			
23	V	2	2	6	2	6	3	2			
24	Cr	2	2	6	2	6	5	1			
25	Mn	2	2	6	2	6	5	2			
26	Fe	2	2	6	2	6	6	2			
27	Co	2	2	6	2	6	7	2			
28	Ni	2	2	6	2	6	8	2			
29	Cu	2	2	6	2	6	10	1			
30	Zn	2	2	6	2	6	10	2			
31	Ga	2	2	6	2	6	10	2	1		
32	Ge	2	2	6	2	6	10	2	2		
33	As	2	2	6	2	6	10	2	3		
34	Se	2	2	6	2	6	10	2	4		
35	Br	2	2	6	2	6	10	2	5		
36	Kr	2	2	6	2	6	10	2	6	—	—

Atomic number	Element symbol	Number of electrons									
		4s	4p	4d	4f	5s	5p	5d	5f	5g	6s
37	Rb	2	6	—	—	1	—	—	—	—	—
38	Sr	2	6	—	—	2	—	—	—	—	—
39	Y	2	6	1	—	2	—	—	—	—	—
40	Zr	2	6	2	—	2	—	—	—	—	—
41	Nb	2	6	4	—	1	—	—	—	—	—
42	Mo	2	6	5	—	1	—	—	—	—	—
43	Tc	2	6	6	—	1	—	—	—	—	—
44	Ru	2	6	7	—	1	—	—	—	—	—
45	Rh	2	6	8	—	1	—	—	—	—	—
46	Pd	2	6	10	—	—	—	—	—	—	—
47	Ag	2	6	10	—	1	—	—	—	—	—
48	Cd	2	6	10	—	2	—	—	—	—	—
49	In	2	6	10	—	2	1	—	—	—	—
50	Sn	2	6	10	—	2	2	—	—	—	—
51	Sb	2	6	10	—	2	3	—	—	—	—
52	Te	2	6	10	—	2	4	—	—	—	—
53	I	2	6	10	—	2	5	—	—	—	—
54	Xe	2	6	10	—	2	6	—	—	—	—
55	Cs	2	6	10	—	2	6	—	—	—	1
56	Ba	2	6	10	—	2	6	—	—	—	2
57	La	2	6	10	—	2	6	1	—	—	2
58	Ce	2	6	10	2	2	6	—	—	—	2
59	Pr	2	6	10	3	2	6	—	—	—	2
60	Nd	2	6	10	4	2	6	—	—	—	2
61	Pm	2	6	10	5	2	6	—	—	—	2
62	Sm	2	6	10	6	2	6	—	—	—	2
63	Eu	2	6	10	7	2	6	—	—	—	2
64	Gd	2	6	10	7	2	6	1	—	—	2
65	Tb	2	6	10	8	2	6	1	—	—	2
66	Dy	2	6	10	9	2	6	1	—	—	2
67	Ho	2	6	10	10	2	6	1	—	—	2
68	Er	2	6	10	11	2	6	1	—	—	2
69	Tm	2	6	10	12	2	6	1	—	—	2
70	Yb	2	6	10	13	2	6	1	—	—	2
71	Lu	2	6	10	14	2	6	1	—	—	2
72	Hf	2	6	10	14	2	6	2	—	—	2

Atomic number	Element symbol	Number of electrons									
		5p	5d	5f	5g	6s	6p	6d	6f	7s	7p
73	Ta	6	3	—	—	2	—	—	—	—	—
74	W	6	4	—	—	2	—	—	—	—	—
75	Re	6	5	—	—	2	—	—	—	—	—
76	Os	6	6	—	—	2	—	—	—	—	—
77	Ir	6	9	—	—	0	—	—	—	—	—
78	Pt	6	9	—	—	1	—	—	—	—	—
79	Au	6	10	—	—	1	—	—	—	—	—
80	Hg	6	10	—	—	2	—	—	—	—	—
81	Tl	6	10	—	—	2	1	—	—	—	—
82	Pb	6	10	—	—	2	2	—	—	—	—
83	Bi	6	10	—	—	2	3	—	—	—	—
84	Po	6	10	—	—	2	4	—	—	—	—
85	At	6	10	—	—	2	5	—	—	—	—
86	Rn	6	10	—	—	2	6	—	—	—	—
87	Fr	6	10	—	—	2	6	—	—	1	—
88	Ra	6	10	—	—	2	6	—	—	2	—
89	Ac	6	10	—	—	2	6	1	—	2	—
90	Th	6	10	—	—	2	6	2	—	2	—
91	Pa	6	10	—	—	2	6	3	—	2	—
92	U	6	10	—	—	2	6	4	—	2	—
93	Np	6	10	5	—	2	6	—	—	2	—
94	Pu	6	10	5	—	2	6	1	—	2	—
95	Am	6	10	6	—	2	6	1	—	2	—
96	Cm	6	10	7	—	2	6	1	—	2	—
97	Bk	6	10	8	—	2	6	1	—	2	—
98	Cf	6	10	9	—	2	6	1	—	2	—
99	—	6	10	10	—	2	6	1	—	2	—
100	—	6	10	11	—	2	6	2	—	2	—

(i) The function given by the above equation may be referred to as the ψ_{200} wave function. Why?

(ii) Determine c_1 from the condition that eqn (4.32) satisfies Schrödinger's equation.

(iii) Find the corresponding energy.

(iv) Determine A from the condition that the total probability of finding the electron somewhere must be unity.

(v) Find the most probable orbit of the electron for the wave function of eqn (4.32) and compare your result with that given by the curve in Fig. 4.3 for $n = 2$.

4.7. Show that ψ_{210} as given by eqn (4.29) satisfies Schrödinger's equation. Find the corresponding energy.

Compare this energy with that obtained for the wave function in the previous example. Can you draw any conclusions from these results concerning the whole $n = 2$ shell?

4.8. Solve Schrödinger's equation for the ground state of helium neglecting the potential term between the two electrons. What is the energy of the ground state calculated this way? The measured value is -24.6 eV. What do you think the difference is caused by? Give an explanation in physical terms.

4.9. Write down the time-independent Schrödinger equation for lithium.

Bonds 5

Striking the electric chain wherewith we are darkly bound
Byron *Childe Harold's Pilgrimage*

Those whom God has joined together,
Let no man put asunder.
Marriage service

5.1 Introduction

As we have seen, an electron and a proton may strike up a companionship, the result being a hydrogen atom. We have found that the energy of the electron is a negative number, that is, the electron in the vicinity of the proton has a lower energy than it would have if it were an infinite distance away, which corresponds to zero energy. The minimum comes about as some sort of compromise between the kinetic and potential energy, but the important thing is that a minimum exists. The electron comes closer to find lower energy.

Can we say the same thing about two hydrogen atoms? Would they too come close to each other in order to reduce the total energy? Yes, they come close; they combine and make up a hydrogen molecule. This combination between atoms is called a *chemical bond,* and the discipline that is concerned with these combinations is chemistry.

You may justifiably ask why we talk of chemistry in a course on electrical properties of materials. Well, in a sense, chemistry is just a branch of the electrical properties of materials. The only way to explain chemical bonds is to use electrical and some specific quantum-mechanical properties. And not only is chemistry relegated to this position: metallurgy, is too. When a large number of atoms conglomerate and make up a solid, the reason is again to be sought in the behaviour of electrons. Thus, all the mechanical properties of solids, including their very solidity, spring from the nature of their electrical components.

This is true in principle but not quite true in practice. We know the fundamental laws, and so we could work out everything (the outcome of all chemical reactions, the strength of all materials) if only the mathematical problems could be overcome. Bigger computers and improved techniques of numerical analysis might one day make such calculations feasible, but for the moment it is not practicable to go back to first principles. So we are not going to solve the problems of chemistry and metallurgy here. Nevertheless, we need to understand the nature of the chemical bond to proceed further. The bond between hydrogen atoms leads to the bond between the atoms of heavier elements. We shall encounter among others germanium and silicon, their band structure, how they can be doped and how they can serve as the basis upon which most of our electronic devices are built. Is it worth starting with the fundamentals? It is far from obvious. Unless you find these glimpses behind the scenes fascinating in themselves you might come to the conclusion that the labour to be expended is just too much. But try not to think in too narrow terms. Learning something about the foundations may help you later when confronting wider problems.

5.2   General mechanical properties of bonds

Before classifying and discussing particular bond types, we can make a few common-sense deductions about what sort of forces must be involved in a bond. First, there must be an attractive force. An obvious candidate for this role is the Coulomb attraction between unlike charges, which we have all met many times, giving a force proportional to r^{-2}, where r is the separation.

We know that sodium easily mislays its outer-ring (often called valence) electron, becoming Na^+, and that chlorine is an avid collector of a spare electron. So, just as we mentioned earlier with lithium and fluorine, the excess electron of sodium will fill up the energy shell of chlorine, creating a positively charged sodium ion with a negatively charged chlorine ion. These two ions will attract each other; that is obvious. What is less obvious, however, is that NaCl crystallizes into a very definite structure with the Na and Cl ions 0.28 nm apart. What stops them getting closer? Surely the Coulomb forces are great at 0.28 nm. Yes, they are great, but they are not the only forces acting. When the ions are very close to each other and start becoming distorted, new forces arise that tend to re-establish the original undistorted separate state of the ions. These repulsive forces are of short range. They come into play only when the interatomic distance becomes comparable with the atomic radius. Thus, we have two opposing forces that balance each other at the equilibrium separation, r_0.

It is possible to put this argument into graphical and mathematical form. If we plot the total energy of two atoms against their separation, the graph must look something like Fig. 5.1. The 'common-sense' points about this diagram are as follows:

1. The energy tends to zero at large distances—in other words, we define zero energy as the energy in the absence of interaction.
2. At large distances the energy is negative and increases with increasing distance. This means that from infinity down to the point r_0 the atoms attract each other.
3. At very small distances the energy is rising rapidly, that is, the atoms repel each other up to the point r_0.
4. The curve has a minimum value at r_0 corresponding to an equilibrium position. Here the attractive and repulsive forces just balance each other.

In the above discussion we have regarded r as the distance between two atoms, and r_0 as the equilibrium distance. The same argument applies, however, if we think of a solid that crystallizes in a cubic structure. We may then interpret r as the interatomic distance in the solid.

Let us now see what happens when we compress the crystal, that is, when we change the interatomic distance by brute force. According to our model, illustrated in Fig. 5.1, the energy will increase, but when the external influence is removed, the crystal will return to its equilibrium position. In some other

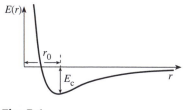

Fig. 5.1

The essential general appearance of the energy versus separation curve if two atoms are to bond together. The equilibrium separation is r_0 and the bond energy is E_c.

branches of engineering this phenomenon is known as elasticity. So if we manage to obtain the $E(r)$ curve, we can calculate all the elastic properties of the solid. Let us work out as an example the bulk elastic constant. We shall take a cubical piece of material of side a (Fig. 5.2) and calculate the energy changes under isotropic compression.

If we regard $E(r)$ as the energy per atom, the total energy of the material is $N_a a^3 E(r_0)$ in equilibrium, where N_a is the number of atoms per unit volume. If the cube is uniformly compressed, the interatomic distance will decrease by Δr, and the total energy will increase to $N_a a^3 E(r_0 - \Delta r)$. Expanding $E(r_0 - \Delta r)$ into a Taylor series and noting that $(\partial E/\partial r)_{r=r_0} = 0$, we get

$$E(r_0 - \Delta r) = E(r_0) + \frac{1}{2}\left(\frac{\partial^2 E}{\partial r^2}\right)_{r=r_0}(\Delta r)^2 + \cdots \tag{5.1}$$

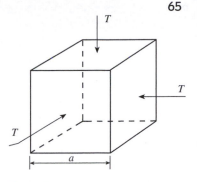

Fig. 5.2
A cube of material, side a, is isotropically compressed.

Hence, the net increase in energy is equal to

$$\frac{1}{2}N_a a^3\left(\frac{\partial^2 E}{\partial r^2}\right)_{r=r_0}(\Delta r)^2. \tag{5.2}$$

This increase in energy is due to the work done by moving the six faces of the cube. The total change in linear dimensions is $(a/r_0)\Delta r$; thus we may say that each face has moved by a distance $(a/2r_0)\Delta r$. Hence, while the stress is increasing from 0 to T, the total work done on the piece of material is

$$6 \times \frac{1}{2}Ta^2\frac{a\Delta r}{2r_0}. \tag{5.3}$$

From the equality of eqns (5.2) and (5.3) we get

$$\frac{3}{2}T\frac{a^3}{r_0}\Delta r = \frac{1}{2}\frac{a^3}{r_0^3}\left(\frac{\partial^2 E}{\partial r^2}\right)_{r=r_0}(\Delta r)^2, \tag{5.4}$$

whence

$$T = \frac{1}{3r_0}\left(\frac{\partial^2 E}{\partial r^2}\right)_{r=r_0}\frac{\Delta r}{r_0}. \tag{5.5}$$

Defining the bulk elastic modulus by the relationship of stress to the volume change caused, that is

$$T = c\frac{\Delta a^3}{a^3} \cong c\frac{3\Delta a}{a} = c\frac{3\Delta r}{r_0}, \tag{5.6}$$

we can obtain c with the aid of eqn (5.5) in the form

$$c = \frac{1}{9r_0}\left(\frac{\partial^2 E}{\partial r^2}\right)_{r=r_0}. \tag{5.7}$$

So we have managed to obtain both Hooke's law and an expression for the bulk elastic modulus by considering the interaction of atoms. If terms higher than second order are not negligible, we have a material that does *not* obey Hooke's law.

It is worth noting that most materials do obey this Hooke's law for small deformations, but not for large ones. This is in line with the assumptions we have made in the derivation.

For the purpose of making some rough calculations, the characteristic curve of Fig. 5.1 may be approximated by the following simple mathematical expression,

$$E(r) = \frac{A}{r^n} - \frac{B}{r^m},$$
(5.8)

where the first term on the right-hand side represents repulsion and the second term attraction. By differentiating eqn (5.8) we can get E_c the minimum of the $E(r)$ curve at the equilibrium distance $r = r_0$, in the form

$$E_c = \frac{B}{r_0^m}\left(\frac{m}{n} - 1\right).$$
(5.9)

The repulsive force has a higher index than the attractive one.

For a stable bond, $E_c < 0$, which can be satisfied only if

$$m < n.$$
(5.10)

5.3 Bond types

There is no sharp distinction between the different types of bonds. For most bonds, however, we may say that one or the other mechanism dominates. Thus, a classification is possible; the four main types are: (i) ionic, (ii) metallic, (iii) covalent, and (iv) van der Waals.

5.3.1 Ionic bonds

A typical representative of an *ionic* crystal is NaCl, which we have already discussed in some detail. The crystal structure is regular and looks exactly like the one shown in Fig. 1.1. We have negatively charged Cl ions and positively charged Na ions. We may now ask the question, what is the cohesive energy of this crystal? Cohesive energy is what we have denoted by E_c in Fig. 5.1, that is the energy needed to take the crystal apart. How could we calculate this? If the binding is due mainly to electrostatic forces, then all we need to do is to sum the electrostatic energy due to pairs of ions.

Let us start with an arbitrary Na ion. It will have six Cl ions at a distance, a, giving the energy,

$$-\frac{e^2}{4\pi\epsilon_0}\frac{6}{a}.$$
(5.11)

There are then 12 Na ions at a distance $a\sqrt{2}$ contributing to the energy by the amount

$$\frac{e^2}{4\pi\epsilon_0}\frac{12}{a\sqrt{2}}.$$
(5.12)

Next come eight chlorine atoms at a distance $a\sqrt{3}$, and so on. Adding up the contributions from all other ions, we have an infinite sum (well, practically infinite) of the form

$$-\frac{e^2}{4\pi\epsilon_0}\left(\frac{6}{a} - \frac{12}{a\sqrt{2}} + \frac{8}{a\sqrt{3}} \cdots\right).$$
(5.13)

We have to add together sums such as eqn (5.13) for every Na and Cl ion to get the cohesive energy. It would actually be twice the cohesive energy, because we counted each pair twice, or we may say that it is the cohesive energy per NaCl unit.

The infinite summations look a bit awkward, but fortunately there are mathematicians who are fond of problems of this sort; they have somehow managed to sum up all these series, not only for the cubical structure of NaCl, but for the more complicated structures of some other ionic crystals as well. Their labour brought forth the formula

$$\text{Electrostatic energy} = -M\frac{e^2}{4\pi\epsilon_0 a}. \tag{5.14}$$

M is called the *Madelung constant*. For a simple cubic structure its value is 1.748.

Taking $a = 0.28$ nm and putting the constants into eqn (5.14) we get for the cohesive energy,

$$E = 8.94\,\text{eV}, \tag{5.15}$$

which is about ten per cent above the experimentally observed value. There are other types of energies involved as well (as, e.g. the energy due to the slight deformation of the atoms) but, as the numerical results show, they must be of lesser significance. We have thus confirmed our starting point that NaCl may be regarded as an ionic bond.

5.3.2 Metallic bonds

Having studied the construction of atoms, we are now in a somewhat better position to talk about metals. Conceptually, the simplest metal is a monovalent alkali metal, where each atom contributes one valence electron to the common pool of electrons. So we are, in fact, back to our very first model, when we regarded a conductor as made up of lattice ions and charged billiard balls bouncing around.

We may now ask the question: how is a piece of metal kept together? 'By electrostatic forces', is the simplest, though not quite accurate, answer. Thus, the *metallic* bond is similar to the ionic bond in the sense that the main role is played by electrostatic forces, but there is a difference as far as the positions of the charges are concerned. In metals the carriers of the negative charge are highly mobile; thus we may expect a bond of somewhat different properties. Since electrons whizz around and visit every little part of the metal, the electrostatic forces are ubiquitous and come from all directions. So we may regard the electrons as a glue that holds the lattice together. It is quite natural, then, that a small deformation does not cause fracture. Whether we compress or try to pull apart a piece of metal the cohesive forces are still there and acting vigorously. This is why metals are so outstandingly ductile and malleable.

5.3.3 The covalent bond

So far we have discussed two bonds, which depend on the fact that unlike charges attract—a familiar, old but nevertheless true, idea. But why should atoms like carbon or silicon hang together? It is possible to purify silicon, so that its resistivity is several ohm metres—there can be no question of a lot of free electrons swarming around, nor is there an ionic bond. Carbon in its diamond form is the hardest material known. Not only must it form strong bonds, but they must also be exceptionally precise and directional to achieve this hardness.

The properties of the covalent bond, also called the *valence* or *homopolar* bond on occasions, is the most important single topic in chemistry, yet its mechanism was completely inexplicable before the rise of quantum mechanics.

The exact mathematical description is immensely difficult, even for people with degrees, so in an undergraduate course we must be modest. The most we can hope for is to get a good physical picture of the bond mechanism, and perhaps an inkling of how a theoretical physicist would start solving the problem.

The simplest example of the covalent bond is the hydrogen molecule, where two protons are kept together by two electrons. The bond comes about because both electrons orbit around both atoms. Another way of describing the bond is to appeal to the atoms' desire to have filled shells. A hydrogen atom needs two electrons (of opposite spin) to fill the 1s shell, and lacking any better source of electrons, it will consider snatching that extra electron from a fellow hydrogen atom. Naturally the other hydrogen atom will resist, and at the end they come to a compromise and share both their electrons. It is as if two men, each anxious to secure two wives for himself, were to agree to share wives.*

Another example is chlorine, which has five 3p electrons and is eagerly awaiting one more electron to fill the shell. The problem is again solved by sharing an electron pair with another chlorine atom. Thus, each chlorine atom for some time has the illusion that it has managed to fill its outer shell.

Good examples of covalent bonds in solids are carbon, silicon, and germanium. Their electron configurations may be obtained from Table 4.1. They are as follows:

$$\text{C:}\quad 1s^2,\ 2s^2,\ 2p^2$$

$$\text{Si:}\quad 1s^2,\ 2s^2,\ 2p^6,\ 3s^2,\ 3p^2$$

$$\text{Ge:}\quad 1s^2,\ 2s^2,\ 2p^6,\ 3s^2,\ 3p^6,\ 3d^{10},\ 4s^2,\ 4p^2.$$

It can be easily seen that the common feature is two s and two p electrons in the outer ring. The s shells (2s, 3s, 4s, respectively) are filled; so one may expect all three substances to be divalent, since they have two extra electrons in the p shells. Alas, all of them are tetravalent. The reason is that because of interaction (which occurs when several atoms are brought close together), the spherical symmetry of the outer s electrons is broken up, and they are persuaded to join the p electrons in forming the bonds. Hence, for the purpose of bonding, the atoms of carbon, silicon, and germanium may be visualized with four dangling electrons at the outside. When the atoms are brought close to each other, these electrons establish the bonds by pairing up. The four electrons are arranged symmetrically in space, and the bonds must therefore be tetrahedral, as shown in Fig. 5.3.

In covalent bonds all the available electrons pair up and orbit around a pair of atoms; none of them can wander away to conduct electricity. This is why carbon in the form of diamond is an insulator.† The covalent bonds are weaker in silicon and germanium, and some of the electrons might be 'shaken off' by the thermal vibrations of the crystal. This makes them able to conduct electricity to a certain extent. They are not conductors; we call them *semiconductors*.

* The analogy also works, as a feminist friend has pointed out, if two women, each anxious to secure two husbands for herself, were to agree to share husbands.

† Incidentally, carbon can have another type of bond as well. In its graphite form, consisting of layers on top of each other, it is a fairly good conductor.

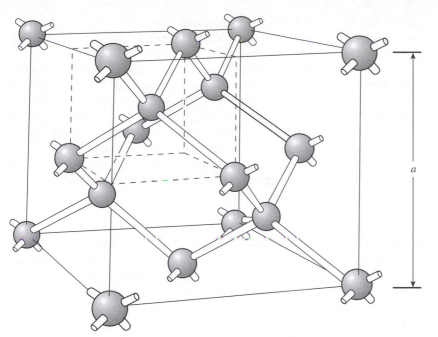

Fig. 5.3
The diamond structure. Notice that each atom is symmetrically surrounded in an imaginary cube by its four nearest neighbours. These are covalently bonded, indicated by tubular connections in the figure.

Table 5.1 *Mohs hardness scale (modified)*

Hardness number	Material
1	Talc $Mg_3Si_4O_{10}(OH)_2$
2	Gypsum $CaSO_4 \cdot 2H_2O$
3	Calcite $CaCO_3$
4	Fluorite CaF_2
5	Apatite $Ca_5(PO_4)_3(OH, F, Cl)$
6	Orthoclase $KAlSi_3O_8$
7	Vitreous Silica SiO_2
8	Quartz, Stellite SiO_2
9	Topaz $Al_2SiO_3(OH, F)_2$
10	Garnet $ZnAl_2O_4$
11	Fused Zirconia ZrO_2
12	Fused Alumina Al_2O_3
13	Silicon Carbide SiC
14	Boron Carbide BC
15	Diamond C

We have already remarked on the hardness of diamond, which is a measure of its resistance to deformation, whether it will crush, scratch, stretch, or dent. It is difficult to quantify precisely—the engineers' rule of thumb is called the Mohs scale, measured by the dent caused by a standard probe. On this scale diamond was initially given the top rating of 10, but to include more hard materials, the scale was uprated to 15, as is shown in Table 5.1. The softest material, rated 1, is the familiarly soft talcum powder. Common abrasive materials include silicon carbide, a group IV compound with a similar diamond type covalent bond, but less symmetry; and tungsten carbide, whose bonds are partly ionic. The diamond structure also gives a high value of bulk elastic modulus [eqn (5.7)].

Diamond has two other superlatives. It has a much greater thermal conductivity than metals at room temperature, conducting energy by lattice vibrations rather than free electrons. It is thus a good heat sink for electronic devices where heat can be removed without prejudicing the electrical behaviour. As well as using 'industrial' rather than gem grade diamonds for this purpose, it is possible to grow suitable plane layers of diamond by molecular beam epitaxy and related techniques described in a little more detail in Section 8.11.

The second superlative that takes us out of Science and into high finance is the 'sparkle' of diamond in jewellery. But see Table 10.1; it is the high refractive index which makes sparklers.

The sale of diamonds of jewel quality is controlled by the Central Selling Organisation (CSO) of de Beers who in 1948 hired a New York advertising agency to create a slogan to stimulate their industry. The young lady assigned to this task thought long and hard, weeks passed with no idea. She thought she would work on this forever, so said "diamonds are forever". It was a great success.

Carbon dating is an important impact of science on ancient history and archaeology, removing some of the luxury of literary speculation from its practitioners. Atmospheric CO_2, which is the source of carbon in living organisms, contains 1 atoms of C^{14} in 7.8×10^{11} atoms of stable C^{12}. C^{14}, decays with a half life on 5700 years emitting an electron. These extreme numbers result in appreciable radioactivity, giving 15 disintegrations per gram per minute. When the organism dies the intake of atmospheric CO_2 stops and the C^{14} within it decays exponentially. Thus, a count of radioactivity in dead bones or wood etc. will give a dating for the time of death.

In the two decades after 1950, scientific archaeology matured rapidly with the advent of the nuclear physics-based methods of carbon and thermoluminescent dating. Samples of diamonds from ancient jewels and numerous samples mined in the past two centuries were dated. They were all found to have been formed within an order of magnitude of 10^9 years ago. On a human scale, 10^9 years is pretty close to 'forever'. So the slogan is more accurate than the average advertisement.

Diamonds are not always forever. Their one vice is that at 700°C in air they burn to carbon dioxide. This rules out diamond for large-scale cutting of steels and other hard metals. It has stimulated research for hard compounds with better temperature stability and hardness perhaps even greater than 15 on the Mohs scale. Compounds of C with N, B, and Si have shown promise, the possible winner is C_3N_4; but not yet.

5.3.4   The van der Waals bond

If the outer shell is not filled, atoms will exert themselves to gain some extra electrons, and they become bonded in the process. But what happens when the shell is already filled, and there are no electrostatic forces either, as for example in argon? How will argon solidify? For an explanation some quantum-mechanical arguments are needed again.

We have described the atoms as consisting of a positive nucleus and the electrons around the nucleus, with the electrons having certain probabilities of being in certain places. Since the electrons are sometimes here and sometimes there, there is no reason why the centres of positive and negative charge should always be coincident. Thus, we could regard atoms as fluctuating dipoles. If atom A has a dipole moment, then it will induce an opposite dipole moment on atom B. On average there will be an attractive force, since the tendency described leads always to attraction, never to repulsion.

This attraction is called a *van der Waals* bond. Such bonds are responsible for the formation of organic crystals.

The forces in van der Waals bonds are fairly weak (and may be shown to vary with the inverse seventh power of distance); consequently these materials have low melting and boiling points.

Searching for an anthropomorphic analogy once more (it's good because it aids the memory) we could look at a dipole as a permanent bond between a man and a woman established by mutual attraction. Now would two such dipoles attract each other? To facilitate the discussion let us introduce the notation m_1 and m_2, and w_1 and w_2 for the two men and two women in dipoles 1 and 2 respectively. For an attractive force to develop between two dipoles all we need is that the attraction between m_1 and w_2, and m_2 and w_1 should be stronger than the repulsions between m_1 and m_2 and w_1 and w_2. In a modern society this is indeed the likely thing to happen. The attraction can be there without the need to break the bond.

5.3.5   Mixed bonds

In most practical cases the bonds are of course not any of these pure types. An example of a mixed bond is that in carbon steel in which the presence of both metallic and ionic bonds leads to a material with considerably more strength than that of iron on its own.

Mixed bonds of particular significance to the semiconductor industry are some III–V and II–VI compounds (where the Roman numbers refer to the respective columns in the periodic table of Fig. 4.5) as for example GaAs or ZnSe. They have a combination of ionic and covalent bonds. We shall discuss their properties in more detail in Section 8.6.

5.3.6   Carbon again

It may be worth noting that the diamond structure is not the only one in which carbon can crystallize. Another form is graphite, which consists of arrays of hexagons stuck together in flat sheets. Interestingly, and rather unexpectedly, lots of further crystalline forms of carbon have been discovered in the last two decades. Their significance in engineering is not obvious as yet, but they are certainly fun to look at. We shall show here only the one discovered earliest, which gave the name of fullerenes to the family after Buckminster Fuller, a US architect who originated the geodesic dome of similar shape. It comes about by removing an atom from some of the hexagons. The sheet may then fold up into a configuration of 60 atoms containing 12 pentagons and 20 hexagons as shown in Fig. 5.4. It resembles a football, which would be a better name for it. Alas, the architects got there first.

Another interesting configuration is the tube which has a thin wall no more than a few atoms thick and may be a hundred atoms long. The family, called

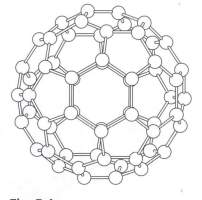

Fig. 5.4
A view of C_{60} containing pentagonal and hexagonal structure.

nanotubes, may lead to a new type of transistor as discussed briefly later in Section 9.25 concerned with nanoelectronics.

A more serious but less picturesque aspect of carbon bonding is the existence of double and triple bonds, which play a crucial role in organic and polymer chemistry. The single covalent bond, which can be expressed in terms of molecular orbitals, occurs when two dangling bonds pair up so that their electron orbitals merge to form a cylindrical two electron cloud shared between the two atoms. Since this looks similar to a pair of s orbitals (quantum number $l = 0$, Section 4.2) it is called a σ bond. When two carbon atoms are so paired, they can each have three spare bonds. These can be combined with other elements (e.g. H), or a pair perpendicular to the plane of the σ bond can form a weaker bond by sharing electrons with a similar structure of two p orbitals ($l = 1$, see Fig. 4.4) called a π bond.

Going even further, a gas such as acetylene (C_2H_2) has the carbons joined to each other and to hydrogen by σ bonds, leaving two pairs of 2p electrons perpendicular to the plane of the carbon σ bond, which form two π bonds. These three bond types hugely contribute to the complexity and versatility of organic chemistry.

5.4   Feynman's coupled mode approach

We are now going to discuss a more mathematical theory of the covalent bond, or rather of its simplest case, the bonding of the hydrogen molecule. We shall do this with the aid of Feynman's coupled modes. This approach proved amazingly powerful in Feynman's hands, enabling him to explain, besides the hydrogen molecule, such diverse phenomena as the nuclear potential between a proton and a neutron, and the change of the K° particle into its own antiparticle. There is in fact hardly a problem in quantum mechanics that Feynman could not treat by the technique of coupled modes. Of necessity, we shall be much less ambitious and discuss only a few relatively simple phenomena.

I should really start by defining the term 'coupled mode'. But to define is to restrict, to put a phenomenon or a method into a neat little box in contradistinction to other neat little boxes. I am a little reluctant to do so in the present instance because I am sure I would then exclude many actual or potential applications. Not being certain of the limitations of the approach, I would rather give you a vague description, just a general idea of the concepts involved.

The coupled mode approach is concerned with the properties of coupled oscillating systems like mechanical oscillators (e.g. pendulums), electric circuits, acoustic systems, molecular vibrations, and a number of other things you might not immediately recognize as oscillating systems. The approach was quite probably familiar to the better physicists of the last century but has become fashionable only recently. Its essence is to divide the system up into its components, investigate the properties of the individual components in isolation, and then reach conclusions about the whole system by assuming that the components are weakly coupled to each other. Mathematicians would call it a perturbation solution because the system is perturbed by introducing the coupling between the elements.

First of all we should derive the equations. These, not unexpectedly, turn out to be coupled linear differential equations. Let us start again with Schrödinger's equation [eqn (3.4)], but put it in the operator form, that is

$$\left(-\frac{\hbar^2}{2m}\nabla^2 + V\right)\Psi = i\hbar\frac{\partial\Psi}{\partial t}. \tag{5.16}$$

The operator in parentheses is usually called the *Hamiltonian operator* and denoted by H.

We may also write Schrödinger's equation in the simple and elegant form

$$H\Psi = i\hbar\frac{\partial\Psi}{\partial t}. \tag{5.17}$$

We have attempted [eqn (3.7)] the solution of this partial differential equation before by separating the variables,

$$\Psi = w(t)\psi(\mathbf{r}). \tag{5.18}$$

Let us try to do the same thing again but in the more general form,

$$\Psi = \sum_j w_j(t)\psi_j(\mathbf{r}), \tag{5.19}$$

where a number of solutions (not necessarily finite) are superimposed.

Up to now we have given all our attention to the spatial variation of the wave function. We have said that if an electron is in a certain state, it turns up in various places with certain probabilities. Now we are going to change the emphasis. We shall not enquire into the spatial variation of the probability at all. We shall be satisfied with asking the much more limited question: what is the probability that the electron (or more generally a set of particles) is in state j at time t? We do not care what happens to the electron in state j as long as it is in state j. We are interested only in the *temporal* variation, that is, we shall confine our attention to the function $w(t)$.

We shall get rid of the spatial variation in the following way. Let us substitute eqn (5.19) into eqn (5.17)

$$\sum_j w_j H\psi_j = i\hbar \sum_j \psi_j \frac{\mathrm{d}w_j}{\mathrm{d}t}, \tag{5.20}$$

then multiply both sides by ψ_k and integrate over the volume. We then obtain

$$\sum_j w_j \int \psi_k H\psi_j \, \mathrm{d}v = i\hbar \sum_j \frac{\mathrm{d}w_j}{\mathrm{d}t} \int \psi_j\psi_k \, \mathrm{d}v, \tag{5.21}$$

where dv is the volume element.

Now ψ_j and ψ_k are two solutions of the time-independent Schrödinger equation, and they have the remarkable property (I have to ask you to believe this) of being *orthogonal* to each other. You may have met simple examples of orthogonality of functions before, if at no other place than in the derivation of the coefficients of a Fourier series. The condition can be simply stated in the following form

$$\int \psi_k \psi_j \, dv = \begin{cases} C_{kj} & \text{if } k = j, \\ 0 & \text{if } k \neq j. \end{cases} \tag{5.22}$$

Multiplying the wave function with judiciously chosen constants, C_{kj} can be made unity, and then the wave functions are called *orthonormal*. Assuming that this is the case and introducing the notation,

$$H_{kj} = \int \psi_k H \psi_j \, dv, \tag{5.23}$$

we get the following differential equations:*

$$i\hbar \frac{dw_k}{dt} = \sum_j H_{kj} w_j \tag{5.24}$$

for each value of k.

This is the equation we sought. It is independent of the spatial variables and depends only on time. It is therefore eminently suitable for telling us how the probability of being in a certain state varies with time.

You may quite justifiably worry at this point about how you can find the wavefunction, how you can make them orthonormal, and how you can evaluate integrals looking as complex as eqn (5.23). The beauty of Feynman's approach is that neither the wave function nor H_{kj} need be calculated. It will suffice to guess H_{kj} on purely physical grounds.

We have not so far said anything about the summation. How many wave functions (i.e. states) are we going to have? We may have an infinite number, as for the electron in a rigid potential well, or it may be finite. If, for example, only the spin of the electron matters, then we have two states and no more. The summation should run through $j = 1$ and $j = 2$. Two is of course the minimum number. In order to have coupling, one needs at least two components, and it turns out that two components are enough to reach some quite general conclusions about the properties of coupled systems. So the differential equations we are going to investigate look as follows:

$$i\hbar \frac{dw_1}{dt} = H_{11} w_1 + H_{12} w_2, \tag{5.25}$$

$$i\hbar \frac{dw_2}{dt} = H_{21} w_1 + H_{22} w_2. \tag{5.26}$$

If $H_{12} = H_{21} = 0$ the two states are not coupled. Then the differential equation for state (1) is

$$i\hbar \frac{dw_1}{dt} = H_{11} w_1, \tag{5.27}$$

which has a solution

$$w_1 = K_1 \exp\left(-i\frac{H_{11}}{\hbar} t\right). \tag{5.28}$$

The probability of being in state (1) is thus

$$|w_1(t)|^2 = |K_1|^2. \tag{5.29}$$

* The derivation would be analogous if, instead of one electron, a set of particles was involved. Schrödinger's equation would then be written in terms of a set of spatial variables and there would be multiple integrals instead of the single integral here. The integrations would be more difficult to perform, but the final form would still be that of eqn (5.24).

This is not a very exciting solution, but it is at least consistent. If there is no coupling between the states, then the probability of being in state (1) does not vary with time—once in state (1), always in state (1). The same is true, of course, for state (2). In the absence of coupling nothing changes.

Before solving the coupled differential equations, let us briefly discuss the physical concepts of uncoupled states and the meaning of coupling. What do we mean exactly by coupling? We can explain this with our chosen example, the hydrogen molecule, or better still the even simpler case, the hydrogen molecular ion.

The hydrogen molecular ion consists of a hydrogen atom to which a proton is attached. We may then imagine our uncoupled states as shown in Fig. 5.5. We choose for state (1) the state when the electron is in the vicinity of proton 1 and occupying the lowest energy (ground) level, and proton 2 is just alone with no electron of its own. State (2) represents the alternative arrangement when the electron is attached to proton 2 and proton 1 is bare.

When we say that we consider only these two states, we are not denying the existence of other possible states. The electron could be in any of its excited states around the proton, and the whole configuration of three particles may vibrate, rotate, or move in some direction. We are going to ignore all these complications. We say that as far as our problem is concerned, only the two states mentioned above are of any significance.

What do we mean when we say that these two states are uncoupled? We mean that if the electron is at proton 1 in the beginning, it will always stay there. Similarly, if the electron is at proton 2 in the beginning, it will always stay at proton 2. Is this complete separation likely? Yes, if the protons are far from each other, this is the only thing that can happen. What can we expect when the protons are brought closer to each other? Classically, the electron that is in the vicinity of proton 1 should still remain with proton 1 because this is energetically more favourable. The electron cannot leave proton 1 because it faces an adverse potential barrier. According to the laws of quantum mechanics, this is no obstacle, however. The electron may tunnel through the potential barrier and arrive at proton 2 with energy unchanged. Thus, as the two protons approach each other, there is an increasing probability that the electron jumps over from proton 1 to proton 2 and vice versa. And this is what we mean by coupling. The two states are not entirely separate.

What do we mean by *weak* coupling? It means that even in the presence of coupling, it is still meaningful to talk about one or the other state. The states influence each other but may preserve their separate entities.

Let us return now to the solution of equations (5.25) and (5.26). As we are going to investigate symmetric cases only, we may introduce the simplifications

$$H_{11} = H_{22} = E_0, \quad H_{12} = H_{21} = -A, \qquad (5.30)$$

leading to

$$i\hbar \frac{dw_1}{dt} = E_0 w_1 - A w_2, \qquad (5.31)$$

$$i\hbar \frac{dw_2}{dt} = -A w_1 + E_0 w_2. \qquad (5.32)$$

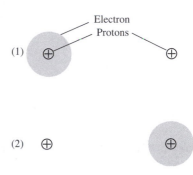

Fig. 5.5
The two basic states of the hydrogen molecular ion. The shaded area represents the electron in its ground state. It is attached either to proton 1 or to proton 2.

When the electron jumps from one proton to the other proton, it introduces coupling between the two states.

Following the usual recipe, the solution may be attempted in the form

$$w_1 = K_1 \exp\left(-i\frac{E}{\hbar}t\right), \quad w_2 = K_2 \exp\left(-i\frac{E}{\hbar}t\right). \quad (5.33)$$

Substituting eqn (5.33) into equations (5.31) and (5.32) we get

$$K_1 E = E_0 K_1 - A K_2 \quad (5.34)$$

and

$$K_2 E = -A K_1 + E_0 K_2, \quad (5.35)$$

which have a solution only if

$$\begin{vmatrix} E_0 - E & -A \\ -A & E_0 - E \end{vmatrix} = 0. \quad (5.36)$$

Expanding the determinant we get

$$(E_0 - E)^2 = A^2, \quad (5.37)$$

whence

$$E = E_0 \pm A. \quad (5.38)$$

If there is no coupling between the two states, then $E = E_0$; that is, both states have the same energy. If there is coupling, the energy level is split. There are two new energy levels $E_0 + A$ and $E_0 - A$. This is a very important phenomenon that you will meet again and again. Whenever there is coupling, the energy splits.

The energies $E_0 \pm A$ may be defined as the energies of so-called stationary states obtainable from linear combinations of the original states. For our purpose it will suffice to know that we can have states with energies $E_0 + A$ and $E_0 - A$.

How will these energies vary with d, the distance between the protons? What is A anyway? A has come into our equations as a coupling term. The larger A, the larger the coupling, and the larger the split in energy. Hence A must be related to the tunnelling probability that the electron may get through the potential barrier between the protons. Since tunnelling probabilities vary exponentially with distance—we have talked about this before when solving Schrödinger's equation for a tunnelling problem—A must vary roughly in the way shown in Fig. 5.6.

Now what is E_0? It is the energy of the states shown in Fig. 5.5. It consists of the potential and kinetic energies of the electron and of the potential energies of the protons (assumed immobile again). When the two protons are far away, their potential energies are practically zero, and the electron's energy, since it is bound to a proton, is a negative quantity. Thus, E_0 is negative for large interproton distances but rises rapidly when the separation of the two protons is less than the average distance of the fluctuating electron from the protons. A plot of E_0 against d is also shown in Fig. 5.6.

We may now obtain the energy of our states by forming the combinations $E_0 \pm A$. Plotting these in Fig. 5.7, we see that $E_0 - A$ has a minimum, that is, at that particular value of d a stable configuration exists. We may also argue in

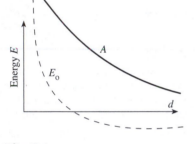

Fig. 5.6
The variation of E_0 and A with the interproton separation, d. E_0 is the energy when the states shown in Fig. 5.5 are uncoupled. A is the coupling term.

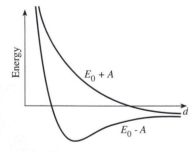

Fig. 5.7
Summing the quantities in Fig. 5.6 to get $E_0 + A$ and $E_0 - A$. The latter function displays all the characteristics of a bonding curve.

terms of forces. Decreasing energy means an attractive force. Thus, when the protons are far away, and we consider the state with the energy $E_0 - A$, there is an attractive force between the protons. This will be eventually balanced by the Coulomb repulsion between the protons, and an equilibrium will be reached.

Thus, in order to explain semi-quantitatively the hydrogen molecular ion, we have had to introduce a number of new or fairly new quantum-mechanical ideas.

5.5 Nuclear forces

Feynman in his *Lectures on Physics* goes on from here and discusses a large number of phenomena in terms of coupled modes. Most of the phenomena are beyond what an engineering undergraduate needs to know; so with regret we omit them. (If you are interested you can always read Feynman's book.) But I cannot resist the temptation to follow Feynman in saying a few words about nuclear forces. With the treatment of the hydrogen molecular ion behind us, we can really acquire some understanding of how forces between protons and neutrons arise.

It is essentially the same idea that we encountered before. A hydrogen atom and a proton are held together owing to the good services of an electron. The electron jumps from the hydrogen atom to the proton converting the latter into a hydrogen atom. Thus, when a reaction

$$\text{H}, \text{p} \rightarrow \text{p}, \text{H} \tag{5.39}$$

takes place, a bond is formed.

Yukawa* proposed in the middle of the 1930s that the forces between nucleons may have the same origin. Let us take the combination of a proton and a neutron. We may say again that a reaction

$$\text{p}, \text{n} \rightarrow \text{n}, \text{p} \tag{5.40}$$

takes place, and a bond is formed. 'Something' goes over from the proton to the neutron which causes the change, and this 'something' is called a positively charged π-meson. Thus, just as an electron holds together two protons in a hydrogen molecular ion, in the same way a positively charged π-meson holds together a proton and a neutron in the nucleus.

5.6 The hydrogen molecule

The hydrogen molecule differs from the hydrogen molecular ion by having one more electron. So we may choose our states as shown in Fig. 5.8. State (1) is when electron a is with proton 1 and electron b with proton 2, and state (2) is obtained when the electrons change places.

How do we know which electron is which? Are they not indistinguishable? Yes, they are, but we may distinguish them by assigning opposite spins to them.

We may now explain the bond of the hydrogen molecule in a manner analogous to that of the hydrogen molecular ion, but instead of a single electron jumping to and fro, we now have two electrons changing places. Thus, we may argue again that owing to symmetry, the energies of the two states are identical.

* If you permit us a digression in a digression, I should like to point out that Yukawa, a Japanese, was the first non-European ever to make a significant contribution to theoretical physics. Many civilizations have struck independently upon the same ideas, as for example the virgin births of gods or the commandments of social conduct, invented independently useful instruments like the arrow or the wheel, and developed independently similar judicial procedures and constitutions, but, interestingly, no civilization other than the European one bothered about theoretical physics.

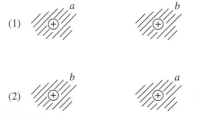

Fig. 5.8

The two basic states of the hydrogen molecule. Each electron can be attached to either proton leading to a coupling between the states.

The coupling between the states—due to the exchange of electrons—splits the energy levels, one becoming somewhat higher, the other somewhat lower. Having the chance to lower the energy results in an attractive force which is eventually balanced by the repulsive force between the protons. And that is the reason why the hydrogen molecule exists.

It is interesting to compare this picture with the purely intuitive one described earlier, based on the atoms' 'desire' to fill the energy shells. In the present explanation we are saying that the bond is due to the *exchange* of electrons; previously we said the bond was due to *sharing* of the electrons. Which is it; is it sharing or swapping? It is neither. Both explanations are no more than physical pictures to help the imagination.

We could equally well have said that the hydrogen molecule exists because it comes out mathematically from our basic premises, that is the spin and Pauli's principle added to Schrödinger's equation. The problem is a purely mathematical one, which can be solved by numerical methods. There is no need, whatsoever, for a physical picture. This argument would hold its ground if numerical solutions were always available. But they are *not* available. Computers are not powerful enough, not as yet and will not be for a long time to come. So we need mathematical approximations based on a simplified physical picture and then we must strive to build up a new, more sophisticated physical picture from the mathematical solution obtained, and then attempt for a better mathematical approximation based on the new physical picture, and so on, and so on. It seems a tortuous way of doing things, but that is how it is.

It is a lot easier in classical physics. Our physical picture is readily acquired in conjunction with our other faculties. We do not need to be taught that two bricks cannot occupy the same place: we know they cannot.

In studying phenomena concerned with extremely small things beyond the powers of direct observation, the situation is different. The picture of an atom with filled and unfilled energy shells is not a picture acquired through personal experience. It has come about by solving a differential equation. But once the solution is obtained, a physical picture starts emerging. We may visualize little boxes, or concentric spheres, or rows of seats in the House of Commons filling up slowly with MPs. The essential thing is that we *do* form some kind of picture of the energy shells. And once the shell picture is accepted, it helps us find an explanation for the next problem, the bond between the atoms.

So you should not be unduly surprised that many alternative explanations are possible. They reflect attempts to develop intuition in a discipline where intuition does not come in a natural way.

Whenever confronted with new problems, one selects from this store of physical pictures the ones likely to be applicable. If one of the physical pictures does turn out to be applicable, it is a triumph both for the picture and for the person who applied it. If all attempts fail, then either a new physical picture or a brighter person is needed to tackle the problem.

5.7   An analogy

One of the most important conclusions of the foregoing discussion was that 'whenever there is coupling, the energy levels split'. This is a very important relationship in quantum mechanics, but it could also be regarded as a

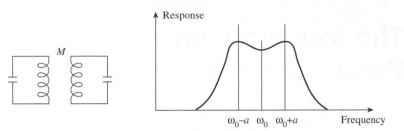

Fig. 5.9
The coupled circuit analogy. Two resonant circuits tuned to ω_0 when far apart (no coupling between them), have their resonant frequency split to $\omega_0 \pm a$ (cf. $E_0 \pm A$) when coupled.

simple mathematical consequence of the mathematical formulation. If we have coupled differential equations, something will always get split somewhere. The example we are all familiar with is that of coupled electric resonant circuits shown in Fig. 5.9. If the two circuits are far away from each other, that is they are uncoupled, both of them have resonant frequencies ω_0. When the circuits are coupled, there are two resonant frequencies $\omega_0 \pm a$, that is, we may say the resonant frequencies are split.

Exercises

5.1. Discuss qualitatively the various mechanisms of bonding. Give examples of materials for each type of bond and also materials that do not have a clear single bond type.

5.2. Show that the force between two aligned permanent dipoles, a distance r apart, is attractive and varies as r^{-4}.

5.3. The interaction energy between two atoms may be phenomenologically described by eqn (5.8). Show that the molecule will break up when the atoms are pulled apart to a distance

$$r_b = \left(\frac{n+1}{m+1}\right)^{1/(n-m)} r_0,$$

where r_0 is the equilibrium distance between the atoms. Discuss the criterion of breaking used to get the above result.

5.4. For the KCl crystal the variation of energy may also be described by eqn (5.8), but now r means the interatomic distance in the cubic crystal, and the energy is for an ion pair. Take $m = 1, n = 9, B = 1.75e^2/4\pi\epsilon_0$. The bulk modulus of elasticity is $1.88 \times 10^{10} \, \text{N m}^{-2}$. Calculate the separation of the $K^+ - Cl^-$ ions in the ionic solid.

5.5. Show with the aid of eqns (5.7)–(5.9) that the bulk modulus may be obtained in the form

$$c = -\frac{mn}{9r_0^3} E_c.$$

5.6. Calculate the energy of a negative ion in a linear chain of equally spaced ions, which carry alternative positive and negative charges.

5.7. For a symmetrical coupled system, the decrease in energy (in respect to the uncoupled case) is A, as shown by eqn (5.38). Show that, for an unsymmetrical system ($H_{11} \neq H_{22}$) with the same coupling ($H_{12} H_{21} = A^2$), the decrease in energy is less than A.

6 The free electron theory of metals

Struggling to be free, art more engaged
Hamlet

Much have I travelled in the realms of gold,
And many goodly states and kingdoms seen.
Keats *On First Looking in Chapman's Homer*

6.1 Free electrons

The electrical and magnetic properties of solids are mainly determined by the properties of electrons in them. Protons can usually be relegated to subordinate roles, like ensuring charge neutrality. Neutrons may sometimes need to be considered, as for example in some superconducting materials, in which the critical temperature depends on the total mass of the nucleus, but on the whole, the energy levels of electrons hold the key to the properties of solids.

The mathematical problem is not unlike the one we met in the case of individual atoms. How can we determine the energy levels of electrons in a solid? Take a wave function depending on the coordinates of 10^{25} electrons; write down the Coulomb potential between each pair of electrons, between electrons and protons; and solve Schrödinger's equation. This is an approach which, as you have probably guessed, we are not going to try. But what can we do instead? We can take a much simpler model, which is mathematically soluble, and hope that the solution will make sense.

Let us start our search for a simple model by taking a piece of metal and noting the empirical fact (true at room temperature) that there are no electrons beyond the boundaries of the metal. So there is some mechanism keeping the electrons inside. What is it? It might be an infinite potential barrier at the boundaries. And what about inside? How will the potential energy of an electron vary in the presence of that enormous number of nuclei and other electrons? Let us say it will be uniform. You may regard this a sweeping assumption (and, of course, you are absolutely right), but it works. It was introduced by Sommerfeld in 1928, and has been known as the 'free electron' model of a metal.

The electrons inside the metal (more correctly, the valence electrons, which occupy the outer ring) are entirely free to roam around, but they are not allowed to leave the metal.

You may recognize that the model is nothing else but the potential well we met before. There we obtained the solution for the one-dimensional case in the following form:

$$E = \frac{\hbar^2 k^2}{2\,m} = \frac{h^2}{8\,m}\frac{n^2}{L^2}.$$

(6.1)

If we imagine a cube of side L containing the electrons, then we get for the energy in the same manner

$$E = \frac{\hbar^2}{2m}(k_x^2 + k_y^2 + k_z^2) = \frac{h^2}{8mL^2}(n_x^2 + n_y^2 + n_z^2). \qquad (6.2)$$

n_x, n_y, n_z are integers.

6.2 The density of states and the Fermi–Dirac distribution

The allowed energy, according to eqn (6.2), is an integral multiple of $h^2/8mL^2$. For a volume of 10^{-6}m^3 this unit of energy is

$$E_{unit} = \frac{(6.62 \times 10^{-34})^2}{8 \times 9.1 \times 10^{-31} \times 10^{-4}} = 0.6 \times 10^{-33} \text{ J} = 3.74 \times 10^{-15} \text{ eV}. \qquad (6.3)$$

This is the energy difference between the first and second levels, but since the squares of the integers are involved, the difference between neighbouring energy levels increases at higher energies. Let us anticipate the result obtained in the next section and take for the maximum energy $E = 3$ eV, which is a typical figure. Taking $n_x^2 = n_y^2 = n_z^2$, this maximum energy corresponds to a value of $n_x \cong 1.64 \times 10^7$. Now an energy level just below the maximum energy can be obtained by taking the integers $n_x - 1$, n_x, n_x. We get for the energy difference

$$\Delta E \cong 1.22 \times 10^{-7} \text{ eV}, \qquad (6.4)$$

We can therefore say that, in a macroscopically small energy interval dE, there are still many discrete energy levels. So we can introduce the concept of density of states, which will simplify our calculations considerably.

The next question we ask is how many states are there between the energy levels E and $E + $ dE. It is convenient to introduce for this purpose the new variable n with the relationship

Even at the highest energy, the difference between neighbouring energy levels is as small as 10^{-7} eV.

$$n^2 = n_x^2 + n_y^2 + n_z^2. \qquad (6.5)$$

Thus n represents a vector to a point n_x, n_y, n_z in three-dimensional space. In this space every point with integer coordinates specifies a state; that is, a unit cube contains exactly one state. Hence, the number of states in any volume is just equal to the numerical value of the volume. Thus, in a sphere of radius n, the number of states is

$$\frac{4n^3\pi}{3}. \qquad (6.6)$$

Since n and E are related, this is equivalent to saying that the number of states having energies less than E is

$$\frac{4n^3\pi}{3} = \frac{4\pi}{3}K^{3/2}E^{3/2} \quad \text{with } K = \frac{8mL^2}{h^2} \qquad (6.7)$$

Similarly, the number of states having energies less than $E + $ dE is

$$\frac{4\pi}{3}K^{3/2}(E + \text{d}E)^{3/2}. \qquad (6.8)$$

So the number of states having energies between E and $E + dE$ is equal to

$$Z(E)dE = \frac{4\pi}{3}K^{3/2}\{(E + dE)^{3/2} - E^{3/2}\}$$
$$\cong 2\pi K^{3/2}E^{1/2}dE. \tag{6.9}$$

This is not the end yet. We have to note that only positive values of n_x, n_y, n_z are permissible; therefore we have to divide by a factor 8. Allowing further for the two values of spin, we have to multiply by a factor 2. We get finally

$$Z(E)dE = CE^{1/2}dE \quad \text{with } C = \frac{4\pi L^3(2m)^{3/2}}{h^3}. \tag{6.10}$$

Equation (6.10) gives us the number of states, but we would also like to know the number of *occupied* states, that is, the number of states that contain electrons. For that we need to know the probability of occupation, $F(E)$. This function can be obtained by a not-too-laborious exercise in statistical mechanics. One starts with the Pauli principle (that no state can be occupied by more than one electron) and works out the most probable distribution on the condition that the total energy and the total number of particles are given. The result is the so-called Fermi–Dirac distribution*

$$F(E) = \frac{1}{\{\exp(E - E_F)/KT\} + 1}, \tag{6.11}$$

where E_F is a parameter called the Fermi level. It has the easily memorized property that at

$$E = E_F, \qquad F(E) = \tfrac{1}{2}, \tag{6.12}$$

that is, at the Fermi level the probability of occupation is $\tfrac{1}{2}$.

As may be seen in Fig. 6.1, $F(E)$ looks very different from the classical distribution $\exp(-E/kT)$. Let us analyse its properties in the following cases:

1. At $T = 0$.

$$\begin{matrix} F(E) = 1 \\ F(E) = 0 \end{matrix} \quad \text{for} \quad \begin{matrix} E < E_F \\ E > E_F. \end{matrix} \tag{6.13}$$

Thus, at absolute zero temperature, all the available states are occupied up to E_F, and all the states above E_F are empty. But remember, $Z(E)\,dE$ is the number of states between E and $E + dE$. Thus, the total number of states is

$$\int_0^{E_F} Z(E)\,dE, \tag{6.14}$$

which must equal the total number of electrons NL^3, where N is the number of electrons per unit volume. Thus, substituting eqn (6.10) into eqn (6.14),

* If we use the assumption that a state may contain any number of particles, the so-called Bose–Einstein distribution is obtained. It turns out that all particles belong to one or the other of these distributions and are correspondingly called fermions or bosons. For this book, it is of great importance that electrons are fermions and they obey the Fermi–Dirac distribution. The properties of bosons (e.g. quantized electromagnetic waves and lattice waves) are of somewhat less significance. We occasionally need to refer to them as photons and phonons but their statistics is usually irrelevant for our purpose. The Bose–Einstein distribution, and the so-called boson condensation does come into the argument in Section 12.4, where we talk briefly about atom lasers and in Chapter 14 concerned with superconductivity, but we shall not need any mathematical formulation of the distribution function.

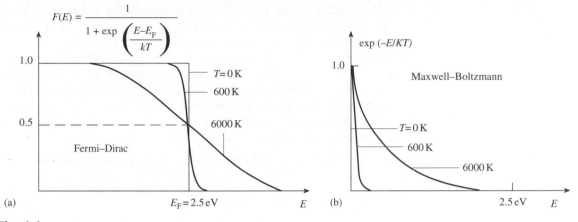

Fig. 6.1
(a) The Fermi–Dirac distribution function for a Fermi energy of 2.5 eV and for temperatures of 0 K, 600 K, and 6000 K.
(b) The classical Maxwell–Boltzmann distribution function of energies for the same temperatures.

the following equation must be satisfied:

$$(4\pi L^3 (2m)^{3/2}/h^3) \int_0^{E_F} E^{1/2} dE = NL^3. \qquad (6.15)$$

Integrating and solving for E_F, we get

$$E_F = \frac{h^2}{2m} \left(\frac{3N}{8\pi} \right)^{2/3}. \qquad (6.16)$$

E_F, calculated from the number of free electrons, is shown in Table 6.1. Thus, even at absolute zero temperature, the electrons' energies cover a wide range. This is strongly in contrast with classical statistics, where at $T = 0$, all electrons have zero energy.

2. For electron energies above the Fermi level, so that

$$E - E_F \gg kT, \qquad (6.17)$$

the term unity in eqn (6.11) may be neglected, leading to

$$F(E) \cong \exp \left\{ -\frac{(E - E_F)}{kT} \right\}, \qquad (6.18)$$

which you may recognize as the classical Maxwell–Boltzmann distribution. That is, for sufficiently large energies the Fermi–Dirac distribution is reduced to the Maxwell–Boltzmann distribution, generally referred to as the 'Boltzmann tail'.

3. For electron energies below the Fermi level, so that

$$E_F - E \gg kT, \qquad (6.19)$$

eqn (6.11) may be approximated by

$$F(E) \cong 1 - \exp \frac{(E - E_F)}{kT}. \qquad (6.20)$$

Table 6.1 *Fermi levels of metals calculated from eqn (6.16)*

Li	4.72 eV
Na	3.12 eV
K	2.14 eV
Rb	1.82 eV
Cs	1.53 eV
Cu	7.04 eV
Ag	5.51 eV
Al	11.70 eV

The probability of a state being occupied is very close to unity.

It is sometimes useful to talk about the probability of a state *not* being occupied and use the function $1 - F(E)$. We may say then, for the present case, that the probability of non-occupation varies exponentially.

4. In the range $E \approx E_F$ the distribution function changes rather abruptly from nearly unity to nearly zero. The rate of change depends on kT. For absolute zero temperature the change is infinitely fast, for higher temperatures (as can be seen in Fig. 6.1) it is more gradual. We may take this central region (quite arbitrarily) as between $F(E) = 0.9$ and $F(E) = 0.1$. The width of the region comes out then [by solving eqn (6.11)] to about $4.4\,kT$.

Summarizing, we may distinguish three regions for finite temperatures: from $E = 0$ to $E = E_F - 2.2\,kT$, where the probability of occupation is close to unity, and the probability of non-occupation varies exponentially; from $E = E_F - 2.2\,kT$ to $E = E_F + 2.2\,kT$, where the distribution function changes over from nearly unity to nearly zero; and from $E = E_F + 2.2\,kT$ to $E = \infty$, where the probability of occupation varies exponentially.

6.3 The specific heat of electrons

Classical theory, as we have mentioned before (Section 1.8), failed to predict the specific heat of electrons. Now we can see the reason. The real culprit is not the wave nature of the electron nor Schrödinger's equation but Pauli's principle. Since only one electron can occupy a state, electrons of lower energy are not in a position to accept the small amount of energy offered to them occasionally. The states above them are occupied, so they stay where they are. Only the electrons in the vicinity of the Fermi level have any reasonable chance of getting into states of higher energy; so they are the only ones capable of contributing to the specific heat.

The specific heat at constant volume per electron is defined as

$$c_V = \frac{d\langle E \rangle}{dT}, \tag{6.21}$$

where $\langle E \rangle$ is the average energy of electrons.

A classical electron would have an average energy $3/2\,kT$, which tends to zero as $T \to 0$. Quantum-mechanically, if an electron satisfies the Fermi–Dirac statistics, then the average energy of the electrons is finite and can quite easily be determined (see example 6.6). For the purpose of estimating the specific heat, we may make up a simple argument and claim that only the electrons in the region $E = E_F - 2.2\,kT$ to $E = E_F$ need to be considered as being able to respond to heat, and they can be regarded as if they were classical electrons possessing an energy $(3/2)\,kT$. Hence the average energy of these electrons is

$$\langle E \rangle = \frac{3}{2}kT\frac{2.2\,kT}{E_F}, \tag{6.22}$$

which gives for the specific heat

$$c_V = 6.6\frac{k^2}{E_F}T. \tag{6.23}$$

A proper derivation of the specific heat would run into mathematical difficulties, but it is very simple in principle. The average energy of an electron following a distribution $F(E)$ is given by

$$\langle E \rangle = \frac{1}{N} \int_0^\infty F(E) Z(E) E \, dE, \tag{6.24}$$

which should be evaluated as a function of temperature* and differentiated. The result is

$$c_v = \frac{\pi^2}{2} \frac{k^2}{E_F} T, \tag{6.25}$$

which agrees reasonably well with eqn (6.23) obtained by heuristic arguments. This electronic specific heat is vastly lower than the classical value $(3/2)k$ for any temperature at which a material can remain solid.

*See, for example. F. Seitz. *Modern theory of solids*. McGraw-Hill, New York, 1940, p. 146.

6.4 The work function

If the metal is heated, or light waves are incident upon it, then electrons may leave the metal. A more detailed experimental study would reveal that there is a certain threshold energy the electrons should possess in order to be able to escape. We call this energy (for historical reasons) the *work function* and denote it by ϕ. Thus, our model is as shown in Fig. 6.2. At absolute zero temperature all the states are filled up to E_F, and there is an external potential barrier ϕ.

It must be admitted that our new model is somewhat at variance with the old one. Not long ago, we calculated the energy levels of the electrons on the assumption that the external potential barrier is infinitely large, and now I go back on my word and say that the potential barrier is finite after all. Is this permissible? Strictly speaking, no. To be consistent, we should solve Schrödinger's equation subject to the boundary conditions for a finite potential well. But since the potential well is deep enough, and the number of electrons escaping is relatively small, there is no need to recalculate the energy levels. So I am cheating, but not excessively.

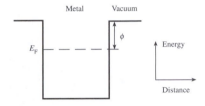

Fig. 6.2

Model for thermionic emission calculation. The potential barrier that keeps the electrons in the metal is above the Fermi energy level by an energy ϕ.

6.5 Thermionic emission

In this section we shall be concerned with the emission of electrons at high temperatures (hence the adjective *thermionic*). As we agreed before, the electron needs at least $E_F + \phi$ energy in order to escape from the metal, and all this, of course, should be available in the form of kinetic energy. Luckily, in the free-electron model, all energy is kinetic energy, since the potential energy is zero and the electrons do not interact; so the relationship between energy and momentum is simply

$$E = \frac{1}{2m}(p_x^2 + p_y^2 + p_z^2). \tag{6.26}$$

A further condition is that the electron, besides having the right amount of energy, must go in the right direction. Taking x as the coordinate perpendicular to the surface of the metal, the electron momentum must satisfy the inequality

$$\frac{p_x^2}{2m} > \frac{p_{x0}^2}{2m} = E_F + \phi. \tag{6.27}$$

However, this is still not enough. An electron may not be able to scale the barrier even if it has the right energy in the right direction. According to the rules of quantum mechanics, it may still suffer reflection. Thus, the probability of escape is $1 - r(p_x)$, where $r(p_x)$ is the reflection coefficient. If the number of electrons having a momentum between p_x and $p_x + dp_x$ is $N(p_x)dp_x$, then the number of electrons arriving at the surface per second per unit area is

$$\frac{p_x}{m} N(p_x)\, dp_x, \tag{6.28}$$

and the number of those escaping is

$$\{1 - r(p_x)\} \frac{p_x}{m} N(p_x)\, dp_x. \tag{6.29}$$

Adding the contributions from all electrons that have momenta in excess of p_{x_0}, we can write for the emission current density

$$J = \frac{e}{m} \int_{p_{x_0}}^{\infty} \{1 - r(p_x)\} p_x N(p_x) dp_x. \tag{6.30}$$

We may obtain the number of electrons in an infinitesimal momentum range in the same way as for the infinitesimal energy range. First, it consists of two factors, the density of states and the probability of occupation. The density of states, $Z(p_x)$, can be easily obtained by noting from eqns (6.2) and (6.26) that

$$p_x = \frac{h}{2} n_x, \quad p_y = \frac{h}{2} n_y, \quad p_z = \frac{h}{2} n_z. \tag{6.31}$$

The number of states in a cube of side one is exactly one. Therefore, the number of states in a volume of sides dn_x, dn_y, dn_z is equal to $dn_x\, dn_y\, dn_z$, which with the aid of eqn (6.31) can be expressed as

$$\left(\frac{2}{h}\right)^3 dp_x\, dp_y\, dp_z. \tag{6.32}$$

Dividing again by 8 (because only positive integers matter) and multiplying by two (because of spin) we get

$$Z(p_x, p_y, p_z) = \frac{2}{h^3}. \tag{6.33}$$

Hence, the number of electrons in the momentum range p_x, $p_x + dp_x$; p_y, $p_y + dp_y$; p_z, $p_z + dp_z$ is

$$N(p_x, p_y, p_z)\, dp_x\, dp_y\, dp_z$$
$$= \frac{2}{h^3} \frac{dp_x\, dp_y\, dp_z}{\exp[\{(1/(2\,m))(p_x^2 + p_y^2 + p_z^2) - E_F\}/kT] + 1}. \tag{6.34}$$

To get the number of electrons in the momentum range p_x, $p_x + dp_x$, the above equation needs to be integrated for all values of p_y and p_z

$$N(p_x)\, dp_x$$
$$= \frac{2}{h^3} dp_x \int_{-\infty}^{\infty} \int_{-\infty}^{\infty} \frac{dp_y\, dp_z}{\exp[\{(1/(2\,m))(p_x^2 + p_y^2 + p_z^2) - E_F\}/kT] + 1}. \tag{6.35}$$

This integral looks rather complicated, but since we are interested only in those electrons exceeding the threshold $\phi(\gg kT)$, we may neglect the unity term in the denominator. We are left then with some Gaussian functions, whose integrals between $\pm\infty$ can be found in the better integral tables (you can derive them for yourself if you are fond of doing integrals). This leads us to

$$N(p_x)\mathrm{d}p_x = \frac{4\pi mkT}{h^3}\mathrm{e}^{E_F/kT}\mathrm{e}^{-p_x^2/2mkT}\mathrm{d}p_x. \tag{6.36}$$

Substituting eqn (6.36) into (6.30) and assuming that $r(p_x) = r$ is independent of p_x, which is not true but gives a good enough approximation, the integration can be easily performed, leading to

$$J = A_0(1 - r)T^2\mathrm{e}^{-\phi/kT}, \tag{6.37}$$

where

$$A_0 = \frac{4\pi emk^2}{h^3} = 1.2 \times 10^6\ \mathrm{A\ m^{-2}\ K^{-2}}. \tag{6.38}$$

The most important factor in eqn (6.37) is $\exp(-\phi/kT)$, which is strongly dependent both on temperature and on the actual value of the work function. Take, for example, tungsten (the work functions for a number of metals are given in Table 6.2), for which $\phi \cong 4.5\,\mathrm{eV}$ and take $T = 2500\,K$. Then, a 10% change in the work function or temperature changes the emission by a factor of 8.

The main merit of eqn (6.37) is to show the exponential dependence on temperature, which is well borne out by experimental results. The actual numerical values are usually below those predicted by the equation, but this is not very surprising in view of the many simplifications we had to introduce. In a real crystal, ϕ is a function of temperature, of the surface conditions, and of the directions of the crystallographic axes, which our simple model did not take into account.

There is one more thing I would like to discuss, which is really so trivial that most textbooks do not even bother to mention it. Our analysis was one for a piece of metal in isolation. The electron current obtained in eqn (6.37) is the current that would start to flow if the sample were suddenly heated to a temperature T. But this current would not flow for long because, as electrons leave the metal, it becomes positively charged, making it more difficult for further electrons to leave. Thus, our formulae are valid only if we have some means of replenishing the electrons lost by emission. That is, we need an electric circuit like the one in Fig. 6.3(a). As soon as an electron is emitted from our piece of metal, another electron will enter from the circuit. The current flowing can be measured by an ammeter.

A disadvantage of this scheme is that the electrons travelling to the electrode will be scattered by air; we should really evacuate the place between the emitter and the receiving electrode, making up the usual cathode–anode configuration of a vacuum tube. This is denoted in Fig. 6.3(b) by the envelope shown. The electrons are now free to reach the anode but also free to accumulate in the vicinity of the cathode. This is bad again because by their negative charge they will compel many of their fellow electrons to interrupt their planned journey to the anode and return instead to the emitter. So again we do not measure the 'natural' current.

In order to prevent the accumulation of electrons in front of the cathode, a d.c. voltage may be applied to the anode [Fig. 6.3(c)]; this will sweep out

Table 6.2 *Work functions of metals*

Metal	Work function (eV)
Li	2.48
Na	2.3
K	2.2
Cs	1.9
Cu	4.45
Ag	4.46
Au	4.9
Mg	3.6
Ca	3.2
Ba	2.5
Al	4.2
Cr	4.6
Mo	4.2
Ta	4.2
W	4.5
Co	4.4
Ni	4.9
Pt	5.3

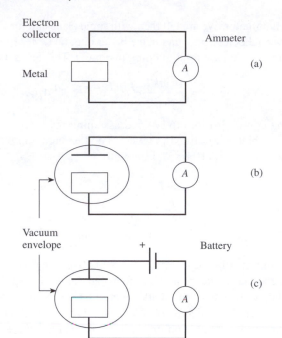

Fig. 6.3

Stages in measuring thermionic emission. (a) A current flows but it is impeded by air molecules. (b) A current flows in a vacuum until it builds up a charge cloud that repels further electrons. The steady-state ammeter reading is much less than the total emission current. (c) By employing a battery all the emission current is measured.

most of the unwanted electrons from the cathode–anode region. This is the arrangement used for measuring thermionic current.

The requirements to be fulfilled by cathode materials vary considerably according to the particular application. The cathodes must have a large emission current for high-power applications, low temperature for low-noise amplification, and long life when the tubes are used at not easily accessible places. All these various requirements have been admirably met by industry, though the feat should not be attributed to the powers of science. To make a good cathode is still an art, and a black art at that.

6.6   The Schottky effect

We are now going to refine our model for thermionic emission further by including (a) image force and (b) electric field.

It is a simple and rather picturesque consequence of the laws of electrostatics that the forces on an electron in front of an infinitely conducting sheet are correctly given by replacing the sheet by the 'mirror' charge (a positively charged particle the same distance behind the sheet as shown in Fig. 6.4). The force between these two charges is

$$F = \frac{e^2}{4\pi\epsilon_0}\frac{1}{(2x)^2},\tag{6.39}$$

and the potential energy is the integral of this force from the point x to infinity:

$$V(x) = \int_x^\infty F(y)\,\mathrm{d}y = -\frac{e^2}{16\pi\epsilon_0 x}.\tag{6.40}$$

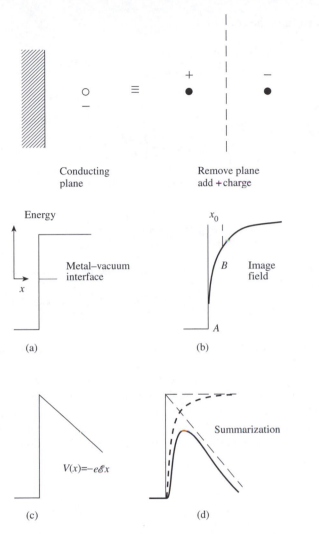

Fig. 6.4
The 'image charge' theorem. The effect of a plane conductor on the static field due to a charged particle is equivalent to a second, oppositely charged, particle in the mirror image position

Fig. 6.5
The Schottky effect. (a) Potential at metal–vacuum interface. (b) Potential changed by image charge field. (c) Potential due to applied anode voltage in vacuum region. (d) Total potential field showing reduction in height of the potential barrier compared with (a).

In the above calculation, we took the potential energy as zero at $x = \infty$ to agree with the usual conventions of electrostatics, but remember that our zero a short while ago was that of a valence electron at rest. Hence, to be consistent, we must redraw the energy diagram inside and outside the metal as shown in Fig. 6.5(a). If we include now the effect of the mirror charges*, then the potential barrier modifies to that shown in Fig. 6.5(b).

In the absence of an electric field this change in the shape of the potential barrier has practically no effect at all. But if we do have electric fields, the small correction due to mirror charges becomes significant.

For simplicity, let us investigate the case when the electric field is constant. Then,

$$V(x) = -e\mathcal{E}x, \qquad (6.41)$$

as shown in Fig. 6.5(c). If both an electric field is present and the mirror charges are taken into account, then the potentials should be added, leading to the potential barrier shown in Fig. 6.5(d). Clearly, there is a maximum that can be

* Note that the curve between A and B does not satisfy eqn (6.40). This is because the concept of a homogeneous sheet is no longer applicable when x is comparable with the interatomic distance. The energy is, however, given for $x = 0$ (an electron resting on the surface must have the same energy as an electron at rest inside the metal); so we simply assume that eqn (6.40) is valid for $x > x_0$ and connect the points A and B by a smooth line.

calculated from the condition

$$\frac{d}{dx}\left(-\frac{e^2}{16\pi\epsilon_0 x} - e\mathcal{E}x\right) = 0, \tag{6.42}$$

leading to

The energy needed to escape from the metal is reduced by $-V_{max}$.

$$V_{max} = -e\left(\frac{e\mathcal{E}}{4\pi\epsilon_0}\right)^{1/2}. \tag{6.43}$$

The effective work function is thus reduced from ϕ to

$$\phi_{eff} = \phi - e\left(\frac{e\mathcal{E}}{4\pi\epsilon_0}\right)^{1/2}. \tag{6.44}$$

Substituting this into eqn (6.37) we get the new formula for thermionic emission

$$J = A_0(1-r)T^2 \exp\left[-\frac{\{\phi - e\sqrt{(e\mathcal{E}/4\pi\epsilon_0)}\}}{kT}\right]. \tag{6.45}$$

The reduction in the effective value of the work function is known as the *Schottky effect*, and plotting $\log J$ against $\mathcal{E}^{1/2}$, we get the so-called *Schottky line*. A comparison with experimental results in Fig. 6.6 shows that above a certain value of the electric field the relationship is quite accurate. Do not be too much impressed, though; in graphs of this sort the constants are generally fiddled to get the theoretical and experimental curves on top of each other. But it certainly follows from Fig. 6.6 that the functional relationship between J and $\mathcal{E}^{1/2}$ is correct.

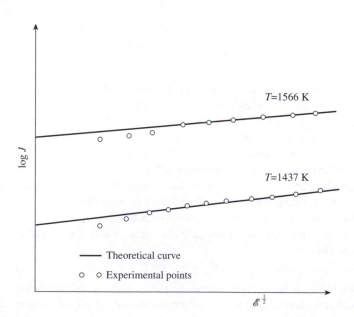

Fig. 6.6
Experimental verification of the Schottky formula [eqn (6.45)].

6.7 Field emission

As we have seen in the previous section, the presence of an electric field increases the emission current because more electrons can escape over the reduced barrier. If we increase the electric field further, towards $10^9\,\mathrm{V\,m^{-1}}$, then a new escape route opens up. Instead of going *over* the potential barrier, the electrons *tunnel across* it. It may be seen in Fig. 6.7 that for high-enough electric fields the barrier is thin, and thus electrons may sneak through. This is called *field emission*, and it is practically independent of temperature.

To derive a theoretical formula for this case, we should consider all the electrons that move towards the surface and calculate their tunnelling probability. It follows from the shape of the potential barrier that electrons with higher energy can more easily slip through, but (at ordinary temperatures) there are few of them; so the main contribution to the tunnelling current comes from electrons situated around the Fermi level. For them the width of the barrier is calculable from the equation (see Fig. 6.7)

$$-\phi = -e\mathcal{E}x_{\mathrm{F}}, \tag{6.46}$$

and the height of the potential barrier they face is ϕ_{eff}. Hence, very approximately, we may represent the situation by the potential profile of Fig. 6.8. It may be shown (see Exercise 3.7) that the tunnelling current varies approximately exponentially with barrier width,

$$J \sim \exp\left(-\frac{(2\,\phi_{\mathrm{eff}})^{1/2}}{\hbar}x_{\mathrm{F}}\right), \tag{6.47}$$

which, with the aid of eqn (6.46) reduces to

$$J \sim \exp\left(-\frac{(2\,m)^{1/2}}{\hbar e}\frac{\phi_{\mathrm{eff}}^{1/2}\phi}{\mathcal{E}}\right). \tag{6.48}$$

The exponential factor in eqn (6.48) represents quite a good approximation to the exact formula, which is unfortunately too long to quote. It may be noted that the role of temperature in equations (6.37) and (6.45) is taken over here by the electric field.

The theory has been fairly well confirmed by experiments. The major difficulty in the comparison is to take account of surface irregularities. The presence of any protuberances considerably alters the situation because the electric field is higher at those places. This is a disadvantage as far as the interpretation of the measurements is concerned, but the existence of the effect made possible the invention of an ingenious device called the field-emission microscope.

6.8 The field-emission microscope

The essential part of a field-emission microscope is a very sharp tip ($\approx 100\,\mathrm{nm}$ in diameter), which is placed in an evacuated chamber (Fig. 6.9). A potential of a few thousand volts is applied between the tip (made usually of tungsten) and the anode, which creates at the tip an electric field high enough to draw out electrons. The emitted electrons follow the lines of force and produce a magnified

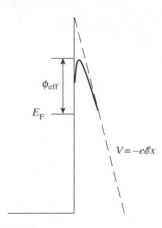

Fig. 6.7
With very high applied electric fields the potential barrier is thin, thus, instead of moving over the barrier, electrons at the Fermi level may tunnel across the barrier.

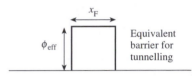

Fig. 6.8
Equivalent barrier, for simplifying the calculation of tunnelling current in Fig. 6.7.

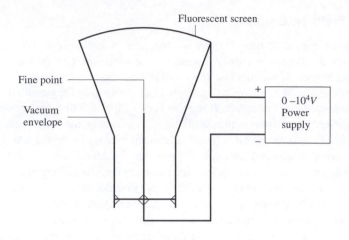

Fluorescent screen

Fine point

Vacuum
envelope

+

0 –10⁴V
Power
supply

–

Fig. 6.9
Sketch showing the principle of
field-emission microscope.

picture (magnification = r_2/r_1, where r_2 = radius of the screen and r_1 = radius
of the tip) on the fluorescent screen. Since the magnification may be as large
as 10^6, we could expect to see the periodic variation in the electron emission
caused by the atomic structure. The failure to observe this is explained by two
reasons: quantum-mechanical diffraction, and deviation from the 'theoretical'
course owing to a random transverse component in the electron velocity when
leaving the metal.

The limitations we have just mentioned can be overcome by introducing
helium into the chamber and reversing the polarity of the applied potential.
The helium atoms that happen to be in the immediate vicinity of the tungsten
tip become ionized owing to the large electric field, thus acquiring a positive charge, and move to the screen. Both the quantum-mechanical diffraction
(remember, the de Broglie wavelength is inversely proportional to mass) and
the random thermal velocities are now smaller, so that the resolution is higher
and individual atoms can indeed be distinguished as shown in Fig. 6.10. This
device, called the *field-ion microscope*, was the first in the history of science to
make individual atoms visible. Thus, just about two and a half millennia after
introducing the concept of atoms, it proved possible to see them.

6.9   The photoelectric effect

Emission of electrons owing to the incidence of electromagnetic waves is called
the *photoelectric effect*. The word *photo* (*light* in Greek) came into the description because the effect was first found in the visible range. Interestingly, it is one
of the earliest phenomena that cast serious doubts on the validity of classical
physics and was instrumental in the birth of quantum physics.

The basic experimental set-up may be seen in Fig. 6.11. When an electromagnetic wave is incident, an electric current starts to flow between the electrodes.
The magnitude of the current is proportional to the input electromagnetic power,
but there is no current unless the frequency is high enough to make

$$\hbar\omega > \phi.$$ (6.49)

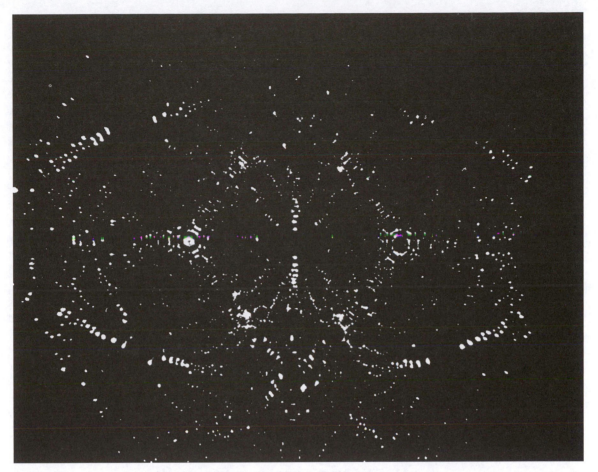

Fig. 6.10
Field-ion micrograph of a tungsten tip. The atoms on the surface can be clearly distinguished (Courtesy of E. W. Muller).

This is probably the best place to introduce photons which are the particle equivalents of electromagnetic waves. Each photon carries an energy of

$$E = \hbar\omega. \tag{6.50}$$

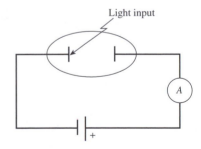

Electrons will be emitted when this energy is larger than the work function. In an electromagnetic wave of power P and frequency ω the number of photons incident per unit time is $N_{\text{phot}} = P/\hbar\omega$.

A detailed calculation of the current is not easy because a photon is under no obligation to give its energy to an electron. One must calculate transition probabilities, which are different at the surface and in the bulk of the material. The problem is rather complex; we shall not go more deeply into the theory. It might be some consolation for you that the first engineers who used and designed photocells (the commercially available device based on the photoelectric effect) knew much less about its functioning than you do.

Fig. 6.11
An experiment showing the photoelectric effect. If the frequency of the light is above a certain threshold value, the incident photons knock out electrons from the cathode. These cause a current in the external circuit by moving to the anode.

6.10   Quartz–halogen lamps

Filaments of tungsten have been used not only as sources of electrons but also as radiating elements in light bulbs. The basic design has changed little during the last hundred years until quite recently when the quartz–halogen lamps were developed. The advantage of the quartz envelope (aided by the judicious use of molybdenum in the sealing process) is that the possible running temperature is higher than with ordinary glass envelopes, and so we can get much better luminous efficiency (light output is proportional to T^4). However, this *alone* is not much good because the tungsten filament has long been a prime example of the universal law of cussedness (things will go wrong if they can) called Sod's or Murphy's law, depending on which side of the Atlantic (and how far from Ireland) you live. What happens is that the filament has a region of cracks or thinning that has higher resistance and thus gets hotter than the rest. Thus, the local rate of evaporation is increased, it gets still thinner, and by a rapidly accelerating process of positive feedback, burnout occurs. Incidentally, the fact that a light seems much brighter for a few seconds before it burns out, even although the electrical power consumed is less, is a qualitative confirmation of the T^4 law. This effect can be overcome by adding some halogen gas, such as chlorine, to the lamp during processing. The tungsten vapour is now converted into chloride, which is sufficiently volatile to leave the hot silica envelope transparent. When chloride molecules strike the much hotter filament, they decompose, depositing tungsten and liberating chlorine to take part in further reactions. The rate of depositing goes up with temperature, so that a 'hot spot' is thickened, and hence cooled. This negative feedback process stabilizes the lamp. So next time you are dazzled by a quartz–halogen headlight, remember that it is an example of the very rare anti-Sod's law.

6.11   The junction between two metals

If two metals of different work functions are brought into contact (Fig. 6.12), the situation is clearly unstable. Electrons will cross from left to right to occupy the lower energy states available. However, as electrons cross over there will be an excess of positive charge on the left-hand side and an excess of negative charge on the right-hand side. Consequently, an electric field is set up with a polarity that hinders the flow of electrons from left to right and encourages the flow of electrons from right to left. A dynamic equilibrium is established when equal numbers of electrons cross in both directions. At what potential difference will this occur? An exact solution of this problem belongs to the

Fig. 6.12
The Fermi levels and work functions of two metals to be brought into contact.

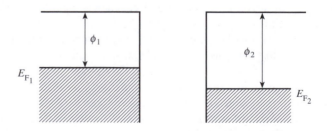

domain of statistical thermodynamics. The solution is fairly lengthy, but the answer, as is so often the case in thermodynamics, could hardly be simpler.

The potential difference between the two metals, called the *contact potential*, is equal to the difference between the two work functions; or, in more general terms, the potential difference may be obtained by equating the Fermi levels of the two media in contact. This is a general law valid for any number of materials in equilibrium at any temperature.

The resulting energy diagram is shown in Fig. 6.13. The potential difference appearing between the two metals is a real one. If we could put an extra electron in the contact region, it would feel a force towards the left. The potential difference is real but, alas, it cannot perform the function of a battery. Why? Because in real life you never get something for nothing and, anyway, extracting power from an equilibrium state is against the second law of thermodynamics.

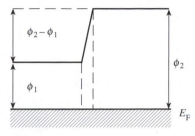

Fig. 6.13
When the two metals are brought into contact there is potential difference $\phi_2 - \phi_1$, between them.

Exercises

6.1. Evaluate the Fermi function for an energy kT above the Fermi energy. Find the temperature at which there is a 1% probability that a state, with an energy 0.5 eV above the Fermi energy, will be occupied by an electron.

6.2. Indicate the main steps in the derivation of the Fermi level and calculate its value for sodium from the data given in example 1.4.

6.3. Ultraviolet light of 0.2 μm wavelength is incident upon a metal. Which of the metals listed in Table 6.2 will emit electrons in response to the input light?

6.4. Determine the density of occupied states at an energy kT above the Fermi level. Find the energy below the Fermi level which will yield the same density of occupied states.

6.5. Use free electron theory to determine the Fermi level in a two-dimensional metal. Take N as the number of electrons per unit area.

6.6. Show that the average kinetic energy of free electrons, following Fermi–Dirac statistics, is $(3/5)E_F$ at $T = 0$ K.

6.7. A tungsten filament is 0.125 mm diameter and has an effective emitting length of 15 mm. Its temperature is measured with an optical pyrometer at three points, at which also the voltage, current, and saturated emission current to a 5 mm diameter anode are measured as given in the table below.

(i) Check that the data obey the Richardson law, and estimate the work function and value of A_0 in eqn (6.37).
(ii) Find a mean value for the temperature coefficient of resistance.
(iii) Find how the heater power varies with temperature, and estimate the Stefan–Boltzmann coefficient.

(iv) If the anode voltage is increased to 2.3 kV by how much will the emission current rise?

Filament temperature (K)	2000	2300	2600
Filament current (A)	1.60	1.96	2.30
Filament voltage (V)	3.37	5.12	7.40
Emission current (mA)	0.131	5.20	91.2

6.8. Light from a He–Ne laser (wavelength 632.8 nm) falls on a photo-emissive cell with a quantum efficiency of 10^{-4} (the number of electrons emitted per incident photon). If the laser power is 2 mW, and all liberated electrons reach the anode, how large is the current? Could you estimate the work function of the cathode material by varying the anode voltage of the photocell?

6.9. Work out the Fermi level for conduction electrons in copper. Estimate its specific heat at room temperature; what fraction of it is contributed by the electrons? Check whether your simple calculation agrees with data on specific heat given in a reference book.

Assume one conduction electron per atom. The atomic weight of copper is 63.5 and its density 9.4×10^3 kg m^{-3}.

6.10. Figure 6.14(a) shows the energy diagram for a metal–insulator–metal sandwich at thermodynamic equilibrium. Take the insulator as representing a high potential barrier. The temperature is sufficiently low for all states above the Fermi level to be regarded as unoccupied. When a voltage U is applied [Fig. 6.14(b)] electrons may tunnel through the insulator from left to right. Assume that the tunnelling current in each energy range dE is proportional to the number of filled

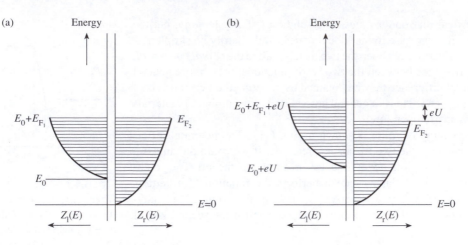

Fig. 6.14

Energy against density of states for a metal–insulator–metal tunnel junction.

states from which tunnelling is possible and to the number of empty states on the other side into which electrons can tunnel.

In the coordinate system of Fig. 6.14(a) the density of states as a function of energy may be written as

$$Z_1(E) = C_1(E - E_0)^{1/2} \quad \text{for } E > E_0$$

and

$$Z_r(E) = C_r E^{1/2} \quad \text{for } E > 0,$$

where C_1, C_r, and E_0 are constants.

Show (i) that the tunnelling current takes the form

$$I \sim \int_{E_{F_2}}^{E_{F_2} + eU} (E - eU - E_0)^{1/2} E^{1/2} \, dE$$

and (ii) that Ohm's law is satisfied for small voltages.

6.11. Assume that the energy levels in a certain system are integral multiples of a basic unit (zero energy being permitted), and each energy level is doubly degenerate, which means that two different states can have the same energies. Assume further that there are only five fermions (particles which obey the rule that only one of them can occupy a state) in the system with a total energy of 12 units.

Find 10 allowed distributions of the particles into the energy levels mentioned.

The band theory of solids

Band of all evils, cradle of causeless care.
Sir Philip Sidney

7.1 Introduction

Most properties of metals can be well explained with the aid of the free electron model, but when we come to insulators and semiconductors the theory fails. This is not very surprising because the term 'free electron', by definition, means an electron free to roam around and conduct electricity; and we know that the main job of an insulator is to insulate; that is, *not* to conduct electricity. It is not particularly difficult to find a model explaining the absence of electrical conductivity. We only need to imagine that the valence electrons cling desperately to their respective lattice ions and are unwilling to move away. So we are all right at the two extremes; free electrons mean high conductivity, tightly bound electrons mean no conductivity. Now what about semiconductors? They are neither good conductors nor insulators; so neither model is applicable. What can we do? Well, we have touched upon this problem before. Silicon and germanium are semiconductors in spite of the covalent bonds between the atoms. The bonding process uses up all the available electrons, so at absolute zero temperature there are no electrons available for conduction. At finite temperatures however, some of the electrons may escape. The lattice atoms vibrate randomly, having *occasionally* much more than the average thermal energy. Thus, at certain instants at certain atoms there is enough energy to break the covalent bond and liberate an electron. This is a possible description of the electrical properties of semiconductors and, physically, it seems quite plausible. It involves no more than developing our physical picture of the covalent bond a little further by taking account of thermal vibrations as well. All we need to do is to put these arguments into some quantitative form, and we shall have a theory of semiconductors. It can be done, but somehow the ensuing theory never caught the engineers' imagination.

The higher the temperature the more likely it is that some electrons escape.

The theory that did gain wide popularity is the one based on the concept of energy bands. This theory is more difficult to comprehend initially, but once digested and understood it can provide a solid foundation for the engineers' flights of fancy.

The job of engineers is to invent. Physicists discover the laws of nature, and engineers exploit the phenomena for some useful (sometimes not so useful) end. But in order to exploit them, the engineer needs to combine the phenomena, to regroup them, to modify them, to interfere with them; that is, to create something new from existing components. Invention has never been an easy task, but at least in the good old days the basic mechanism was simple to understand. It was not very difficult to be wise after the event. It was, for example, an early triumph of engineering to turn the energy of steam into a steam engine but, having accomplished the feat, most people could comprehend that the expanding steam moved a piston, which was connected to a wheel, etc. It needed perhaps a little more abstract thought to appreciate Watt's invention of the separate condenser, but even then any intelligent man willing to devote half an hour of his time to the problem of heat exchange could realize the advantages. Alas, these times have gone. No longer can a layman hope to understand the working principles of complex mechanisms, and this is particularly true in electronics. And most unfortunately, it is true not only for the layman. Even electronic engineers find it hard nowadays to follow the phenomena in an electronic device. Engineers may nowadays be expected to reach for the keyboard of a computer at the slightest provocation, but the fundamental equations are still far too complicated for a direct numerical attack. We need models. The models need not be simple ones, but they should be comprehensive and valid under a wide range of conditions. They have to serve as a basis for intuition. Such a model and the concurrent physical picture are provided by the band theory of solids. It may be said without undue exaggeration that the spectacular advance in solid-state electronic devices in the second half of the 20th century owes its existence to the power and simplicity of the band theory of solids.

Well, after this rather lengthy introduction, let us see what this theory is about. There are several elementary derivations, each one giving a slightly different physical picture. Since our aim is a thorough understanding of the basic ideas involved it is probably the best to show you the three approaches I know.

7.2 The Kronig–Penney model

This model is historically the first (1930) and is concerned with the solution of Schrödinger's equation, assuming a certain potential distribution inside the solid. According to the free electron model, the potential inside the solid is uniform; the Kronig–Penney model, goes one step further by taking into account the variation of potential due to the presence of immobile lattice ions.

If we consider a one-dimensional case for simplicity, the potential energy of an electron is shown in Fig. 7.1. The highest potential is halfway between the ions, and the potential tends to minus infinity as the position of the ions is approached. This potential distribution is still very complicated for a straightforward mathematical solution. We shall, therefore, replace it by a simpler one, which still displays the essential features of the function, namely (i) it has the same period as the lattice; (ii) the potential is lower in the vicinity of the lattice ion and higher between the ions. The potential distribution chosen is shown in Fig. 7.2.

The solution of the time-independent Schrödinger equation

The ions are located at $x=0, a, 2a, \ldots$ etc. The potential wells are separated from each other by potential barriers of height V_0, and width w.

$$\frac{\hbar^2}{2m} \frac{\mathrm{d}^2 \psi}{\mathrm{d}x^2} + \{E - V(x)\}\psi = 0 \qquad (7.1)$$

for the above chosen potential distribution is not too difficult. We can solve it for the $V(x) = V_0/2$ and $V(x) = -V_0/2$ regions separately, match the solutions at the boundaries, and take good care that the solution is periodic. It is all fairly simple in principle; one needs to prove a new theorem followed by

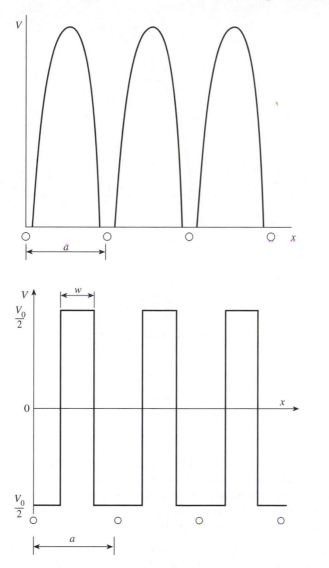

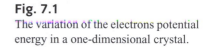

Fig. 7.1
The variation of the electrons potential
energy in a one-dimensional crystal.

Fig. 7.2
An approximation to the potential
energy of Fig. 7.1, suitable for
analytical calculations.

a derivation, which takes the best part of an hour, and then one gets the final
result. We cannot go through the lot, so I shall just say that the wavefunctions
are assumed to be of the form

$u_k(x)$ is a periodic function.

$$u_k(x)e^{ikx}. \tag{7.2}$$

A solution exists if k is related to the energy E by the following equation*

$$\cos ka = P\frac{\sin \alpha a}{\alpha a} + \cos \alpha a, \tag{7.3}$$

where

$$P = \frac{ma}{\hbar^2}V_0 w, \tag{7.4}$$

* There is actually one more mathematical simplification introduced in arriving at eqn (7.3), namely w and V_0 are
assumed to tend to zero and infinity
respectively, with their product $V_0 w$ kept
constant.

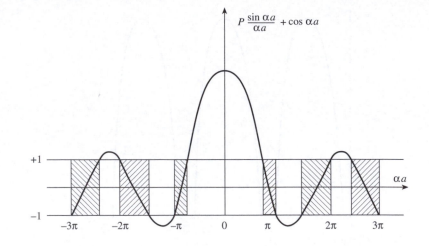

Fig. 7.3

The right-hand side of eqn (7.3) for $P = 3\pi/2$ as a function of αa.

and

$$\alpha = \frac{1}{\hbar}\sqrt{(2mE)}. \tag{7.5}$$

Remember, for a free electron we had the relationship

$$E = \frac{\hbar^2 k^2}{2m}. \tag{7.6}$$

The relationship is now different, implying that the electron is no longer free.

In order to find the $E - k$ curve, we plot the right-hand side of eqn (7.3) in Fig. 7.3 as a function of αa. Since the left-hand side of eqn (7.3) must always be between $+1$ and -1, a solution exists only at those values of E for which the right-hand side is between the same limits; that is, there is a solution for the shaded region and no solution outside the shaded region. Since α is related to E, this means that the electron may possess energies within certain bands but not outside them. This is our basic conclusion, but we can draw some other interesting conclusions from eqn (7.3).

There are allowed and forbidden bands of energy.

1. If $V_0 w$ is large, that is, if P is large, the function described by the right-hand side of eqn (7.3) crosses the $+1, -1$ region at a steeper angle, as shown in Fig. 7.4. Thus, the allowed bands are narrower and the forbidden bands wider. In the limit $P \to \infty$ the allowed band reduces to one single energy level; that is, we are back to the case of the discrete energy spectrum existing in isolated atoms.

For $P \to \infty$ it follows from eqn (7.3) that

$$\sin \alpha a = 0; \tag{7.7}$$

that is, the permissible values of energy are

$$E_n = \frac{\pi^2 \hbar^2}{2ma^2} n^2, \tag{7.8}$$

which may be recognized as the energy levels for a potential well of width a. Accordingly, all electrons are independent of each other, and each one is confined to one atom by an infinite potential barrier.

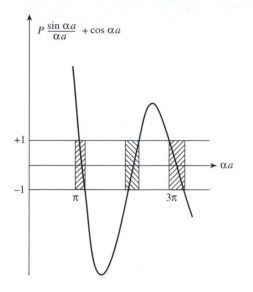

Fig. 7.4
The right-hand side of eqn (7.3) for $P = 6\pi$ as a function of αa.

2. In the limit $P \to 0$, we get

$$\cos \alpha a = \cos ka; \qquad (7.9)$$

that is,

$$E = \frac{\hbar^2 k^2}{2m}, \qquad (7.10)$$

as for the free electron. Thus, by varying P from zero to infinity, we cover the whole range from the completely free electron to the completely bound electron.

3. At the boundary of an allowed band $\cos ka = \pm 1$; that is,

$$k = \frac{n\pi}{a}, \quad n = 1, 2, 3 \ldots \qquad (7.11)$$

Looking at a typical energy versus k plot (Fig. 7.5), we can see that the discontinuities in energy occur at the values of k specified above. We shall say more about this curve, mainly about the discontinuities in energy, but let us see first what the other models can tell us.

7.3 The Ziman model

This derivation relies somewhat less on mathematics and more on physical intuition. We may start again with the assertion that the presence of lattice ions will make the free electron model untenable—at least under certain circumstances.

Let us concentrate now on the wave aspect of the electron and look upon a free electron as a propagating plane wave. Its wavefunction is then

$$\psi_k = e^{ikx}. \qquad (7.12)$$

We know that waves (whether X-rays or electron waves) can easily move across a crystal lattice; so plane waves (i.e. free electrons) may after all represent

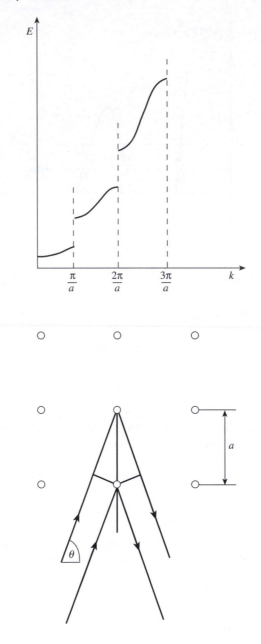

Fig. 7.5
The energy as a function of k. The discontinuities occur at $k = n\pi/a$, $n = 1, 2, 3\ldots$

Fig. 7.6
Geometry of reflection from atomic planes.

the truth. Waves *may* move across a crystal lattice, but not always. There is strong disturbance when the individual reflections add in phase; that is, when

$$n\lambda = 2a \sin\theta, \quad n = 1, 2, 3\ldots, \tag{7.13}$$

as follows from the sketch in Fig. 7.6. This is a well-known relationship (called *Bragg reflection*) for X-rays and, of course, it is equally applicable to electron waves. So we may argue that the propagation of electrons is strongly disturbed whenever eqn (7.13) is satisfied. In one dimension the condition reduces to

$$n\lambda = 2a. \tag{7.14}$$

Using the relationship between wavelength and wave number, the above equation may be rewritten as

$$k = \frac{n\pi}{a}. \tag{7.15}$$

Thus, we may conclude that our free electron model is not valid when eqn (7.15) applies. The wave is reflected, so the wave function should also contain a term representing a wave in the opposite direction

$$\psi_{-k} = \mathrm{e}^{-\mathrm{i}kx}. \tag{7.16}$$

Since waves of that particular wavelength are reflected to and fro, we may expect the forward- and backward-travelling waves to be present in the same proportion; that is, we shall assume wave functions in the form

$$\psi_{\pm} = \frac{1}{\sqrt{2}}(\mathrm{e}^{\mathrm{i}kx} \pm \mathrm{e}^{-\mathrm{i}kx}) = \sqrt{2}\begin{pmatrix} \cos kx \\ \mathrm{i}\sin kx \end{pmatrix}, \tag{7.17}$$

where the constant is chosen for correct normalization.

Let us now calculate the potential energies of the electrons in both cases. Be careful; we are not here considering potential energy in general but the potential energy of the electrons that happen to have the wave functions $\psi_{\pm}$. These electrons have definite probabilities of turning up at various places, so their potential energy* may be obtained by averaging the *actual* potential $V(x)$ weighted by the probability function $|\psi_{\pm}|^2$. Hence,

$$V_{\pm} = \frac{1}{L}\int |\psi_{\pm}|^2 V(x)\,\mathrm{d}x$$

$$= \frac{1}{L}\int \begin{pmatrix} 2\cos^2 kx \\ 2\sin^2 kx \end{pmatrix} V(x)\,\mathrm{d}x. \tag{7.18}$$

Since $k = n\pi/a$, the function $V(x)$ contains an integral number of periods of $|\psi_{\pm}|^2$; it is therefore sufficient to average over one period. Hence,

$$V_{\pm} = \frac{1}{a}\int_0^a \begin{pmatrix} 2\cos^2 kx \\ 2\sin^2 kx \end{pmatrix} V(x)\,\mathrm{d}x$$

$$= \frac{1}{a}\int_0^a \begin{pmatrix} 1 + \cos 2kx \\ 1 - \cos 2kx \end{pmatrix} V(x)\,\mathrm{d}x$$

$$= \pm\frac{1}{a}\int_0^a \cos 2kx\, V(x)\,\mathrm{d}x, \tag{7.19}$$

since $V(x)$ integrates to zero. Therefore,

$$V_{\pm} = \pm V_n. \tag{7.20}$$

* You may also look upon eqn (7.18) as an application of the general formula given by eqn (3.43).

L is the length of the one-dimensional 'crystal' and $V(x)$ is the same function that we met before in the Kronig–Penney model but now, for simplicity, we take $2w = a$.

The integration in eqn (7.19) can be easily performed, but we are not really interested in the actual numerical values. The important thing is that $V_n \neq 0$ and has opposite signs for the wave functions $\psi_\pm$.

Let us go through the argument again. If the electron waves have certain wave numbers [satisfying eqn (7.15)], they are reflected by the lattice. For each value of k two distinct wave functions ψ_+ and ψ_- can be constructed, and the corresponding potential energies turn out to be $+V_n$ and $-V_n$.

The kinetic energies are the same for both wave functions, namely

$$E = \frac{\hbar^2 k^2}{2m}. \tag{7.21}$$

Thus, the total energies are

$$E_\pm = \frac{\hbar^2 k^2}{2m} \pm V_n. \tag{7.22}$$

This is shown in Fig. 7.7 for $k = \pi/a$. The energy of the electron may be

$$\frac{\hbar^2 k^2}{2m} - V_1 \quad \text{or} \quad \frac{\hbar^2 k^2}{2m} + V_1, \tag{7.23}$$

but *cannot be any value in between*. There is an energy gap.

What will happen when $k \neq n\pi/a$? The same argument can be developed further, and a general form may be obtained for the energy.*

It is, however, not unreasonable to assume that apart from the discontinuities already mentioned, the $E-k$ curve will proceed smoothly; so we could construct it in the following manner. Draw the free electron parabola (dotted lines in Fig. 7.8) add and subtract V_n at the points $k = n\pi/a$, and connect the end points with a smooth curve, keeping close to the parabola. Not unexpectedly, Fig. 7.8 looks like Fig. 7.5, obtained from the Kronig–Penney model.

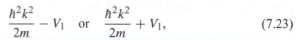

Fig. 7.7
The two possible values of the electron's total energy at $k = \pi/a$.

* J.M. Ziman, *Electrons in metals, a short guide to the Fermi surface*, Taylor and Francis, 1962.

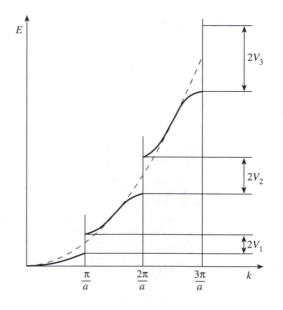

Fig. 7.8
Construction of the $E-k$ curve from the free electron parabola.

7.4 The Feynman model

This is the one I like best, because it combines mathematical simplicity with an eloquent physical picture. It is essentially a generalization of the model we used before to understand the covalent bond—another use of the coupled mode approach.

Remember, the energy levels of two interacting atoms are split; one is slightly higher, the other slightly below the original (uncoupled) energy. What happens when n atoms are brought close together? It is not unreasonable to expect that there will be an n-fold split in energy. So if the n atoms are far away from each other, each one has its original energy levels denoted by E_1 and E_2 in Fig. 7.9(a), but when there is interaction they split into n separate energy levels. Now looking at this cluster of energy levels displayed in Fig. 7.9(b), we are perfectly entitled to refer to allowed energy bands and to forbidden gaps between them.

To make the relationship a little more quantitative, let us consider the one-dimensional array of atoms shown in Fig. 7.10. We shall now put a single electron on atom j into an energy level E_1 and define by this the state (j). Just as we discussed before in connection with the hydrogen molecular ion, there is a finite probability that the electron will jump from atom j to atom $j + 1$, that is from state (j) into state $(j + 1)$. There is of course no reason why the electron should jump only in one direction; it has a chance of jumping the other way too. So the transition from state (j) into state $(j - 1)$ must have equal probability. It is quite obvious that a direct jump to an atom farther away is also possible but much less likely; we shall therefore disregard that possibility.

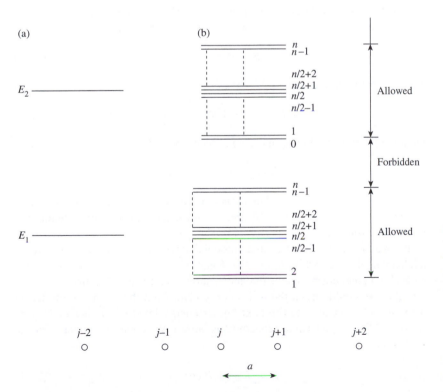

Fig. 7.9
There is an n-fold split in energy when n atoms are brought close to each other, resulting in a band of allowed energies, when n is large.

Fig. 7.10
A one-dimensional array of atoms.

We have now a large number of states so we should turn to eqn (5.24), which looks formidable with j running through all the atoms of the one-dimensional crystal. Luckily only nearest neighbours are coupled (or so we claim), so the differential equation for atom j takes the rather simple form

$$i\hbar \frac{dw_j}{dt} = E_1 w_j - A w_{j-1} - A w_{j+1}, \qquad (7.24)$$

where, as we mentioned before, E_1 is the energy level of the electron in the absence of coupling, and the coupling coefficient is taken again as $-A$. We could write down analogous differential equations for each atom, but fortunately there is no need for it. We can obtain the general solution for the whole array of atoms from eqn (7.24).

Let us assume the solution in the form

E is the energy to be found.

$$w_j = K_j e^{-iEt/\hbar}, \qquad (7.25)$$

Substituting eqn (7.25) into eqn (7.24) we get

$$E K_j = E_1 K_j - A(K_{j-1} + K_{j+1}). \qquad (7.26)$$

Note now that atom j is located at x_j, and its neighbours at $x_j \pm a$, respectively. We may therefore look upon the amplitudes K_j, K_{j+1}, and K_{j-1} as functions of the x-coordinate. Rewriting eqn (7.26) in this new form we get

$$E K(x_j) = E_1 K(x_j) - A\{K(x_j + a) + K(x_j - a)\}. \qquad (7.27)$$

This is called a *difference equation* and may be solved by the same method as a differential equation. We can assume the trial solution

$$K(x_j) = e^{ikx_j}, \qquad (7.28)$$

which, substituted into eqn (7.27) gives

$$E e^{ikx_j} = E_1 e^{ikx_j} - A\{e^{ik(x_j+a)} + e^{ik(x_j-a)}\}. \qquad (7.29)$$

Dividing by $\exp(ikx_j)$ the eqn (7.29) reduces to the final form

$$E = E_1 - 2A \cos ka, \qquad (7.30)$$

which is plotted in Fig. 7.11. Thus, once more, we get the result that energies within a band, between $E_1 - 2A$ and $E_1 + 2A$, are allowed and outside that range are forbidden.

It is a great merit of the Feynman model that we have obtained a very simple mathematical relationship for the $E - k$ curve within a given energy band. But what about other energy bands that have automatically come out from the other models? We could obtain the next energy band from the Feynman model by planting our electron into the next higher energy level of the isolated atom, E_2, and following the same procedure as before. We could then obtain for the next band

The coupling coefficient between nearest neighbours is now taken as $-B$.

$$E = E_2 - 2B \cos ka, \qquad (7.31)$$

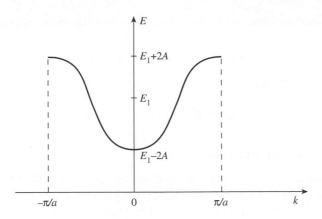

Fig. 7.11
Energy as a function of k obtained
from the Feynman model.

Another great advantage of the Feynman model is that it is by no means restricted to electrons. It could apply to any other particles. But, you may ask, what other particles can there be? Well, we have talked about positively charged particles called holes. They may be represented as deficiency of electrons, and so they too can jump from atom to atom. But there are even more interesting possibilities. Consider, for example, an atom that somehow gets into an excited state, meaning that one of its electrons is in a state of higher energy, and can with a certain probability transfer its energy to the next atom down the line. The concepts are all familiar, and so we may describe this process in terms of a particle moving across the lattice.

Yet another advantage of this model is its easy applicability to three-dimensional problems. Whereas the three-dimensional solution of the Kronig–Penney model would send shudders down the spines of trained numerical analysts, the solution of the same problem, using the Feynman approach, is well within the power of engineering undergraduates, as you will presently see.

In a three-dimensional lattice, assuming a rectangular structure, the distances between lattice points are a, b, and c in the directions of the coordinates axes x, y, and z, respectively. Denoting the probability that an electron is attached to the atom at the point x, y, z by $|w(x, y, z, t)|^2$, we may write down a differential equation analogous to eqn (7.24):

$$i\hbar \frac{\partial w(x, y, z, t)}{\partial t} = E_1 w(x, y, z, t)$$
$$- A_x w(x + a, y, z, t) - A_x w(x - a, y, z, t)$$
$$- A_y w(x, y + b, z, t) - A_y w(x, y - b, z, t)$$
$$- A_z w(x, y, z + c, t) - A_z w(x, y, z - c, t), \quad (7.32)$$

The solution of the above differential equation can be easily guessed by analogy with the one-dimensional solution in the form

$$w(x, y, z, t) = \exp(-iEt/\hbar) \exp\{i(k_x x + k_y y + k_z z)\}, \quad (7.33)$$

which, substituted in eqn (7.32), gives in a few easy steps

$$E = E_1 - 2A_x \cos k_x a - 2A_y \cos k_y b - 2A_z \cos k_z c. \quad (7.34)$$

A_x, A_y, A_z are the coupling coefficients between nearest neighbours in the x, y, z directions respectively.

Thus, in the three-dimensional case, the energy band extends from the minimum energy

$$E_{min} = E_1 - 2(A_x + A_y + A_z) \qquad (7.35)$$

to the maximum energy

$$E_{max} = E_1 + 2(A_x + A_y + A_z). \qquad (7.36)$$

7.5 The effective mass

The mass of an electron in a crystal appears, in general, different from the free electron mass, and is usually referred to as the *effective mass*.

It has been known for a long time that an electron has a well-defined mass, and when accelerated by an electric field it obeys Newtonian mechanics. What happens when the electron to be accelerated happens to be inside a crystal? How will it react to an electric field? We have already given away the secret when talking about cyclotron resonance.

We shall obtain the answer by using a semi-classical picture, which, as the name implies, is 50% classical and 50% quantum-mechanical. The quantum-mechanical part describes the velocity of the electron in a one-dimensional lattice by its group velocity,

$$v_g = \frac{1}{\hbar} \frac{\partial E}{\partial k}, \qquad (7.37)$$

which depends on the actual E–k curve. The classical part expresses dE as the work done by a classical particle travelling a distance, $v_g dt$, under the influence of a force $e\mathcal{E}$ yielding

$$dE = e\mathcal{E} v_g dt$$
$$= e\mathcal{E} \frac{1}{\hbar} \frac{\partial E}{\partial k} dt. \qquad (7.38)$$

We may obtain the acceleration by differentiating eqn (7.37) as follows

$$\frac{dv_g}{dt} = \frac{1}{\hbar} \frac{d}{dt} \frac{\partial E}{\partial k} = \frac{1}{\hbar} \frac{\partial^2 E}{\partial k^2} \frac{dk}{dt}. \qquad (7.39)$$

Expressing now dk/dt from eqn (7.38) and substituting it into eqn (7.39) we get

$$\frac{dv_g}{dt} = \frac{1}{\hbar^2} \frac{\partial^2 E}{\partial k^2} e\mathcal{E}. \qquad (7.40)$$

Comparing this formula with that for a free, classical particle

$$m \frac{dv}{dt} = e\mathcal{E}, \qquad (7.41)$$

we may define

$$m^* = \hbar^2 \left(\frac{\partial^2 E}{\partial k^2} \right)^{-1} \qquad (7.42)$$

as the effective mass of an electron. Thus, the answer to the original question is that an electron in a crystal lattice *does* react to an electric field, but its mass is given by eqn (7.42) in contrast to the mass of a free electron. Let us just check whether we run into any contradiction with our E–k curve for a free electron. Then

$$E = \frac{\hbar^2 k^2}{2m},$$

and thus

$$\frac{\partial^2 E}{\partial k^2} = \frac{\hbar^2}{m},$$

which substituted into eqn (7.42) gives

$$m^* = m.$$

So everything is all right.

For an electron in a one-dimensional lattice we may take E in the form of eqn (7.30), giving

$$m^* = \frac{\hbar^2}{2Aa^2} \sec ka. \tag{7.43}$$

The graphs of energy, group velocity, and effective mass are plotted for this case in Fig. 7.12 as a function of k between $-\pi/a$ and π/a. Oddly enough, m^* may go to infinity and may take on negative values as well.

If an electron, initially at rest at $k = 0$, is accelerated by an electric field, it will move to higher values of k and will become heavier and heavier, reaching infinity at $k = \pi/2a$. For even higher values of k the effective mass becomes negative, heralding the advent of a new particle, the hole, which we have casually met from time to time and shall often meet in the rest of this course.

The definition of effective mass as given in eqn (7.42) is for a one-dimensional crystal, but it can be easily generalized for three dimensions. If the energy is given in terms of k_x, k_y, and k_z, as for example in eqn (7.34), then the effective mass in the x-direction is

$$m_x^* = \hbar^2 \left(\frac{\partial^2 E}{\partial k_x^2} \right)^{-1}$$

$$= \frac{\hbar^2}{2A_x a^2} \sec k_x a. \tag{7.44}$$

In the y-direction it is

$$m_y^* = \hbar^2 \left(\frac{\partial^2 E}{\partial k_y^2} \right)^{-1}$$

$$= \frac{\hbar^2}{2A_y b^2} \sec k_y b. \tag{7.45}$$

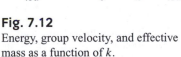

(a)

(b)

(c)

$-\pi/a$ 0 k π/a

Fig. 7.12
Energy, group velocity, and effective mass as a function of k.

A similar formula applies in the z-direction. So, oddly enough, the effective mass may be quite different in different directions. Physically, this means that the same electric field applied in different directions will cause varying amounts of acceleration. This is bad enough, but something even worse may happen. There can be a term like

If you are fond of mathematics you may think of the effective mass (or rather of its reciprocal) as a tensor quantity, but if you dislike tensors just regard the electron in a crystal as an extremely whimsical particle which, in response to an electric field in the (say) z-direction, may move in a different direction.

$$m_{xy}^* = \hbar^2 \left(\frac{\partial^2 E}{\partial k_x \partial k_y} \right)^{-1}. \tag{7.46}$$

With our simple model m_{xy}^* turns out to be infinitely large, but it is worth noting that in general an electric field applied in the x-direction may accelerate an electron in the y-direction. As far as I know there are no electronic devices making use of this effect; if you want to invent something quickly, bear this possibility in mind.

7.6   The effective number of free electrons

Let us now leave the fanciful world of three dimensions and return to the mathematically simpler one-dimensional case. In a manner rather similar to the derivation of effective mass we can derive a formula for the number of electrons available for conduction. According to eqn (7.40).

*Do not mistake this for the rate of change of electric current under stationary conditions to apply this for the steady state, one must take collisions into account as well.

$$\frac{dv_g}{dt} = \frac{1}{\hbar^2} \frac{\partial^2 E}{\partial k^2} e\mathcal{E}. \tag{7.47}$$

We have here the formula for the acceleration of an electron. But we have not only one electron, we have lots of electrons. Every available state may be filled by an electron; so the total effect of accelerating all the electrons may be obtained by a summation over all the occupied states. We wish to sum dv_g/dt for all electrons. Multiplying by the electron charge,

†We have already done it twice before, but since the density of states is a rather difficult concept (making something continuous having previously stressed that it must be discrete), and since this is a slightly different situation, we shall do the derivation again. Remember, we are in one dimension, and we are interested in the number of states in momentum space in an interval dp_x. According to eqn (6.31)

$$p_x = \frac{h}{2} n_x,$$

where n_x is an integer. So for unit length there is exactly one state and for a length dp_x the number of states, dn_x, is $(2/h) \, dp_x$, which is equal to $(1/\pi)dk_x$. We have to divide by 2 because only positive values of n are permitted and have to multiply by 2 because of the two possible values of spin. Thus, the number of states in a dk_x interval remains

$$\frac{1}{\pi} dk_x.$$

$$\sum \frac{d}{dt}(ev_g)$$

is nothing else but the rate of change of electric current that flows initially when an electric field is applied.*Thus,

$$\frac{dI}{dt} = \sum \frac{d}{dt}(ev_g)$$
$$= \frac{1}{\hbar^2} e^2 \mathcal{E} \sum \frac{\partial^2 E}{\partial k^2}, \tag{7.48}$$

or, going over to integration,

$$\frac{dI}{dt} = \frac{1}{\hbar^2} e^2 \mathcal{E} \frac{1}{\pi} \int \frac{d^2 E}{dk^2} dk, \tag{7.49}$$

where the density of states in the range dk is dk/π.†

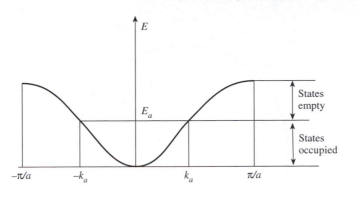

Fig. 7.13
One-dimensional energy band filled
up to k_a at $T = 0\,\mathrm{K}$.

If there were N non-interacting free electrons, we should obtain

$$\frac{\mathrm{d}I}{\mathrm{d}t} = \frac{e^2 \mathcal{E}}{m} N. \tag{7.50}$$

For free electrons eqn (7.50) applies; for electrons in a crystal eqn (7.49) is true. Hence, if we wish to create a mental picture in which the electrons in the crystal are replaced by 'effective' electrons, we may define the number of effective electrons by equating eqn (7.49) with eqn (7.50). Hence

$$N_{\mathrm{eff}} = \frac{1}{\pi} \frac{m}{\hbar^2} \int \frac{\mathrm{d}^2 E}{\mathrm{d}k^2} \mathrm{d}k. \tag{7.51}$$

This, as you may have already guessed, applies only at absolute zero because we did not include the probability of occupation. In this case all the states are occupied up to an energy $E = E_a$, and all the states above E_a are empty.

If E_a happens to be somewhere inside an energy band (as shown in Fig. 7.13) then the integration goes from $k = -k_a$ to $k = k_a$. Performing the integration:

$$N_{\mathrm{eff}} = \frac{1}{\pi} \frac{m}{\hbar^2} \left\{ \left(\frac{\mathrm{d}E}{\mathrm{d}k} \right)_{k=k_a} - \left(\frac{\mathrm{d}E}{\mathrm{d}k} \right)_{k=-k_a} \right\}$$

$$= \frac{2}{\pi} \frac{m}{\hbar^2} \left(\frac{\mathrm{d}E}{\mathrm{d}k} \right)_{k=k_a}. \tag{7.52}$$

This is a very important result. It says that the effective number of electrons capable of contributing to electrical conduction depends on the slope of the E–k curve at the highest occupied energy level.

At the highest energy in the band $\mathrm{d}E/\mathrm{d}k$ vanishes. We thus come to the conclusion that the number of effective electrons for a full band is zero.

If the energy band is filled there is no electrical conduction.

7.7   The number of possible states per band

In order to find the number of states, we must introduce boundary conditions. The simplest one (though physically the least defensible) is the so-called 'periodic boundary condition'.* It is based on the argument that a macroscopic crystal is so large in comparison with atomic dimensions that the detailed nature

* It would be more logical to demand that the wavefunction should disappear at the boundary, but that would involve us only in more mathematics without changing any of the conclusions. So I must ask you to accept the rather artificial boundary condition expressed by eqn (7.53).

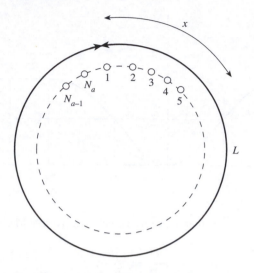

Fig. 7.14

Illustration of periodic boundary condition for a one-dimensional crystal.

of the boundary conditions does not matter, and we should choose them for mathematical convenience.

In the case of the periodic boundary condition, we may simply imagine the one-dimensional crystal biting its own tail. This is shown in Fig. 7.14, where the last atom is brought into contact with the first atom. For this particular configuration, it must be valid that

$$\psi(x + L) = \psi(x) \tag{7.53}$$

Then, with the aid of eqn (7.2), it follows that

$$e^{ik(x+L)}u_k(x + L) = e^{ikx}u_k(x). \tag{7.54}$$

Since u_k is a periodic function repeating itself from atom to atom,

$$u_k(x + L) = u_k(x) \tag{7.55}$$

and, therefore, to satisfy eqn (7.54), we must have

r is a positive or negative integer.

$$kL = 2\pi r. \tag{7.56}$$

It follows from the Kronig–Penney and from the Ziman models that in an energy band (that is in a region without discontinuity in energy) k varies from $n\pi/a$ to $(n + 1)\pi/a$.* Hence

* The Feynman model gives only one energy band at a time, but it shows clearly that the energy is a periodic function of ka, that is the same energy may be described by many values of k. Hence it would have been equally justified (as some people prefer) to choose the interval from $k = 0$ to $k = \pi/a$.

$$k_{max} \equiv (n + 1)\frac{\pi}{a} = \frac{2\pi r_{max}}{L} \tag{7.57}$$

and

$$k_{min} \equiv n(\pi/a) = \frac{2\pi r_{min}}{L}. \tag{7.58}$$

Rearranging, we have

$$r_{max} - r_{min} = \frac{L}{2a}$$

$$= N_a\frac{L}{2}, \tag{7.59}$$

where N_a is the number of atoms per unit length. Since r may have negative values as well, the total number of permissible values of k is

$$2(r_{max} - r_{min}) = N_a L. \qquad (7.60)$$

Now to each value of k belongs a wavefunction; so the total number of wavefunctions is $N_a L$, and thus, including spin, the total number of available states is $2N_a L$.

7.8 Metals and insulators

At absolute zero some materials conduct well, some others are insulators. Why? The answer can be obtained from the formulae we have derived.

If each atom in our one-dimensional crystal contains one electron, then the total number of electrons is $N_a L$, and the band is half-filled. Since dE/dk is large in the middle of the band, this means that there is a high effective number of electrons; that is, high conductivity.

If each atom contains two electrons, the total number of electrons is $2N_a L$; that is, each available state is filled. There is no conductivity: the solid is an insulator.

If each atom contains three electrons, the total number of electrons is $3N_a L$; that is, the first band is filled and the second band is half-filled. The value of dE/dk is large in the middle of the second band; therefore a solid containing atoms with three electrons each (it happens to be lithium) is a good conductor.

It is not difficult to see the general trend. Atoms with even numbers of electrons make up the insulators, whereas atoms with odd numbers of electrons turn out to be metals. This is true in general, but it is not true in every case. All we need to know is the number of electrons, even or odd, and the electric behaviour of the solid is determined. Diamond, with six electrons, must be an insulator and aluminium, with thirteen electrons, must be a metal. Simple, isn't it?

It is a genuine triumph of the one-dimensional model that the electric properties of a large number of elements may be promptly predicted. Unfortunately, it does not work always. Beryllium with four electrons and magnesium with twelve electrons should be insulators. They are not. They are metals; though metals of an unusual type in which electric conduction, evidenced by Hall-effect measurements, takes place both by holes and electrons. What is the mechanism responsible? For that we need a more rigorous definition of holes.

7.9 Holes

We first met holes as positively charged particles that enjoy a carefree existence quite separately from electrons. The truth is that they are not separate entities but merely by-products of the electrons' motion in a periodic potential. There is no such thing as a free hole that can be fired from a hole gun. Holes are artifices but quite lively ones. The justification for their existence is as follows.

Using our definition of effective mass [eqn (7.42)] we may rewrite eqn (7.48) in the following manner:

$$\frac{\mathrm{d}I}{\mathrm{d}t} = e^2 \mathcal{E} \sum_i \frac{1}{m_i^*}, \tag{7.61}$$

where the summation is over the occupied states.

If there is only one electron in the band, then

$$\frac{\mathrm{d}I_e}{\mathrm{d}t} = \frac{e^2 \mathcal{E}}{m^*}. \tag{7.62}$$

If the band is full, then according to eqn (7.52) the effective number of electrons is zero; that is,

$$\frac{\mathrm{d}I}{\mathrm{d}t} = e^2 \mathcal{E} \sum_i \frac{1}{m_i^*} = 0. \tag{7.63}$$

Assume now that somewhere towards the top of the band an electron, denoted by j, is missing. Then, the summation in eqn (7.61) must omit the state j, which we may write as

$$\frac{\mathrm{d}I_h}{\mathrm{d}t} = e^2 \mathcal{E} \sum_{\substack{i \\ i \neq j}} \frac{1}{m_i^*}. \tag{7.64}$$

But from eqn (7.63)

$$e^2 \mathcal{E} \left(\frac{1}{m_j^*} + \sum_{\substack{i \\ i \neq j}} \frac{1}{m_i^*} \right) = 0. \tag{7.65}$$

Equation (7.64) therefore reduces to

$$\frac{\mathrm{d}I_h}{\mathrm{d}t} = -e^2 \mathcal{E} \frac{1}{m_j^*}. \tag{7.66}$$

In the upper part of the band, however, the effective mass is negative; therefore

Hence, an electron missing from the top of the band leads to exactly the same formula as an electron present at the bottom of the band.

$$\frac{\mathrm{d}I_h}{\mathrm{d}t} = \frac{e^2 \mathcal{E}}{|m_j^*|}. \tag{7.67}$$

Now there is no reason why we should not always refer to this phenomenon as a current due to a missing electron that has a negative mass. But it is a lot shorter, and a lot more convenient, to say that the current is caused by a positive particle, called a *hole*. We can also explain the reason why the signs of eqn (7.62) and of eqn (7.67) are the same. In response to an electric field, holes move in an *opposite* direction carrying an *opposite* charge; their contribution to electric current is therefore the same as that of electrons.

7.10   Divalent metals

We may now return to the case of beryllium and magnesium and to their colleagues, generally referred to as divalent metals. One-dimensional theory is unable to explain their electric properties; let us try two dimensions.

The $E - k_x$, k_y surface may be obtained from eqn (7.34) as follows:

The divalent metals, having two valency electrons, are found in groups IIA and IIB of the periodic table (Fig. 4.5)

$$E = E_1 - 2A_x \cos k_x a - 2A_y \cos k_y b. \qquad (7.68)$$

Let us plot now the constant energy curves in the $k_x - k_y$ plane for the simple case when

$$E_1 = 1, \quad A_x = A_y = \tfrac{1}{4}, \quad a = b. \qquad (7.69)$$

It may be seen in Fig. 7.15 that the minimum energy $E = 0$ is at the origin and for higher values of k_x and k_y the energy increases. Note well that the boundaries $k_x = \pm\pi/a$ and $k_y = \pm\pi/a$ represent a discontinuity in energy. (This is something we have proved only for the one-dimensional case, but the generalization to two dimensions is fairly obvious.) There is an energy gap there. If the wave vector changes from point B just inside the rectangle, to point C, just outside the rectangle, the corresponding energy may jump from one unit to (say) 1.5 units.

Let us now follow what happens at $T = 0$ as we fill up the available states with electrons. There is nothing particularly interesting until all the states up to $E = 1$ are filled, as shown in Fig. 7.16(a). The next electron coming has an itch to leave the rectangle; it looks out, sees that the energy outside is 1.5 units, and therefore stays inside. This will go on until all the states are filled up to an energy $E = 1.5$, as shown in Fig. 7.16(b). The remaining states inside our rectangle have energies in excess of 1.5, and so the next electron in its search for lowest energy will go outside. It will go into a higher band because there are lower energy states in that higher band (in spite of the energy gap) than inside the rectangle.

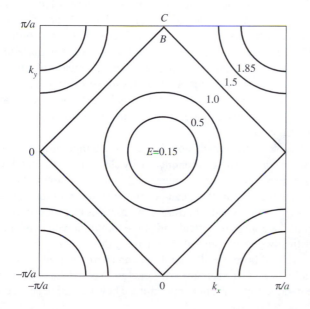

Fig. 7.15
Constant energy contours for a two-dimensional crystal in the $k_x - k_y$ plane on the basis of the Feynman model.

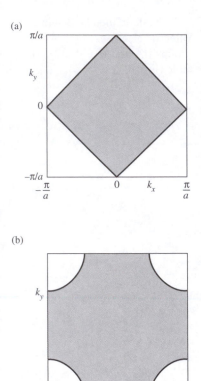

Fig. 7.16

(a) All energy levels filled up to $E = 1$. (b) All energy levels filled up to $E = 1.5$.

At finite temperatures a metal is a metal, but an insulator is no longer an insulator.

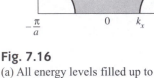

As you can see, there is no profound difference in principle; insulators and semiconductors are distinguished only by the magnitudes of their respective energy gaps.

The detailed continuation of the story depends on the variation of energy with k in the higher band, but one thing is certain: the higher band will not be empty.

For an atom with two electrons, the number of available states is equal to the number of electrons. If the energy gap is large (say, two units instead of half a unit), then all the states in the rectangle would be filled, and the material would be an insulator. If the energy gap is small (half a unit in our example), then some states will remain unfilled in the rectangle, and some states will be filled in the higher band. This means that both bands will contribute to electrical conduction. There will be holes coming from the rectangle and electrons from the higher band. This is how it happens that in some metals holes are the dominant charge-carriers.

7.11   Finite temperatures

All we said so far applies to zero temperature. What happens at finite temperatures? And what is particularly important, what happens at about room temperature, at which most electronic devices are supposed to work?

For finite temperatures it is no longer valid to assume that all states up to the Fermi energy are filled and all states above that are empty. The demarcation line between filled and unfilled states will become less sharp.

Let us see first what happens to a metal. Its highest energy band is about half-filled at absolute zero; at higher temperatures some of the electrons will acquire somewhat higher energies in the band, but that is all. There will be very little change in the effective number of electrons. A metal will stay a metal at higher temperatures.

What will happen to an insulator? If there are many electrons per atom, then there are a number of completely filled bands that are of no interest. Let us concentrate our attention to the two highest bands, called *valence* and *conduction* bands, and take the zero of energy at the top of the valence band. Since at absolute zero the valence band is completely filled, the Fermi level must be somewhere above the top of the valence band. Assuming that it is about half-way between the bands (I shall prove this later), the situation is depicted in Fig. 7.17(a) for zero temperature and in Fig. 7.17(b) for finite temperature. Remember, when the Fermi function is less than 1, it means that the probability of occupation is less than 1; thus, some states in the valence band must remain empty. Similarly, when the Fermi function is larger than 0, it means that the probability of occupation is finite; that is, some electrons will occupy states in the conduction band.

We have come to the conclusion that at finite temperatures, an insulator is no longer an insulator. There is conduction by electrons in the conduction band, and conduction by holes in the valence band. The actual amount of conduction depends on the energy gap. This can be appreciated if you remember that well away from the Fermi level the Fermi function varies exponentially; its value at the bottom of the conduction band and at the top of the valence band therefore depends critically on the width of the energy gap.

For all practical purposes diamond with an energy gap of 5.4 eV is an insulator, but silicon and germanium with energy gaps of 1.1 and 0.65 eV show noticeable conduction at room temperature. They are called semi-conductors.

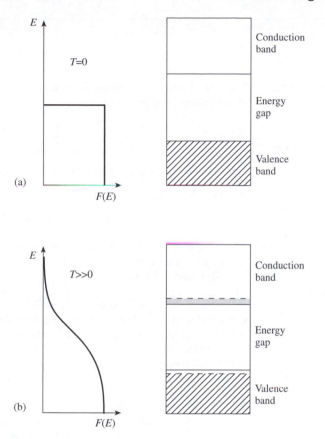

Fig. 7.17
(a) The two highest bands at $T = 0\,\mathrm{K}$.
(b) The two highest bands at
$T \gg 0\,\mathrm{K}$. There are electrons at the
bottom of the conduction band, and
holes at the top of the valence band.

7.12 Concluding remarks

The band theory of solids is not an easy subject. The concepts are a little bewildering at first and their practical utility is not immediately obvious.

You could quite well pass examinations without knowing much about band theory, and you could easily become the head of a big electrical company without having any notion of bands at all. But if you ever want to create something new in solid-state electronic devices, which will be more and more numerous in your professional life, a thorough understanding of band theory is imperative. So my advice would be to go over it again and again until familiarity breeds comprehension.

I would like to add a few more words about the one-dimensional models we use so often. The reason for using one-dimensional models is mathematical simplicity, and you must appreciate that the results obtained are only qualitatively true. The real world is three-dimensional, thus our models must also be three-dimensional if we want to have good agreement between theory and experiment. Having said that I must admit that this is not quite true. When we look at some of the recently invented devices, we find that some are two-dimensional, some are one-dimensional and some have zero dimension.*

*I know it is rather hard to swallow that there can be such things as zero-dimensional devices but all will be clear when we discuss them (they are known as low-dimensional devices) in Section 12.7 among semiconductor lasers.

Exercises

7.1. Classify metals, semiconductors, and insulators on the basis of their band structure. Find out what you can about materials on the border lines of these classifications.

7.2. X-ray measurements show that electrons in the conduction band of lithium have energies up to 4.2 eV. Take this as the highest filled energy level in the band. If you further identified this energy level with the Fermi level, what average effective mass will give you the same result from free electron theory? Assume one free electron per atom. The atomic weight of lithium is 6.94, and its density is 530 kg m^3.

7.3. Show, using the Feynman model, that the effective mass at the bottom of the band is inversely proportional to the width of the band.

7.4. Show from eqns (7.3) and (7.5) that the group velocity of the electron is zero at $k = n\pi/a$.

7.5. In general, the reciprocal of the effective mass is a tensor whose components are given by the mixed derivatives. How would the classical equation of motion

$$m\frac{\mathrm{d}v}{\mathrm{d}t} = \mathbf{F}$$

be modified?

7.6. The reciprocal mass tensor for Bi close to the bottom of the conduction band is of the form

$$\begin{pmatrix} a_{xx} & 0 & 0 \\ 0 & a_{yy} & a_{yz} \\ 0 & a_{yz} & a_{zz} \end{pmatrix}$$

(i) Find the components of the effective mass tensor.

(ii) Find the function $E(k_x, k_y, k_z)$.

(iii) Show that the constant energy surfaces are ellipsoids.

7.7. Using the potential energy distribution of Fig. 7.2, determine, with the aid of the Ziman model, the width of the first forbidden band. Take $w = a/2$.

7.8. Assume, as in the Kronig–Penney model of Section 7.2, that w and V_0 in Fig. 7.2 tend to zero and infinity, respectively, but their product $V_0 w$ is kept constant. Determine with the aid of the Ziman model the widths of the nth allowed band and the nth forbidden band. Can you conclude that the higher the band the wider it is?

7.9. The lowest energy bands in a solid are very narrow because there is hardly any overlap of the wave functions. In general, the higher up the band is, the wider it becomes. Assuming that the collision time is about the same for the valence and conduction bands, which would you expect to have higher mobility, an electron or a hole?

Semiconductors

Whatsoever things are true ...
whatsoever things are pure ...
think on these things.
                    Philippians iv 8

8.1 Introduction

With the aid of band theory we have succeeded in classifying solids into metals, insulators, and semiconductors. We are now going to consider semiconductors (technologically the newest class) in more detail. Metals and insulators have been used for at least as long as we have been civilized, but semiconductors have found application only in the last century, and their more widespread application dates from the 1950s. During this period the electronics industry has been (to use a hackneyed word justifiably) revolutionized, first by the transistor, then by microelectronic circuitry. Each of these in succession, by making circuitry much cheaper and more compact, has led to the wider use of electronic aids, such as computers, in a way that is revolutionary in the social sense too.

Perhaps the key reason for this sudden change has been the preparation of the extremely pure semiconductors, and hence the possibility of controlling impurity; this was a development of the 1940s and 1950s. By crystal pulling, zone-refining, and epitaxial methods it is possible to prepare silicon and germanium with an impurity of only 1 part in 10^{10}. Compare this with long-established engineering materials, such as steel, brass, or copper where impurities of a few parts per million are still virtually unattainable (and for most purposes, it must be admitted, not required). Probably the only other material that has ever been prepared with purity comparable to that of silicon and germanium is uranium, but people seem a little shy of quoting figures.

I shall now try to show why the important electrical properties of semiconductors occur and how they are influenced and controlled by small impurity concentrations. Next, we shall consider what really came first, the preparation of pure material. We shall be ready then to discuss junction devices and integrated circuit technology.

8.2 Intrinsic semiconductors

The aim in semiconductor technology is to purify the material as much as possible and then to introduce impurities in a controlled manner. We shall call the pure semiconductor 'intrinsic' because its behaviour is determined by its intrinsic properties alone, and we shall call the semiconductor 'extrinsic' after external interference has changed its inherent properties. In devices it is mostly extrinsic semiconductors that are used, but it is better to approach our subject gradually and discuss intrinsic semiconductors first.

To be specific, let us think about silicon, although most of our remarks will be qualitatively true of germanium and other semiconductors. Silicon has the diamond crystalline structure; the four covalent bonds are symmetrically arranged. All the four valence electrons of each atom participate in the covalent bonds, as we discussed before. But now, having learned band theory, we may express the same fact in a different way. We may say that all the electrons are in the valence band at 0 K. There is an energy gap of 1.1 eV above this before the conduction band starts. Thus, to get an electron in a state in which it can take up kinetic energy from an electric field and can contribute to an electric current, we first have to give it a package of at least 1.1 eV of energy. This can come from thermal excitation, or by photon excitation quite independently of temperature.

Let us try to work out now the number of electrons likely to be free to take part in conduction at a temperature T. How can we do this? We have already solved this problem for the one-dimensional case: eqn (7.51) gives us the effective number of electrons in a partly filled band; so all we need to do is to include the Fermi function to take account of finite temperature and to generalize the whole thing to three dimensions. It can be done, but it is a bit too complicated. We shall do something else, which is less justifiable on strictly theoretical grounds, but is physically much more attractive. It is really cheating because we use only those concepts of band theory that suit us, and instead of solving the problem honestly, we shall appeal to approximations and analogies. It is a compromise solution that will lead us to easily manageable formulae.

First of all we shall say that the only electrons and holes that matter are those near the bottom of the conduction band and the top of the valence band, respectively. Thus, we may assume that

$$k_x a, \quad k_y b, \quad k_z c \ll 1, \tag{8.1}$$

and we may expand eqn (7.34) to get the energy in the form,

$$E = E_1 - 2A_x \left(1 - \tfrac{1}{2}k_x^2 a^2\right) - 2A_y \left(1 - \tfrac{1}{2}k_y^2 b^2\right) - 2A_z \left(1 - \tfrac{1}{2}k_z^2 c^2\right). \tag{8.2}$$

Using our definition of effective mass, we can easily show from the above equation that

$$m_x^* = \frac{\hbar^2}{2A_z a^2}, \quad m_y^* = \frac{\hbar^2}{2A_y b^2}, \quad m_z^* = \frac{\hbar^2}{2A_z c^2}. \tag{8.3}$$

Substituting the values of $A_x a^2$, $A_y b^2$, and $A_z c^2$ from eqn (8.3) back into eqn (8.2) and condensing the constant terms into a single symbol, E_0, we may now express the energy as

$$E = E_0 + \frac{\hbar^2}{2} \left(\frac{k_x^2}{m_x^*} + \frac{k_y^2}{m_y^*} + \frac{k_z^2}{m_z^*} \right). \tag{8.4}$$

Taking further $E_0 = 0$, and assuming that everything is symmetric, that is

$$m_x^* = m_y^* = m_z^* = m^*, \tag{8.5}$$

we get

$$E = \frac{\hbar^2}{2m^*}\left(k_x^2 + k_y^2 + k_z^2\right). \tag{8.6}$$

The mass in the denominator is not the real mass of an electron but the effective mass. But that is the only difference between eqn (8.6) and the free electron model. Thus, we are going to claim that electrons in the conduction band have a different mass but apart from that behave in the same way as free electrons. Hence the formula derived for the density of states [eqn (6.10)] is also valid, and we can use the same method to determine the Fermi level. So we shall have the total number of electrons by integrating.... Wait, we forgot about holes. How do we include them? Well, if holes are the same sort of things as electrons apart from having a positive charge, then everything we said about electrons in the conduction band should be true for holes in the valence band. The only difference is that the density of states must increase downwards for holes.

Choosing now the zero of energy at the top of the valence band, we may write the density of states in the form

$$Z(E) = C_e(E - E_g)^{1/2}, \quad C_e = 4\pi(2m_e^*)^{3/2}/h^3 \tag{8.7}$$

for electrons, and

$$Z(E) = C_h(-E)^{1/2}, \quad C_h = 4\pi(2m_h^*)^{3/2}/h^3 \tag{8.8}$$

for holes, both of them per unit volume. This is shown in Fig. 8.1, where E is plotted against $Z(E)$. You realize of course that the density of states has meaning only in the allowed energy band and must be identically zero in the gap between the two bands.

Let us return now to the total number of electrons. To obtain that we must take the density of states, multiply by the probability of occupation (getting thereby the total number of occupied states) and integrate from the bottom to the top of the conduction band. So, formally, we have to solve the following integral

$$N_e = \int_{\text{bottom of conduction band}}^{\text{top of conduction band}} (\text{density of states})(\text{Fermi Function})\,dE \tag{8.9}$$

There are several difficulties with this integral:

1. Our solution for the density of states is valid only at the bottom of the band,
2. The Fermi function

$$F(E) = \left\{1 + \exp\left(\frac{E - E_F}{kT}\right)\right\}^{-1} \tag{8.10}$$

is not particularly suitable for analytical integration.
3. We would need one more parameter in order to include the width of the conduction band.

We are saved from all these difficulties by the fact that the Fermi level lies in the forbidden band, and in practically all cases of interest its distance from

This formula is identical to eqn (6.2) obtained from the free electron model—well, nearly identical.

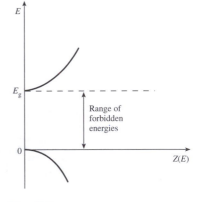

Fig. 8.1
Density of states plotted as a function of energy for the bottom of the conduction (electrons) and top of the valence (holes) bands. See eqns (8.7) and (8.8).

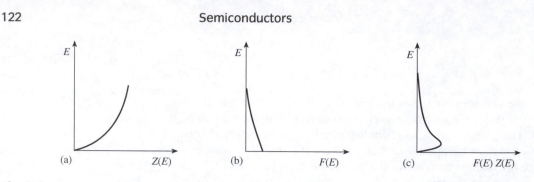

Fig. 8.2
(a) The density of states as a function of energy for the bottom of the conduction band. (b) The Fermi function for the same range of energies. (c) A plot of $F(E)Z(E)$ showing that the filled electron states are clustered together close to the bottom of the conduction band.

the band edge is large in comparison with kT (0.025 eV at room temperature). Hence,

$$E - E_F \gg kT \tag{8.11}$$

and the Fermi function may be approximated by

$$F(E) = \exp\left\{\frac{-(E - E_F)}{kT}\right\}, \tag{8.12}$$

as shown already in eqn (6.18).

If the Fermi function declines exponentially, then the $F(E)Z(E)$ product will be appreciable only near the bottom of the conduction band as shown in Fig. 8.2. Thus, we do not need to know the density of states for higher energies (nor the width of the band) because the fast decline of $F(E)$ will make the integrand practically zero above a certain energy. But if the integrand is zero anyway, why not extend the upper limit up to infinity? We may then come to an integral that is known to mathematicians.

Substituting now eqns (8.7) and (8.12) into eqn (8.9), we get

$$N_e = C_e \int_{E_g}^{\infty} (E - E_g)^{1/2} \exp\left\{\frac{-(E - E_F)}{kT}\right\} dE. \tag{8.13}$$

Introducing now the new variable

$$x = \frac{(E - E_g)}{kT}, \tag{8.14}$$

the integral takes the form

$$N_e = C_e(kT)^{3/2} \exp\left\{\frac{-(E_g - E_F)}{kT}\right\} \int_0^{\infty} x^{1/2} e^{-x} dx. \tag{8.15}$$

According to mathematical tables of high reputation*

* Even better, you could work it out for yourself; it's not too difficult.

$$\int_0^\infty x^{1/2} e^{-x} \, dx = \frac{1}{2}\sqrt{\pi}, \tag{8.16}$$

leading to the final result

$$N_e = N_c \exp\left\{\frac{-(E_g - E_F)}{kT}\right\}, \tag{8.17}$$

where

$$N_c = 2\left(\frac{2\pi m_e^* kT}{h^2}\right)^{3/2}. \tag{8.18}$$

Thus, we have obtained the number of electrons in the conduction band as a function of some fundamental constants, of temperature, of the effective mass of the electron at the bottom of the band, and of the amount of energy by which the bottom of the band is above the Fermi level.

We can deal with holes in an entirely analogous manner. The probability of a hole being present (that is of an electron being absent) is given by the function

$$1 - F(E), \tag{8.19}$$

which also declines exponentially along the negative E-axis. So we can choose the lower limit of integration as $-\infty$, leading to the result for the number of holes in the valence band:

$$N_h = N_v \exp(-E_F/kT), \tag{8.20}$$

where

$$N_v = 2(2\pi m_h^* kT/h^2)^{3/2}. \tag{8.21}$$

For an intrinsic semiconductor each electron excited into the conduction band leaves a hole behind in the valence band. Therefore, the number of electrons should be equal to the number of holes (this would actually follow from the condition of charge neutrality too); that is

$$N_e = N_h. \tag{8.22}$$

Substituting now eqns (8.17) and (8.20) into eqn (8.22), we get

$$N_c \exp\{-(E_g - E_F)/kT\} = N_v \exp(-E_F/kT), \tag{8.23}$$

from which the Fermi level can be determined. With a little algebra we get

$$E_F = \frac{E_g}{2} + \frac{3}{4}kT \log_e \frac{m_h^*}{m_e^*}. \tag{8.24}$$

Since kT is small, and the effective masses of electrons and holes are not very much different, we can say that the Fermi level is roughly halfway between the valence and conduction bands.

We may now ask the question how carrier concentration varies with temperature. Strictly speaking, the energy gap is also a function of temperature for the reason that it depends on the lattice constant, which does vary with temperature. That is, however, a small effect in the normally used temperature range, so we are nearly always entitled to disregard it. Substituting eqn (8.24) into eqns (8.17) and (8.20), we find that both N_e and N_h are proportional to $\exp(-E_g/2kT)$, an important relationship.

We know now everything we need about intrinsic semiconductors. Let us now look at the effect of impurities.

8.3   Extrinsic semiconductors

We shall continue to consider silicon as our specific example, but now with a controlled addition of a group V impurity (this refers to column five in the periodic table of elements) as, for example, antimony (Sb), arsenic (As), or phosphorus (P). Each group V atom will replace a silicon atom and use up four of its valence electrons for covalent bonding [Fig. 8.3(a)]. There will, however, be a spare electron. It will no longer be so tightly bound to its nucleus as in a free group V atom, since the outer shell is now occupied (we might look at it this way) by eight electrons, the number of electrons in an inert gas; so the dangling spare electron cannot be very tightly bound. However, the impurity nucleus still has a net positive charge to distinguish it from its neighbouring silicon atoms. Hence, we must suppose that the electron still has some affinity for its parent atom. Let us rephrase this somewhat anthropomorphic picture in terms of band theory. We have said the energy gap represents the minimum energy required to ionize a silicon atom by taking one of its valenc electrons. The electron belonging to the impurity atom clearly needs far less energy than this to become available for conduction. We would expect its energy level to be something like E_D in Fig. 8.3(b). $(E_g - E_D)$ is typically of the order of 10^{-2} eV (see Table 8.1). At absolute zero temperature an electron does occupy this energy level, and it is *not* available for conduction. But at finite temperatures this electron needs no more than about

If the impurity is less than, say, 1 in 10^6 silicon atoms, the lattice will be hardly different from that of a pure silicon crystal.

Fig. 8.3

(a) The extra electron 'belonging' to the group V impurity is much more weakly bound to its parent atom than the electrons taking part in the covalent bond. (b) This is equivalent to a *donor level* close to the conduction band in the band representation.

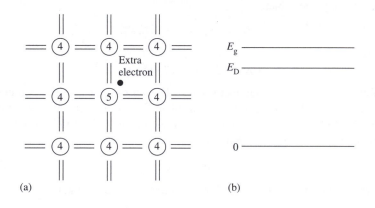

Table 8.1 *Energy levels of donor (group V) and acceptor (group III) impurities in Ge and Si. The energies given are the ionization energies, that is, the distance of the impurity level from the band edge (in electron volts)*

	Impurity	Ge	Si
Donors	Antimony (Sb)	0.0096	0.039
	Phosphorus (P)	0.0120	0.045
	Arsenic (As)	0.0127	0.049
Acceptors	Indium (In)	0.0112	0.160
	Gallium (Ga)	0.0108	0.065
	Boron (B)	0.0104	0.045
	Aluminium (Al)	0.0102	0.057

10^{-2} eV of energy to put it into the conduction band. This phenomenon is usually referred to as an electron *donated* by the impurity atom. For this reason E_D is called the *donor* level. The energy difference between the donor levels and the bottom of the conduction band is shown in Table 8.1 for several impurities.

Interestingly, a very rough model serves to give a quantitative estimate of the donor levels. Remember, the energy of an electron in a hydrogen atom [given by eqn (4.18)] is

$$E = -me^4/8\epsilon_0^2 h^2. \tag{8.25}$$

We may now argue that the excess electron of the impurity atom is held by the excess charge of the impurity nucleus; that is, the situation is like that in the hydrogen atom, with two minor differences.

1. The dielectric constant of free space should be replaced by the dielectric constant of the material.
2. The free electron mass should be replaced by the effective mass of the electron at the bottom of the conduction band.

Thus, this model leads to the following estimate

$$E_g - E_D = m^* e^4/8\epsilon^2 h^2, \tag{8.26}$$

which for silicon, with a relative dielectric constant of 12 and effective mass about half the free electron mass, gives a value of about 0.05 eV for ($E_g - E_D$), which is not far from what is actually measured. Note, however, that the parameters in eqn (8.26) depend only on the properties of the host material, so this model cannot possibly say anything on how $E_g - E_D$ varies with the type of dopant.

If instead of a group V impurity we had some group III atoms, as, for example, indium (In), aluminium (Al), or boron (B), there would be an electron missing from one of the covalent bonds (see Fig. 8.4). If one electron is missing, there must be a hole present.

Before going further, let me say a few words about holes. You might have been slightly confused by our rather inconsistent references to them. To clear this point—there are three equivalent representations of holes, and you can always (or nearly always) look at them in the manner most convenient under the circumstances.

You may think of a hole as a full-blooded positive particle moving around in the crystal, or as an electron missing from the top of the valence band, or as the actual physical absence of an electron from a place where it would be desirable to have one.

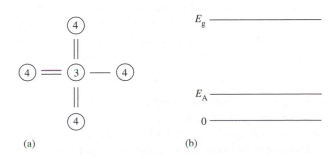

(a)   (b)

Fig. 8.4
(a) In the case of a group III impurity one bonding electron is missing—there is a 'hole' which any valence electron with a little surplus energy can fall into. (b) This shows in the band representation as an *acceptor level* just above the valence band edge.

In the present case, the third interpretation is the most convenient one—to start with. A group III atom has three valence electrons. So when it replaces a silicon atom at a certain atomic site, it will try to contribute to bonding as much as it can. However, it possesses only three electrons for four bonds. Any electron wandering around would thus be welcomed to help out. More aggressive impurity nuclei might even consider stealing an electron from the next site.

The essential point is that a low energy state is available for electrons—not as low an energy as at the host atom but low enough to come into consideration when an electron has acquired some extra energy and feels an urge to jump somewhere. Therefore (changing now to band theory parlance), the energy levels due to group III impurities must be just above the valence band. Since these atoms accept electrons so willingly, they are called *acceptors* and the corresponding energy levels are referred to as *acceptor levels*.

A real material will usually have both electron donors and acceptors present (not necessarily group V and group III elements; these were chosen for simplicity of discussion and because they are most often used in practice). However, usually one type of impurity exceeds the other, and we can talk of impurity semiconductors as n- (negative carrier) or p- (positive carrier) types according to whether the dominant charge carriers are electrons or holes. If we had some silicon with 10^{20} atoms per cubic metre of trivalent indium, it would be a p-type semiconductor. If we were somehow to mix in 10^{21} atoms per cubic metre of pentavalent phosphorus, the spare phosphorus electrons would not only get to the conduction band but would also populate the acceptor levels, thus obliterating the p characteristics of the silicon and turning it into an n-type semiconductor.

Let us calculate now the number of electrons and holes for an extrinsic semiconductor. We have in fact already derived formulae for them, as witnessed by eqns (8.17) and (8.20); but they were for intrinsic semiconductors. Did we make any specific use of the fact that we were considering intrinsic semiconductors? Perhaps not. We said that only electrons at the bottom of the band matter, and we also said that the bottom of the band is many times kT away from the Fermi level in energy but this could all be equally valid for extrinsic semiconductors. It turns out that these approximations are valid, apart from certain exceptional cases (we shall meet one exception when we consider devices).

But how can we determine the Fermi level? It is certainly more difficult for an extrinsic semiconductor. We have to consider now all the donors and acceptors. The condition is that the crystal must be electrically neutral, that is, the net charge density must be zero. Let us take a count of what sort of charges we may meet in an extrinsic semiconductor. There are our old friends, electrons and holes; then there are the impurity atoms that donated an electron to the conduction band, and are left with a positive charge; and finally there are the acceptor atoms that accepted an electron from the valence band and thus have a negative charge. Hence the formula for overall charge neutrality is

N_A^- is the number of *ionized* acceptor atoms, which accepted an electron from the valence band, and N_D^+ is the number of *ionized* donor atoms, which donated an electron to the conduction band.

$$N_e + N_A^- \leftrightharpoons N_h + N_D^+, \qquad (8.27)$$

We have written eqn (8.27) with the $\leftrightharpoons$ sign used by chemists to show that it is a dynamic equilibrium, rather than a once and for all equation.

How can we find the number of ionized impurities, N_A^- and N_D^+, from the actual number of impurity atoms N_A and N_D? Looking at Fig. 8.3(b) you may recollect that the N_D donor electrons live at the level E_D (at 0 K) all ready and willing to become conduction electrons if somehow they can acquire $(E_g - E_D)$ joules of energy. N_D^+ is hence a measure of how many of them have gone. Therefore,

$$N_D^+ = N_D\{1 - F(E_D)\}. \tag{8.28}$$

If we multiply N_D by the probability of an electron *not* being at E_D we should get N_D^+.

For acceptor atoms the argument is very similar. The probability of an electron occupying a state at the energy level E_A is $F(E_A)$. Therefore the number of ionized acceptor levels is

$$N_A^- = N_A F(E_A). \tag{8.29}$$

We are now ready to calculate the position of the Fermi level in any semiconductor whose basic properties are known, that is, if we know N_A and N_D, the energy gap, E_g, and the effective masses of electrons and holes. Substituting for N_e, N_h, N_A^-, N_D^+ from eqns (8.17), (8.20), (8.28), and (8.29) into eqn (8.27) we get an equation that can be solved for E_F. It is a rather cumbersome equation but can always be solved with the aid of a computer. Fortunately, we seldom need to use all the terms, since, as mentioned above, the dominant impurity usually swamps the others. For example, in an n-type semiconductor, usually $N_e \gg N_h$ and $N_D^+ \gg N_A^-$ and eqn (8.27) reduces to

$$N_e \cong N_D^+. \tag{8.30}$$

This, of course, implies that all conduction electrons come from the donor levels rather than from host lattice bonds. Substituting eqns (8.17) and (8.28) into eqn (8.30) we get*

$$N_c \exp\left(-\frac{E_g - E_F}{kT}\right) \cong N_D \left(1 + \exp\frac{E_F - E_D}{kT}\right)^{-1}. \tag{8.31}$$

For a particular semiconductor eqn (8.31) is easily solvable, and we may plot E_F as a function of N_D or of temperature. Let us first derive a formula for the simple case when $(E_F - E_D)/kT$ is a large negative number. Equation (8.31) then reduces to

$$(\text{constant}) \exp\frac{E_F}{kT} \cong N_D. \tag{8.32}$$

* Equation (8.31) is not, however, valid at the limit of *no* impurity, because holes cannot then be neglected; nor is it valid when N_D is very large, because some of the approximations [e.g. eqn (8.11)] are then incorrect, and in any case many impurity atoms getting close enough to each other will create their own impurity band.

So we have already learned that the position of the Fermi level moves upwards, and varies rather slowly with impurity concentration.

Let us consider now a slightly more complicated situation where the above approximation does not apply. Take silicon at room temperature with the data

$$E_g = 1.15\,\text{eV}, \quad E_g - E_D = 0.049\,\text{eV}, \quad N_D = 10^{22}\,\text{m}^{-3}. \tag{8.33}$$

We may get easily the solution by introducing the notation

$$x = \exp\frac{E_F}{kT}, \tag{8.34}$$

reducing thereby eqn (8.31) to the form

$$Ax = \frac{N_D}{(1 + Bx)}, \tag{8.35}$$

E_F increases with the logarithm of N_D.

which leads to a quadratic equation in x, giving finally for the Fermi level, $E_F = 0.97 \, \text{eV}$. Thus, in the practical case of an extrinsic semiconductor the Fermi level is considerably above the middle of the energy gap.

Let us consider now the variation of Fermi level with temperature. This is somewhat complicated by the fact that E_g and E_D are dependent on lattice dimensions, and hence both change with temperature, but we shall ignore this effect for the moment.

At very low temperatures (a few degrees absolute), the chance of excitation across the gap is fantastically remote compared with the probability of ionization from a donor level (which is only remote). Try calculating this for a 1 eV value of E_g and a 0.05 eV value of $E_g - E_D$. You should find that with 10^{28} lattice atoms per cubic metre, of which only 10^{21} are donors, practically all the conduction electrons are from the latter. Thus, at low temperature the material will act like an intrinsic semiconductor whose energy gap is only $E_g - E_D$. So we can argue that the Fermi level must be about halfway within this 'gap', that is

$$E_F \cong \tfrac{1}{2}(E_g + E_D). \qquad (8.36)$$

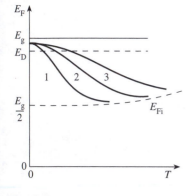

Fig. 8.5

The variation of the Fermi level as a function of temperature for an n-type semiconductor. The curves 1, 2, and 3 correspond to increasing impurity concentrations. E_{Fi} is the intrinsic Fermi level (plotted from eqn (8.24)) for $m_h > m_e$ to which all curves tend at higher temperatures.

Is this analogy so close that we can generalize the relationship obtained before for the temperature variation of intrinsic carriers? Can we claim that it will now be $\exp[-(E_g - E_D)/2kT]$ instead of $\exp(-E_g/2kT)$? Yes, this is true under certain conditions. It can be derived from eqn (8.31) (see example 8.14).

At the other extreme, at very high temperatures, practically all the electrons from the impurity atoms will be ionized, but because of the larger reservoir of valency electrons, the number of carriers in the conduction band will be much greater than N_D. In other words, the material (now a fairly good conductor) will behave like an intrinsic semiconductor with the Fermi level at about $E_g/2$. For larger impurity concentration the intrinsic behaviour naturally comes at a higher temperature. Thus, a sketch of E_F against temperature will resemble Fig. 8.5 for an n-type semiconductor. The relationship for a p-type semiconductor is entirely analogous and is shown in Fig. 8.6.

There is just one further point to note about the variation of energy gap with temperature. We have seen in the Ziman model of the band structure that the interband gap is caused by the interaction energy when the electrons' de Broglie half-wavelength is equal to the lattice spacing. It is reasonable to suppose that this energy would be greater at low temperatures for the following reason. At higher temperatures the thermal motion of the lattice atoms is more vigorous; the lattice spacing is thus less well defined and the interaction is weaker. This, qualitatively, is the case; for example, in germanium the energy gap decreases from about 0.75 eV at 4 K to 0.67 eV at 300 K.

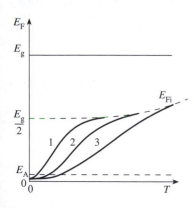

Fig. 8.6

The variation of the Fermi level as a function of temperature for a p-type semiconductor. The curves 1, 2, and 3 correspond to increasing impurity concentrations. E_{Fi} is the intrinsic Fermi level to which all curves tend at higher temperatures.

8.4 Scattering

Having learned how to make n-type and p-type semiconductors and how to determine the densities of electrons and holes, we now know quite a lot about semiconductors. But we should not forget that so far we have made no statement about the electrical conductivity, more correctly nothing beyond that at the beginning of the course [eqn (1.10)], where we produced the formula

$$\sigma = \frac{e^2}{m} \tau N_e.$$

(8.37)

An obvious modification is to put the effective mass in place of the actual mass of the electron. But there is still τ, the mean free time between collisions. What will τ depend on and how?

We have now asked one of the most difficult questions in the theory of solids. As far as I know no one has managed to derive an expression for τ starting from first principles (i.e., without the help of experimental results).

Let us first see what happens at absolute zero temperature. Then all the atoms are at rest;* so the problem seems to be: how long can an electron travel in a straight line without colliding with a stationary atom? Well, why would it collide with an atom at all? The electron hasn't really any desire to bump into an atom. As we know from the Feynman model, the electron does something quite different. It sits in an energy level of a certain atom, then tunnels through the adverse potential barrier and takes a seat at the next atom, and again at the next atom—so it just walks across the crystal without any collision whatsoever. The mean free path is the length of the crystal—so it is *not* the presence of the atoms that causes the collisions. What then? The imperfections? If the crystal were perfect, we should have nice periodic solutions [as in eqn (7.33)] for the wavefunction, and there would be equal probability for an electron being at any atom. It could thus start at any atom and could wriggle through the crystal to appear at any other atom. But the crystal is not perfect. The ideal periodic structure of the atoms is upset, partly by the thermal motion of the atoms, and partly by the presence of impurities, to mention only the two most important effects. So, strictly speaking, the concept of collisions, as visualized for gas molecules, makes little sense for electrons. Strictly speaking, there is no justification at all for clinging to the classical picture. Nevertheless, as so often before, we shall be able to advance some rough classical arguments, which lead us into the right ballpark.

First notice [eqns (1.11) and (1.13)] that the mean free path may be written with good approximation as proportional to

$$l \sim \tau T^{1/2}.$$

(8.38)

Arguing now that the mean free path is inversely proportional to the scattering probability, and the scattering probability may be taken to be proportional to the energy of the lattice wave (i.e. to T), we obtain for the collision time

$$\tau_{\text{thermal}} \sim l T^{-1/2} \sim T^{-3/2}.$$

(8.39)

The argument for ionized impurities is a little more involved. We could say that no scattering will occur unless the electron is so close to the ion that the electrostatic energy [given by eqn (4.2) if e^2 is replaced by Ze^2] is comparable

* For the purpose of the above discussion we can assume that the atoms are at rest, but that can never happen in an actual crystal. If the atoms were at rest then we would know both their positions and velocities at the same time, which contradicts the uncertainty principle. Therefore, even at absolute zero temperature, the atoms must be in some motion.

with the thermal energy

$$\frac{Ze^2}{4\pi\epsilon r_s} = \frac{3}{2}kT \tag{8.40}$$

leading to a radius,

$$r_s = \frac{Ze^2}{6\pi\epsilon kT}. \tag{8.41}$$

Next, we argue that the scattering power of the ion may be represented at the radius by a scattering cross-section

$$S_s = r_s^2 \pi = \frac{1}{\pi}\left(\frac{Ze^2}{6\epsilon kT}\right)^2. \tag{8.42}$$

Finally, assuming that the mean free path is inversely proportional to the scattering cross-section, we obtain the relationship

$$\tau_{\text{ionized impurity}} = lT^{-1/2} \sim S_s^{-1} T^{-1/2} \sim T^{3/2}. \tag{8.43}$$

When both types of scattering are present the resultant collision time may be obtained from the equation

If a high value of τ is required, then one should use a very pure material and work at low temperatures.

$$\frac{1}{\tau} = \frac{1}{\tau_{\text{thermal}}} + \frac{1}{\tau_{\text{ionized impurity}}}. \tag{8.44}$$

A high value of τ means high mobility and hence high average velocity for the electrons.

* Two present applications are the Gunn effect and very high frequency transistors, both to be discussed in Chapter 9.

We do not know yet whether high electron velocities in crystals will have many useful applications,* but since fast electrons in vacuum give rise to interesting phenomena, it might be worth while making an effort to obtain high carrier velocities in semiconductors.

How would mobility vary as a function of impurity density? It is bound to decline. Instead of a mathematical model that is quite complicated, we are going to give here actual measured curves for electron and hole mobilities at $T = 300\,\text{K}$ for Ge and Si (see Fig. 8.7).

Note that electron mobilities are higher than hole mobilities due to the fact that in both materials the effective mass of electrons is smaller than that of holes.

When both electrons and holes are present the conductivities add

$$\sigma = \frac{e^2 \tau_e N_e}{m_e^*} + \frac{e^2 \tau_h N_h}{m_h^*}. \tag{8.45}$$

8.5 A relationship between electron and hole densities

Let us now return to eqns (8.17) and (8.20). These were derived originally for intrinsic semiconductors, but they are valid for extrinsic semiconductors as well. Multiplying them together, we get

$$N_e N_h = 4\left(\frac{2\pi kT}{h^2}\right)^3 (m_e^* m_h^*)^{3/2} \exp\left(-\frac{E_g}{kT}\right). \tag{8.46}$$

It is interesting to note that the Fermi level has dropped out, and only the 'constants' of the semiconductor are contained in this equation. Thus, for a given

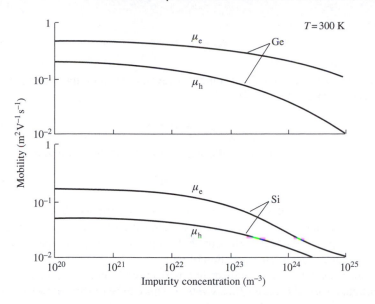

Fig. 8.7
Electron and hole mobilities in Ge
and Si as a function of impurity
concentration.

semiconductor (i.e. for known values of m_e^*, m_h^*, and E_g) and temperature we
can define the product $N_e N_h$ exactly, *whatever the Fermi energy and hence
whatever the impurity density*. In particular, for an intrinsic material, where
$N_e = N_h = N_i$, we get

$$N_e N_h = N_i^2. \tag{8.47}$$

Let us think over the implications. We start with an intrinsic semiconductor;
so we have equal numbers of electrons and holes. Now add some donor atoms.
The number of electrons must then increase, but according to eqn (8.47) the
product must remain constant. At first this seems rather odd. One would think
that the number of electrons excited thermally from the valence band into the
conduction band (and thus the number of holes left behind) would depend on
temperature only, and be unaffected by the presence of donor atoms. This is
not so. By increasing the concentration of donors, the total number of electrons
in the conduction band is increased, but the number of electrons excited across
the gap is decreased (not only in their relative proportion but in their absolute
number too). Why?

If the number of electrons increases, the number of holes must decrease.

We can obtain a qualitative answer to this question by considering the
'dynamic equilibrium' mentioned briefly before. It means that electron–hole
pairs are constantly created and annihilated and there is equilibrium when the
rate of creation equals the rate of annihilation (the latter event is more usually
referred to as 'recombination').

Now it is not unreasonable to assume that electrons and holes can find each
other more easily if there are more of them present. For an intrinsic material
we may write

The rate of recombination must be proportional to the densities of holes and electrons.

$$r_{\text{intrinsic}} = a N_i^2, \quad g_{\text{intrinsic}} = a N_i^2, \tag{8.48}$$

where a is a proportionality constant, and r and g are the rates of recombination
and creation, respectively.

Now we may argue that by adding a small amount of impurity, neither the rate
of creation nor the proportionality constant should change. So for an extrinsic

semiconductor

$$g_{\text{extrinsic}} = aN_i^2 \tag{8.49}$$

is still valid.

The rate of recombination should, however, depend on the actual densities of electrons and holes, that is

$$r_{\text{extrinsic}} = aN_e N_h. \tag{8.50}$$

From the equality of eqns (8.49) and (8.50) we get the required relationship

$$N_i^2 = N_e N_h. \tag{8.51}$$

So we may say that as the density of electrons is increased above the intrinsic value, the density of holes must decrease below the intrinsic value in order that the rate of recombination of electron–hole pairs may remain at a constant value equal to the rate of thermal creation of pairs.

Those of you who have studied chemistry may recognize this relationship as a particular case of the *law of mass action*. This can be illustrated by a chemical reaction between A and B, giving rise to a compound AB, viz.

$$A + B \rightleftharpoons AB. \tag{8.52}$$

If we represent the molecular concentration of each component by writing its symbol in square brackets, the quantity

$$[A][B][AB]^{-1} \tag{8.53}$$

is a constant at a given temperature. Now our 'reaction' is

$$\text{electron} + \text{hole} \rightleftharpoons \text{bound electron}. \tag{8.54}$$

As the number of bound electrons (cf. [AB]) is constant, this means that

$$[\text{electron}][\text{hole}] = N_e N_h \tag{8.55}$$

will also be constant.

8.6   III–V and II–VI compounds

In our examples up to now we have referred to germanium and silicon as typical semiconductors, and indeed they are typical, their technology was certainly the earliest mastered. They are both tetravalent, so they can be found in column IVB of our periodic table shown in Fig. 4.5. There are, of course, many other semiconductors. In this section we shall be concerned with two further types which are compounds of elements from columns IIIB, VB, IIB, and VIB, respectively.

Let us talk first about the III–V compounds. Why are they semiconductors? We can say the same thing about them as about germanium and silicon. They are insulators at low temperatures because all the electrons participate in the bonding process: none of them is available for conduction. At higher temperatures, however, the electronic bond can be broken by thermal excitations, that is electrons can be excited into the conduction band. The only difference relative

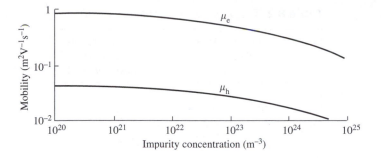

Fig. 8.8
Electron and hole mobilities in GaAs
as a function of impurity
concentration.

to Ge and Si is that III–V compounds have an ionic contribution to the bonding as well. This is not particularly surprising. We mentioned in Chapter 5 that the ionic bond of NaCl comes about because the one outer electron of sodium is happy to join the seven outer electrons of chlorine to make up a completed ring—so it is easy to see that, with Ga having three outer electrons and As having five outer electrons, they will also strike up a companionship in order to complete the ring.

Which are the most important III–V materials? The oldest one, and technologically the best developed, is GaAs, to which serious attention has been paid since the middle of the 1950s and which has been the preferred material for a host of devices. Why? One might expect that arsenic was the last thing anyone in any laboratory would have wanted to work on. However, some of the rivals, for example AlSb, fell by the wayside because of a tendency to decompose quickly, and, most importantly, GaAs was the material that offered high mobility relative to Ge and Si. For measured curves of electron and hole mobilities against impurity concentration, see Fig. 8.8.

What can we say about the energy gap of GaAs? It is, 1.40 eV, a lot higher than that of Ge, the element between them in the periodic table. The energy gap is higher which means that it is more difficult to break a bond in GaAs than in Ge. Why? Because of the presence of ionic bonding. We can also say something definite about the energy gaps of some other compounds relative to that of GaAs. If we combine with Ga the element in the periodic table *above* As, we obtain GaP with an energy gap of 2.25 eV. If we choose for the other element in the compound Sb, which is *below* As in the periodic table, then we obtain GaSb with an energy gap of 0.7 eV. The same is true if we combine various elements from column III with As. The energy gap of AlAs is 2.1 eV, whereas the energy gap declines to 0.35 eV for InAs. The general rule is easy to remember: the lower you go in the periodic table in your choice of the elements the smaller is the gap. What is the reason? The farther down the columns of the periodic table (Fig. 4.2.) the higher are both the nuclear charge (Ze) and the number of filled electronic inner shells. Hence, the valency electrons are farther from the nucleus and so more loosely bound. Thus, the bonding force between atoms is weaker (lower melting point) and the energy to stimulate electrons into the conduction band is less (lower energy gap). This is shown in Table 8.2 where also we have omitted the heaviest group III–V elements thallium (atomic number 81) and bismuth (83) as their compounds have very narrow gap and so are almost metallic.

Note, however, that a III–V ionic bond is weaker than a I–VII ionic bond.

Table 8.2 *Size of atoms in tetrahedral bonds*

Element	Atomic number	Atomic weight (AMU)	Atomic radius in tetrahedral covalent bonds Å
IIB			
Zn	30	68.38	1.31
Cd	48	112.4	1.48
Hg	80	200.59	1.48
IIIB			
B	5	10.81	0.88
Al	13	26.98	1.26
Ga	31	69.72	1.26
In	49	114.82	1.44
IVB			
C	6	12.01	0.77
Si	14	28.09	1.17
Ge	32	72.59	1.22
Sn	50	118.69	1.40
VB			
N	7	14.007	0.70
P	15	30.97	1.10
As	33	74.92	1.18
Sb	51	121.7	1.36
VIB			
O	8	16.0	0.66
S	16	32.06	1.04
Se	34	78.96	1.14
Te	52	127.60	1.32

How can we make a III–V material, say GaAs n-type or p-type? The answer is easy in principle. If there is excess Ga it will be p-type, if there is excess As it will be n-type. Or we can try as a dopant a column IV material, for example silicon. It acts as a donor if it replaces a Ga atom, and as an acceptor if it replaces an As atom. In practice, of course, it is not so easy to produce any of these materials to a given specification.

Most of the semiconductors we consider in this section crystallize like GaAs in the zinc blende structure. This is very like our diamond picture (Fig. 5.3) with the C atoms replaced alternately with a III and V or II and VI. So all bonds are between unlike atoms. We can also visualize this structure as each sub-set of atoms arranged in a face centred cubic (FCC) structure. The two sub-sets are displaced from each other in three dimensions by a half lattice spacing, that is one fourth of the FCC sub-lattice cubic separation. This results (back to Fig. 5.3) in a tetragonal arrangement, each atom being bonded symmetrically to four unlike atoms. In Table 8.2 we list the basic size of atoms, including the atomic radius that they have in tetragonal bonding. One can calculate the bond length of atomic separation in, say, GaP by adding the Ga and P radii. Then by fairly simple geometry we can obtain the lattice parameter by multiplying by 2.31.

All the compounds mentioned so far have been binary compounds, that is they consist of two elements. In principle there are no difficulties in growing ternary crystals consisting of three elements. Al for example is quite happy to occupy a Ga site, so there are no difficulties in producing a GaAlAs crystal. What will be the effect of adding Al to GaAs? According to the argument given above, the energy gap should increase. The more Al is added the larger will be the energy gap. Is that good for something? Yes. The sources for optical communications depend crucially on our ability to tailor the energy gap. Let me give you an example. The preferred material for blue-green semiconductor lasers is GaN, but its energy gap puts it into the ultraviolet region. This can be adjusted by replacing some of the gallium with indium, the heavier element reducing the gap. We shall say a lot more about this in Section 12.7.

There are a few obvious things we can say about II–VI materials. They have more ionic bonding than III–V materials and less ionic bonding than I–VII materials. For more accurate figures of the contribution of ionic forces to bonding see Table 8.3. What about the energy gaps? Just as GaAs has a higher energy gap than Ge, we would also expect ZnSe (look it up in the periodic table: Zn is to the left of Ga, and Se is to the right of As) to have a higher energy gap

Table 8.3 *Covalent and ionic bonds*

Semiconductor	Energy gap eV	Melting point K	Ionic % of bond	Lattice spacing Å
Group IV				
C	5.4	382	0	3.56
Si	1.11	1680	0	5.43
Ge	0.67	1210	0	5.66
SiC	2.9		18	3.08, 5.05
Group III–V				
Al N	6.28	3070		3.11, 4.98
Al P	3.34	1770		5.45
Al As	2.2	1870		5.66
Al Sb	1.6	1330		6.15
Ga N	3.42	2770		3.19, 5.18
Ga P	2.24	1730		5.45
Ga As	1.42	1520	31	5.65
Ga Sb	0.67	980	26	6.10
In N	1.89	2475		3.54, 5.70
In P	1.27	1330	42	5.80
In As	0.36	1215	36	6.06
In Sb	0.17	798	32	6.48
Group II–VI				
Zn O	3.20	2248	62	4.63
Zn S	3.67	1925	62	5.41
Zn Se	2.58	1790	63	5.67
Zn Te	2.26	1658	61	6.10
Cd O	2.5	2020	79	
Cd S	2.42	1750	69	5.58
Cd Se	1.74	1512	70	6.05
Cd Te	1.50	1368	67	6.48

than GaAs. It is actually 2.6 eV. The same rule, as for III–V materials, applies again for going up and down in the periodic table. ZnS (S is above Se) has an energy gap of 3.6 eV, whereas that of ZnTe (Te is below Se) is 2.26 eV.

Zinc blende is of course the name of the ore from which ZnS is obtained. It is a little unfortunate that ZnS is one of the II–VI compounds which also crystallizes in another form called wurtzite. Here the crystals have similar bond lengths, the formal difference is that alternate (111) plane layers are rotated 180° about the 111 axis. This gives a hexagonal atomic arrangement. It is reminiscent of the deviations of diamond to hexagonal planes in graphite and C_{60}. It is not too serious a problem having two versions of ZnS, as well as ZnO, ZnSe and CdS. The bond length and density are the same, and electrical properties practically identical. The bonding is only slightly different because of different distances of third nearest atoms. This is characterized by the Madelung constant (Section 5.3.1) which is 1.638 for zinc blende and 1.641 for wurtzite. So we do not have to record different band gaps and melting points for the two crystal types. Most of the III–V compounds crystallize in the zinc blende structure but the nitrides have the wurtzite form. A clue to why this happens is that the third nearest atoms are unlike hence attractive. The nitrogen atom with its small size and high electro-negativity is prone to take a more ionic form of crystal.

The main problems with II–VI materials used to arise from the fact that some compounds could be made p-type, some others n-type, but no compound could be made both types. The change for the better came with the advent of molecular beam epitaxy (see Section 8.10) which makes it possible to produce junctions from II–VI materials.

8.7   Non-equilibrium processes

In our investigations so far, the semiconductor was always considered to be in thermal equilibrium. Let us look briefly at a few cases where the equilibrium is disturbed.

The simplest way of disturbing the equilibrium is to shine electromagnetic waves (in practice these are mostly in the visible range) upon the semiconductor. As a result *photoemission* may occur, as in metals, but more interestingly, the number of carriers available for conduction may significantly increase. This case is called *photoconduction*.

The three possible processes of producing carriers for conduction are shown in Fig. 8.9; (i) creating an electron–hole pair, that is exciting an electron from the valence band into the conduction band; (ii) exciting an electron from an impurity level into the conduction band; (iii) exciting an electron from the top of the valence band into an impurity level, and thus leaving a hole behind.

The extra carriers are available for conduction as long as the semiconductor is illuminated. What happens when the light is switched off? The number of carriers must fall gradually to the equilibrium value. The time in which the extra density is reduced by a factor e is called the *lifetime* of the carrier and is generally denoted by τ (and is thus quite often confused with the collision time). It is an important parameter in the design of many semiconductor devices.

Assume now that only part of the semiconductor is illuminated; we shall then have a region of high concentration in connection with regions of lower concentration. This is clearly an unstable situation, and by analogy with gases,

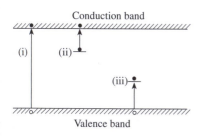

Fig. 8.9

Three models of obtaining free carriers by illumination: (i) band-to-band transitions yielding an electron–hole pair, (ii) ionization of donor atoms, (iii) ionization of acceptor atoms.

we may expect the carriers to move away from places of high concentration towards places of lower concentration. The analogy is incidentally correct; this motion of the carriers has been observed, and can be described mathematically by the usual *diffusion equation*

$$J = eD\nabla N. \tag{8.56}$$

Equation (8.56) is equally valid for holes and electrons, though in a practical case, the signs should be chosen with care.

D is the diffusion coefficient. This equation is quite plausible physically; it means that if there is a density gradient, a current must flow.

8.8 Real semiconductors

All our relationships obtained so far have been based on some idealized model. Perhaps the greatest distortion of reality came from our insistence that all semiconductors crystallize in a simple cubical structure. They do not. The most commonly used semiconductors, silicon and germanium, crystallize in the diamond structure, and this makes a significant difference.

Plotting the E vs k_x curve (Fig. 8.10) for the conduction band of silicon, for example, it becomes fairly obvious that it bears no close resemblance to our simple $E = E_0 - 2A_x \cos k_x a$ curve, which had its minimum at $k_x = 0$. Even worse, the $E(k_y)$ curve would be very different from the one plotted. The surfaces of constant energy in k-space are *not spheres*.

The situation is not much better in the valence band, where the constant energy surfaces are nearly spheres but—I regret to say—there are three different types of holes present. This is shown in Fig. 8.10, where the letters h, l, and s stand for *heavy, light*, and *split-off* bands respectively. What does it mean to have three different types of holes? Well, just think of them as holes painted red, blue, and green. They coexist peacefully, though occasionally, owing to collisions, a hole may change its complexion.

Does this mean that all we have said so far is wrong? Definitely not. Does it mean that considerable modifications are needed? No, for most purposes not

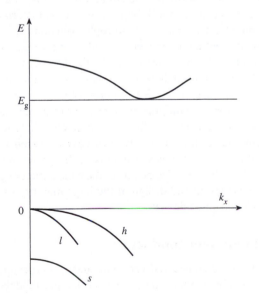

Fig. 8.10
$E - k$ curve for silicon in a particular direction. Note that the minimum of the conduction band is not at $k = 0$ and there are three different types of holes. The situation is similar in germanium, which is also an indirect-gap (minimum of conduction band not opposite to maximum of valence band) semiconductor.

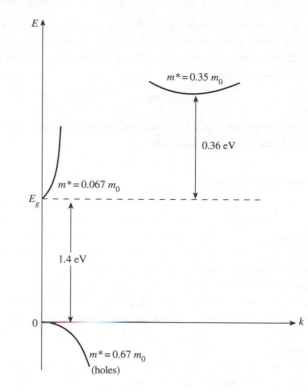

Fig. 8.11

Energy band diagram for gallium arsenide showing subsidiary valley in conduction band.

even that. We can get away with our simple model because, in general, only average values are needed. It is nice to know that there are three different types of holes in the valence band of silicon but for device operation only the average effective mass and some sort of average collision time are needed.

The picture is not as black as it seems. In spite of the anisotropy in the $E(k)$ curves the grand total is isotropic. I mean that by performing all the relevant averaging processes in silicon (still for a single-crystal material), the final result is an isotropic effective mass and isotropic collision time. Measuring the conductivity in different directions will thus always give the same result.*

Another important deviation from the idealized band structure occurs in a number of III–V compounds, where a subsidiary valley appears in the conduction band (shown for GaAs in Fig. 8.11). For most purposes (making a gallium arsenide transistor, for example) the existence of this additional valley can be ignored but it acquires special significance at high electric fields. It constitutes the basis for the operation of a new type of device, which we shall discuss later among semiconductor devices. You must realize, however, that this is the exception rather than the rule. The details of the band structure generally do not matter. For the description and design of the large majority of semiconductor devices our model is quite adequate.

* This is not true for graphite, for example, which has very different conductivities in different directions, but fortunately single crystal graphite is not widely used in semiconductor devices. Polycrystalline graphite has, of course, been used ever since the birth of the electrical and electronics industries (for brushes and microphones) but the operational principles of these devices are so embarrassingly simple that we cannot possibly discuss them among the more sophisticated semiconductor devices.

8.9 Amorphous semiconductors

In this chapter we have considered very pure and regular crystal forms of Ge, Si, and other semiconductors, and have used simple models based on this symmetry to derive basic properties like energy gap and conductivity. What happens if the material is not a single crystal—suppose one evaporates a film of Si on to a suitable substrate? The film consists of randomly oriented clusters of crystallites (small crystals). The structure within the crystallites leads to an energy gap. Remember, in all our three models of band structure (Chapter 7) band edges form when the electrons' de Broglie wave interferes with the lattice spacing. A small array of lattice points gives a less sharp interference pattern, so the band edges are not so well defined. We still get optical absorption and photoconduction, but with a less sharp spectral variation than shown in Fig. 8.18(b). Another effect of small crystallites is that the covalent bonds break off where the orientation changes, so that there are 'dangling bonds', where the unpaired electrons can act as traps for both itinerant electrons and holes. Mobility is greatly reduced; and also doping is much less effective, as the carriers from the dopant are trapped. This effect ties the Fermi level near the middle of the energy gap, as it is in intrinsic materials and insulators.

If the amorphous layer is formed in a gas discharge containing hydrogen, the 'dangling bond' is dramatically reduced. The H atoms neutralize the unpaired electrons, and the atomic nucleus has a minimal effect on lattice behaviour. In these circumstances amorphous Si can be doped into both p- and n-type.

Why should we use amorphous semiconductors when we can have them in superior single crystal form? The reason is purely economical. When we need them in large areas, as in solar cells (Section 13.2) or in xerographic applications (Section 10.14), we use the amorphous variety.

8.10 Measurement of semiconductor properties

The main properties to be measured are (i) mobility, (ii) Hall coefficient, (iii) effective mass, (iv) energy gaps (including the distance of any impurity layer from the band edge), and (v) carrier lifetime.

8.10.1 Mobility

This quantity was defined as the carrier drift velocity for unit field:

$$\mu = v_\text{D}/\mathcal{E}. \tag{8.57}$$

The most direct way of measuring mobility is to measure the drift velocity caused by a known d.c. electric field.

Since the electric field is constant in a conductor (and in a semiconductor too), it can be deduced from measurements of voltage and distance. How can we measure the drift velocity? Well, in the same way we always measure velocity; by measuring the time needed to get from point A to C. But can we follow the passage of carriers? Not normally. When, in the circuit of Fig. 8.12, we close the switch the electrons acquire an ordered motion *everywhere*. Those that happen to be at point C when the switch is closed will arrive at point A some time later, but we have no means of learning when. From the moment the switch is closed, the flow of electrons is uniform at both points, C and A. What we need is a circuit in which carriers can be launched at one point and detected at another point. A circuit that can do this was first described by Haynes and Shockley and is shown in Fig. 8.13(a). When S is open, there is a certain current flowing across the resistor R. At t_1 the switch is closed and according to the well-known laws of Kirchhoff there is a sudden increase of

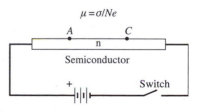

Fig. 8.12
Current flow in an n-type semiconductor.

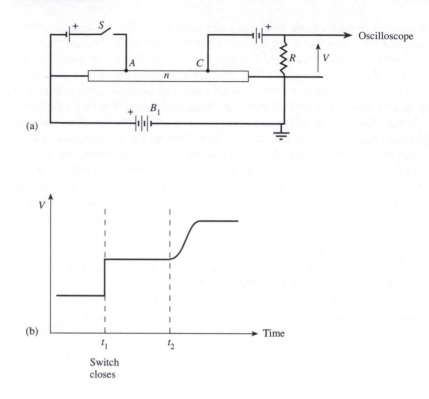

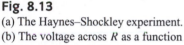

Fig. 8.13
(a) The Haynes–Shockley experiment.
(b) The voltage across R as a function
of time. The switch S is closed at
$t = t_1$. The holes drift from A to C in
a time $t_2 - t_1$.

current [and voltage, as shown in Fig. 8.13(b)] through R. But that is not all.
The contact between the metal wire and the n-type semiconductor is a rather
special one. It has the curious property of being able to *inject holes*. We shall
say more about injection later, but for the time being please accept that holes
appear at point A, when S is closed. Under the influence of the battery B_1 the
holes injected at A will move towards C. When they arrive at C (say at time
t_2), there is a new component of current that must flow across R. The rise in
current (and in voltage) will be gradual because some holes have a velocity
higher than the average, but after a while a steady state develops. Now we
know the distance between points A and C, and we know fairly accurately the
time needed by the holes to get from A to C; the drift velocity can thus be
determined. The electric field can easily be obtained, so we have managed to
measure, the mobility (Table 8.4).

A more modern version of the Haynes–Shockley experiment is to use a narrow light beam for exciting the extra carriers. The physics is then a lot more
complicated. As many as three separate phenomena take place simultaneously:
drift in the applied field, diffusion due to the nonuniform distribution of the created carriers, and recombination as the excess carriers relax back to equilibrium.
It is then a little more difficult to work out the mobility, but the basic principles
are the same.

A less direct way of determining the mobility is to measure the conductivity
and use the relationship

$$\mu = \sigma/Ne. \tag{8.58}$$

Table 8.4 *Effective masses and mobilities (at T = 293 K)*

Semiconductor	Effective electron m_e	Mass of holes m_h	Mobility cm^2 V^{-1} s^{-1} electron μ_e	holes μ_h
Group IV				
C	0.2	0.25	1800	1400
Si	0.58	1.06	1450	500
Ge	0.35	0.56	3800	1820
SiC			300	50
Group III–V				
Al N	0.33			
Al P			80	
Al As			1200	420
Al Sb	0.09	0.4	200	550
Ga N	0.22		1350	13
Ga P	0.35	0.5	300	150
Ga As	0.068	0.5	8800	400
Ga Sb	0.050	0.23	4000	400
In N	0.11			
In P	0.067	2.0	4600	150
In As	0.022	1.2	33 000	460
In Sb	0.014	0.4	78 000	750
Group II–VI				
Zn O	0.38	1.5	180	
Zn S			180	5
Zn Se			540	28
Zn Te			340	100
Cd O	0.10		120	
Cd S	0.165	0.8	400	50
Cd Se	0.13	1.0	450	
Cd Te	0.14	0.35	1200	50

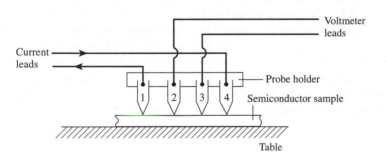

Fig. 8.14
The four-point probe. The probes are sharply pointed and held rigidly in a holder which can be pressed with a known force on to the semiconductor. A typical spacing is 1 mm between probes.

A method often used in practice is the so-called 'four-point probe' arrangement shown in Fig. 8.14. The current is passed from contact 1 to 4, and the voltage drop is measured with a voltmeter of very high impedance between points 2 and 3. Since the current flow between the probes is not laminar, some further calculations must be performed. For equally spaced probes, *d* apart, on a

Mobility can be calculated from this equation if we know the carrier concentration.

semiconductor of much greater thickness than d, the relationship obtained is

$$\sigma = I/2\pi V d. \tag{8.59}$$

It is important to realize that in some applications mobility is a function of field. Since practically everything obeys Ohm's law at low enough fields we may define the *low field mobility* as a constant. For high fields the differential mobility

$$\mu_{\text{diff}} = \frac{\mathrm{d}v_{\mathrm{D}}}{\mathrm{d}\mathcal{E}} \tag{8.60}$$

is usually the important quantity in device applications.

8.10.2　Hall coefficient

The Hall coefficient [eqn (1.20)] is a measure of the charge density, and hence it can be used to relate conductivity to mobility.

For this measurement four contacts have to be made so as to measure the voltage at right angles to the current flow. The basic measurement was described in Chapter 1. However, geometrical factors also come into this. If the distance between voltage probes is greater than that between the current probes the Hall voltage is reduced. Again, this reduction factor is calculable by detailed consideration of the patterns of current flow.

8.10.3　Effective mass

These measurements are commonly made in the microwave region (10^{10} Hz) at liquid helium temperature (about 4 K) or in the infrared (about 10^{13} Hz) at liquid nitrogen temperature (77 K).

The standard method of measuring effective mass uses the phenomenon of cyclotron resonance absorption discussed in Chapter 1. It is essentially an interaction of an electromagnetic wave with charge carriers, which leads to an absorption of the wave when the magnetic field causes the electron to vibrate at the same frequency as that of the applied electric field. For the resonant absorption to be noticeable the electron must travel an appreciable part of the period without collisions; thus a high-frequency electric field, a high-intensity magnetic field, and low temperatures are used.

In the apparatus for a microwave measurement, shown diagrammatically in Fig. 8.15, the sample is enclosed in a waveguide in a Dewar flask filled with liquid helium, which is placed between the poles of a large electromagnet. The microwave signal is fed in through a circulator.* Thus the signal entering arm (2) is reflected by the reflecting plate at the end of the waveguide having passed through the semiconductor in each direction and ends up in the receiver connected to arm (3). Employing a wave of fixed frequency and a variable magnetic field; the effective mass is given [eqn (1.69)] by

* The circulator has the following magical properties: a signal fed into arm (1) goes out entirely by arm (2) and a signal fed into arm (2) leaves the circulator by arm (3).

B is the magnetic field corresponding to an absorption of signal.

$$m^* = eB/\omega_{\mathrm{c}}, \tag{8.61}$$

There will generally be several absorption peaks, corresponding to the various holes and electrons present. The experimentally obtained absorption curve for germanium is shown in Fig. 8.16 for a certain orientation between the magnetic field and the crystal axes. As may be seen, there are two types of holes, light and heavy. A third resonance peak for the holes in the split-off band is missing because there are hardly any holes so much below the band edge. The two resonance peaks for electrons indicate that something sinister is going on in the

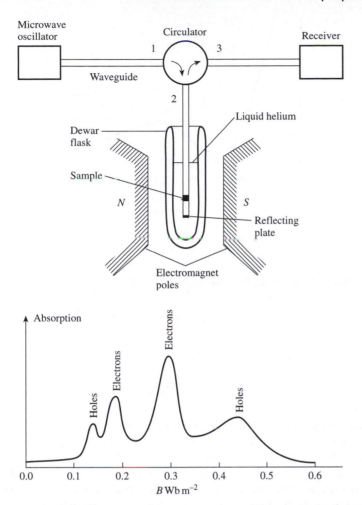

Fig. 8.15
Schematic representation of the cyclotron resonance experiment.

Fig. 8.16
Result of a cyclotron resonance experiment for germanium.

conduction band as well. As a matter of fact, these measurements, repeated in various directions, are just the tools for unravelling what the real $E - k$ curve looks like. In addition, from the amplitude and width of the peaks, information about the density and collision times of the various carriers can be obtained.

8.10.4 Energy gap

A simple way to measure the energy gap between the valence and conduction bands is to see how the conductivity varies with temperature. For any semiconductor the conductivity is given by

$$\sigma = (N_e \mu_e + N_h \mu_h)e, \tag{8.62}$$

which is the same as eqn (8.45). For an intrinsic material, we have from eqn (8.46)

$$N_e = N_h = N_i = \text{constant} \times T^{3/2} \exp\left(\frac{-E_g}{2kT}\right). \tag{8.63}$$

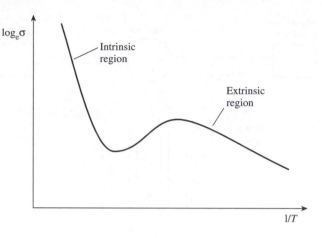

Fig. 8.17

Typical log conductivity–reciprocal temperature curve for an extrinsic semiconductor.

We shall ignore the $T^{3/2}$ variation, which will almost always be negligible compared with the exponential temperature variation. Hence a plot of $\log_e \sigma$ versus $1/T$ will have a slope of $-E_g/2k$, which gives us E_g. Also in eqn (8.64) we have ignored the variation of E_g with temperature.

Combining eqn (8.62) with eqn (8.63) we get

$$\sigma = \text{constant} \times e(\mu_e + \mu_h)T^{3/2} \exp\left(-\frac{E_g}{2kT}\right)$$

$$= \sigma_0 \exp\left(-\frac{E_g}{2kT}\right). \qquad (8.64)$$

Let us now consider what happens with an impurity semiconductor. We have discussed the variation of the Fermi level with temperature and concluded that at high temperatures semiconductors are intrinsic in behaviour, and at low temperatures they are pseudo-intrinsic with an energy gap equal to the gap between the impurity level and the band edge. Thus, we would expect two definite straight-line regions with greatly different slopes on the plot of $\log_e \sigma$ against $1/T$, as illustrated in Fig. 8.17. In the region between these slopes the temperature is high enough to ionize the donors fully but not high enough to ionize an appreciable number of electrons from the host lattice. Hence, in this middle temperature range the carrier density will not be greatly influenced by temperature, and the variations in mobility and the $T^{3/2}$ factor that we neglected will determine the shape of the curve.

An even simpler method of measuring the energy gap is to study optical transmission. The light is shone through a thin slice of semiconductor [Fig. 8.18(a)] and the amount of transmission is plotted as a function of wavelength. If the wavelength is sufficiently small (i.e. the frequency is sufficiently large), the incident photons have enough energy to promote electrons from the valence into the conduction band. Most of the photons are then absorbed, and the transmission is close to zero. As the wavelength increases, there will be a particular value ($\lambda = c/f = hc/E_g$) when band-to-band transitions are no longer possible. The absorption then suddenly declines, and correspondingly, transmission sharply increases as shown in Fig. 8.18(b) for a thin GaAs sample. The point where the sudden rise starts may be estimated from the figure as about 880 nm, which corresponds to an energy gap of 1.41 eV, which is just about right. Fig. 8.18(b) is typical for the so-called direct-gap semiconductors which have an $E - k$ energy band structure [illustrated in Fig. 8.11 and again in Fig. 8.19(a)] where

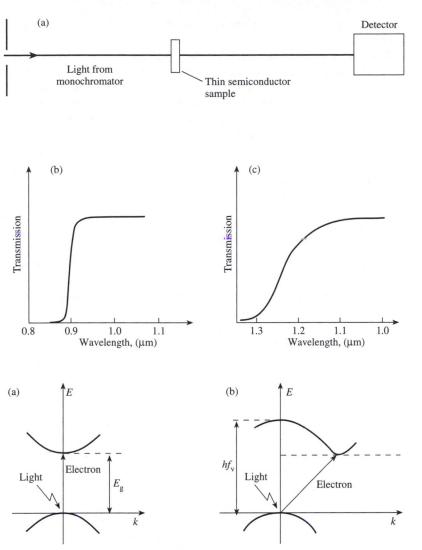

Fig. 8.18
(a) General arrangement of an optical transmission measurement and the result for (b) GaAs and (c) Si.

Fig. 8.19
Photon absorption by (a) a direct and (b) an indirect-gap semiconductor.

the maximum of the valence band is at the same k value as the minimum of the conduction band.

Silicon and germanium are indirect-gap semiconductors as shown in Fig. 8.10 and also in Fig. 8.19(b). The measured transmission as a function of wavelength for a thin Si sample is plotted in Fig. 8.18(c). The transmission may be seen to vary much more gradually with the wavelength.

How does an electron make a transition when excited by a photon? We have not yet studied this problem in any detail. All I have said so far is that if an electron receives the appropriate amount of energy, it can be excited to a state of higher energy. This is not true in general because, as in macroscopic collision processes, not only energy but momentum, as well, should be conserved.

There is no way out now. If we wish to explain the excitation of electrons by light in an indirect-gap semiconductor, as shown in Fig. 8.19(b), we have

There are quite a number of direct-gap semiconductors. In fact, most of the III–V and II–VI compounds belong to that family.

Semiconductors

to introduce phonons and have to consider the momentum of our quantized particles.

Phonons are the quantum-mechanical equivalents of lattice vibrations. For a wave of frequency ω the energy of the phonon is $\hbar\omega$, analogous to the energy of a photon. What can we say about momentum? In Chapter 3 we talked about the momentum operator, and subsequently we showed that the energy of a free electron is equal to $E = \hbar^2 k^2 / 2m$, where k comes from the solution of the wave equation. It is the equivalent of the wavenumber of classical waves.

The momentum of a free electron is $\hbar k$. When an electron is in a lattice, its energy and momentum are no longer related to each other by the simple quadratic expression. The relationship is then given by the $E-k$ curve, but the momentum is still $\hbar k$. The momenta of photons and phonons are given again by $\hbar k$, but now $k = \omega/v$, where v is the velocity of the wave. For an electromagnetic wave it is $v = c = 3 \times 10^8 \, \text{m s}^{-1}$. The velocity of a lattice wave, which may also be called a sound wave, is smaller by four or five orders of magnitude. Hence, for the same frequency the momentum of a phonon is much higher than that of a photon.

Let us next do a simple calculation for the momentum of an optical wave. With a wavelength of $\lambda = 1 \, \mu\text{m}$, the corresponding frequency is $3 \times 10^{14} \, \text{Hz}$, the energy is $1.24 \, \text{eV}$, and $k = 6.28 \times 10^6 \, \text{m}^{-1}$. In our simple models of Chapter 7 the zone boundary came to π/a. With $a = 3 \, \text{nm}$, the corresponding k is about $10^{10} \, \text{m}^{-1}$, that is much larger than that of the photon. Hence, when we talk about electron excitation in a direct-gap semiconductor, we can regard the transition as practically vertical. The picture of a two-particle interaction is permissible: a photon gives all its energy to an electron.

> It may be seen that the energy of the photon is comparable with the gap energies of semiconductors, but its k value is relatively small.

In order to excite an electron in an indirect-gap semiconductor, the photon still has enough energy, but its momentum is insufficient. It needs the good services of a third type of particle, which can provide the missing momentum. These particles are phonons, which are always present owing to the finite temperature of the solid. They can provide high enough momentum to satisfy momentum conservation. Nevertheless, this is a three-particle interaction between an electron, a photon, and a phonon, which is much less likely than a two-particle interaction. Hence, as the wavelength decreases, the transmission will not suddenly increase as in Fig. 8.18(b). It will instead gently rise and reach saturation when the frequency is large enough [f_v in Fig. 8.19(b)] to affect direct transitions between the valence and conduction bands. Whether or not a material is a direct-gap material is of increasing importance in the development of optical semiconductor devices—semiconductor lasers are all direct-gap materials.

8.10.5 Carrier lifetime

We are usually interested in *minority* carrier lifetime. The reason is simply that, owing to injection or optical generation, the minority carrier density may be considerably above the thermal equilibrium value, whereas the change in the density of majority carriers is generally insignificant. Consider, for example, silicon with 10^{22} fully ionized impurities per cubic metre. Then, as N_i for silicon is about $10^{16} \, \text{m}^{-3}$ at room temperature [eqn (8.63)], N_h will be about $10^{10} \, \text{m}^{-3}$. Now suppose that in addition 10^{15} electron–hole pairs per cubic metre are

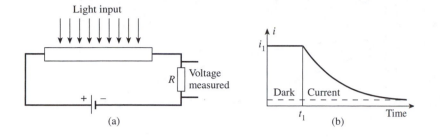

Fig. 8.20
(a) Photoconduction experiment in a semiconductor. (b) When the light is switched off the current decays to its dark current value.

created by input light. The hole density in the silicon will then increase by a huge factor, 10^5, but the change in electron density will be an imperceptible $10^{-5}\%$. Thus, to 'see' the change of hole current is relatively simple; the only trick is to make a junction that lets through the holes but restricts the electron flow to a low value. (This again is something we shall discuss later.) Thus, the current flowing in the circuit of Fig. 8.20(a) consists mainly of holes created by the input light. If the light is switched off at $t = t_1$, the current (and so the voltage) across the resistance R declines exponentially as $\exp(-t/\tau_p)$. By measuring the decay of the current [Fig. 8.20(b)] τ_p can be determined. How does the exponential decay come about? The differential equation can be easily derived (see example 8.17) on the basis of the physical picture developed in Section 8.5. The rate of change of carriers may always be written as the rate of creation minus the rate of recombination.

τ_p is the lifetime of the holes.

8.11 Preparation of pure and controlled-impurity single-crystal semiconductors

8.11.1 Crystal growth from the melt

This is the simplest way of preparing a single crystal. The material is purified by chemical means, perhaps to an impurity concentration of a few parts per million, then melted in a crucible of the shape shown in Fig. 8.21. The crucible is slowly cooled down. As the pointed end tends to cool slightly faster than the bulk of the material, the crystal 'seeds' at the bottom, then grows through the melt. If conditions are well controlled, a single crystal growth is obtained. It is found that the impurity concentration is no longer constant throughout the crystal, but there is a definite concentration gradient, usually with the purest material at the bottom.

To understand the reason for this we have to consider the *metallurgical phase diagram* for the semiconductor and the impurity. You have probably come across the phase diagram for copper and zinc, stretching from 100% copper, 0% zinc to 0% copper, 100% zinc, with brasses in the middle, and curves representing liquidus and solidus lines, with temperature as the ordinate. We do not need to consider such a range of composition, since we are considering only a minute amount of impurity in silicon. We need only look at the region close to pure Si, where there will be no complications of eutectics, but only the liquidus and solidus lines, shown diagrammatically in Fig. 8.22. The temperature separations of these lines will be only a few degrees.

Suppose there is initially an impurity content C_a (Fig. 8.22). As the melt cools down it stays liquid until it reaches the temperature T_a. At this temperature

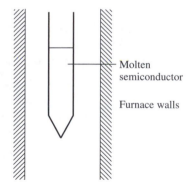

Fig. 8.21
A form of crucible for melt-grown single crystals.

Molten semiconductor

Furnace walls

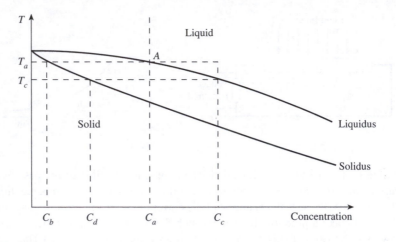

Fig. 8.22
Illustration of change of composition
on freezing.

there can exist liquid of composition C_a and solid of composition C_b. Solid of
the latter composition is the first to crystallize out. As this is purer material and
becomes lost to the rest of the melt once it solidifies, the remainder has a higher
impurity concentration, say C_c. Thus, no more solidification occurs until the

The impurity concentration of the
solid, still starting at C_b smoothly
increases up the crystal.

temperature T_c is reached, when more solid impurity concentration C_d comes
out. And so it goes on. Of course, if the cooling is slow, this is a continuous
process.

It is usual to describe this process in terms of the *distribution coefficient*
k, defined as the concentration of impurity in the solid phase divided by the
concentration in the liquid phase, both measured close to the phase boundary.
For the case of Fig. 8.22

$$k = C_b/C_a. \tag{8.65}$$

If it is assumed that k does not
change during solidification, it is
a simple matter to find the impurity concentration gradient of the
crystal.

8.11.2 Zone refining

The different concentrations of impurity in the solid and liquid phase can be
exploited in a slightly different way. We start with a fairly uniform crystal, melt
a slice of it, and arrange for the *molten zone* to travel along the crystal length.
This can be done by putting it in a refractory boat and dragging it slowly through
a furnace, as shown in Fig. 8.23. At any point, the solid separating out at the
back of the zone will be k times as impure as the melted material which, as
$k < 1$, is an improvement. By a fairly simple piece of algebra it can be shown
that the impurity concentration in the solid, $C_s(x)$, after the zone has passed
down the crystal (of length l) once is

Since k is typically 0.1, it is possible to drive most of the impurity
to a small volume at the far end with
relatively few passes.

$$C_s(x) = C_0\{1 - (1 - k)\exp(-kx/z)\}, \tag{8.66}$$

where C_0 is the initial concentration and z is the length of the molten zone.
Clearly, at the end of the crystal that is melted first, the value of impurity
concentration will be

$$C_s(0) = k^n C_0 \tag{8.67}$$

if this process is repeated n times.

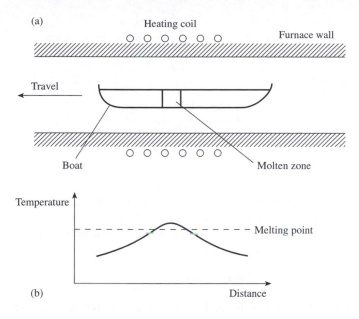

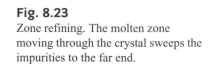

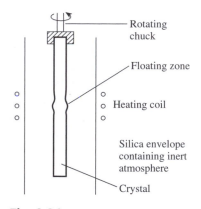

Fig. 8.23
Zone refining. The molten zone moving through the crystal sweeps the impurities to the far end.

This very simple idea is the basis of the great success of semiconductor engineering. As we have said before impurities can be reduced to few parts in 10^{10}, and then they are usually limited by impurities picked up from reactions with the boat.

8.11.3 Floating zone purification

This latter problem showed up rather strongly when the semiconductor industry went over from germanium (melting point 937 °C) to silicon (melting point 1958 °C). The solution was the *floating zone method*, which dispensed with the boat altogether. In this method the crystal is held vertically in a rotating chuck (Fig. 8.24). It is surrounded at a reasonable distance by a cool silica envelope, so that it can be kept in an inert atmosphere, then outside this is a single-turn coil of water-cooled copper tubing. A large high-frequency current (several MHz) is passed through the coil, and the silicon crystal is heated to melting point by the *eddy currents* induced in it. The coil is slowly moved up the crystal so that the molten zone passes along its length. This technique can be used only for fairly small crystals because the weight has to be supported by the surface tension of the molten zone.

8.11.4 Epitaxial growth

The process of growing and refining single crystals made possible the advent of the transistor in the 1950s. The next stage has been the *planar* technique, starting in about 1960, that have led to the development of integrated circuits to be discussed in the next chapter. I shall just describe here the epitaxial growth method of material preparation, which is eminently compatible with the manufacture of integrated circuits.

'Epitaxial' is derived from a Greek word meaning 'arranged upon'.* There are several ways in which such growth can be carried out. To deposit silicon

Fig. 8.24
Floating zone refining.

* My friends who speak ancient Greek tell me that *epitactic* should be the correct adjective. Unfortunately, *epitaxial* has gained such a wide acceptance among technologists having no Greek-speaking friends that we have no alternative but to follow suit.

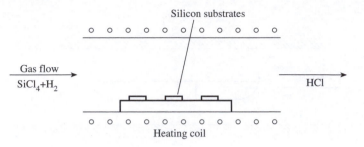

Fig. 8.25
Vapour phase epitaxy. The Si forms on the single crystal substrates at a temperature of about 1200°C in the furnace.

epitaxially from the *vapour phase*, the arrangement of Fig. 8.25 can be used. Wafers of single-crystal silicon are contained in a tube furnace at (typically) 1250 °C. Silicon tetrachloride vapour in a stream of hydrogen is passed through the furnace and the chemical reaction

$$SiCl_4 + 2H_2 \rightleftharpoons Si + 4HCl \qquad (8.68)$$

takes place. The Si is deposited on the silicon wafers as a single crystal layer following the crystal arrangement of the substrate. Sometimes the silane reaction

$$SiH_4 \rightleftharpoons Si + 2H_2 \qquad (8.69)$$

is preferred, since it gives no corrosive products.

The epitaxial layer can be made very pure by controlling the purity of the chemicals; or more usefully it can be deliberately doped to make it n- or p-type by bubbling the hydrogen through a weak solution of (for example) phosphorus trichloride or boron trichloride, respectively, before it enters the epitaxy furnace. In this way epitaxial layers of about 2–20 μm thick can be grown to a known dimension and a resistivity that is controllable to within 5% from batch to batch.

Liquid Phase Epitaxy (LPE) has also been used, mainly with compound semiconductors. The substrate crystal is held above the melt on a quartz plate and dipped into the molten semiconductor (Fig. 8.26). By accurately controlling the cooling rate a single-crystal layer can be grown epitaxially on the crystal.

In recent years liquid phase epitaxy has been the workhorse in growing semiconductors for lasers (semiconductor lasers will be discussed in Section 12.7). It is simple and quite fast, and it coped heroically with the problem of putting upon each other semiconductors of differing bandgaps when there was no alternative, but it cannot really produce the sharply defined layers needed for the latest devices. Some new techniques were bound to come. They are represented by Molecular Beam Epitaxy (MBE) and Metal–Organic Chemical Vapour Deposition (MOCVD).

8.11.5 Molecular beam epitaxy

This is probably the best and most versatile method. Each material (various semiconductors which make up the desired compound plus the dopants) sits in a little box of its own in which it is heated by an oven to a temperature

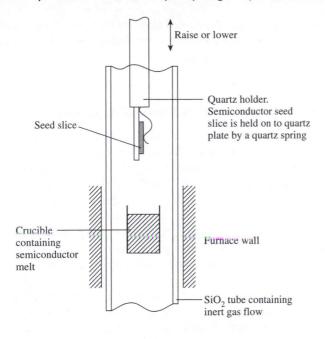

Raise or lower

Quartz holder.
Semiconductor seed
slice is held on to quartz
plate by a quartz spring

Seed slice

Crucible
containing
semiconductor
melt

Furnace wall

SiO$_2$ tube containing
inert gas flow

Fig. 8.26
Liquid phase epitaxy. The
semiconductor slice is held on the
plate by a quartz spring clip and
lowered into the molten
semiconductor alley. By correct
cooling procedures the pure
semiconductor is encouraged to
precipitate onto the surface of the
slice.

usually above its melting point, all in ultra-high vacuum. The atoms evaporated from the surface of the heated material are ready to move to the substrate, provided they can get out of the box, that is the shutter is open. The duration for which each shutter is open and the temperature of the oven will then determine the flux of each element toward the substrate, and thus the composition of the growth material. The growth can be monitored by electron scattering from the surface.

Altogether this is a very precise method capable of excellent composition control, once all the sources have been experimentally calibrated. It was used to explore ternary and quaternary compounds such as GaAlAs but gave way to cheaper liquid or vapour phase epitaxy once these techniques were established. A further use has been for compounds such as many of the II–VIs, for example ZnS, ZnSe and CdS where it proved difficult to get both n and p types, usually p was the stumbling block. Whilst the p type impurities could be established in the crystal lattice, the process of chemical equilibrium on the heated substrate, essential for a good crystalline quality caused the hole centres to be neutralized by electrons. MBE can avoid thermodynamic equilibrium and maintain the p centres active, provided temperature and the ambient atmosphere are controlled.

Growing InGaN has been proved successful too after overcoming the difficulty of a nitrogen source. Initially NH$_3$ was used, but the decomposition is only moderately efficient at temperatures below 800°C, so copious streams of NH$_3$ are needed in what is ideally a high vacuum system. Nitrogen plasma sources have evolved based on either an RF discharge at 13.56 MHz or an electron cyclotron resonance breakdown at 2.45 GHz. This allows higher quality growth at about $1\,\mu m\,h^{-1}$. MOCVD is preferred for commercial production, but MBE has produced higher quality with GaN, for example an electron mobility of $1200\,cm^2\,V^{-1}\,s^{-1}$ has been reported for an epilayer grown by MBE on a

MOCVD template. The usually quoted mobility for 'typical' GaN is about half this. The combined use of two techniques to produce better material properties is perhaps the way ahead for InGaN.

8.11.6 Metal–organic chemical vapour deposition

This process is more suited to be adopted by manufacturers. As the name implies, it is a chemical process. The required elements are introduced into the growth tube as compounds bound to organic substances, and they are then chemically released to be deposited upon the substrate. One of the advantages of the technique is that the otherwise poisonous gases are relatively harmless in organic compounds. Another advantage is that the reaction can take place over substrates that have a large area and, in addition, the process is quite fast. Its disadvantage is that it is done at about atmospheric pressure which is in general detrimental to accurate deposition and, among others, it prevents the use of an electron beam to monitor thickness.

What was largely an effort to pacify safety officers and stop the use of large quantities of arsine and phosphine, came into its own with the advent of nitrides. The nitrogen source dimethylhydrazine was used, but it was found that using a similar R.F. discharge source to that used in MBE was preferable. In 1992, several laboratories started work on dilute N additions to semiconductors such as GaAs, InAs and GaP.

Initially the fact that N added to GaAs reduced the band gap, although GaN has a much wider gap surprised researchers, but soon the large bowing of the conduction band became well known. A large number of long wavelength devices have been made including lasers in the $1.3\,\mu$m range (GaInNAs) and solar cells. In 2002 a 10 Gbit s^{-1} MOVPE grown VCSEL went onto the market. This work has been done in parallel with MBE and MOCVD usually in the same labs, the former is for research and quality, the latter is preferred for production, especially multilayered structures.

The great success of this technique in the past decade has been in the difficult area of producing reasonable quality InGaN for LEDs. There are no nitride substrate crystals, so growth has usually been on sapphire (Al_2O_3). The reactants are trimethyl gallium and trimethyl indium and ammonia (NH_3) as a source of nitrogen. The two flow MOCVD method used at Nichia Chemical Industries first heats the sapphire substrate to $1050°$C in a stream of H_2. Then the temperature is lowered to $510°$C to grow a buffer layer of GaN approximately 30 nm thick. Then the main growth of GaN film of approximately $2\,\mu$m at a temperature of $1020°$C takes about 30 min. At a reduced temperature of about $800°$C the InGaN active layer is grown $0.3\,\mu$m in about 60 min. The active layer has a pn detailed structure. InGaN is naturally n-type but the conductivity can be controlled by silicon impurity via a stream of silane (SiH_4) in the reactant gas. Then there is a very thin layer of InGaN with the In content controlled so that band edge recombination gives the required LED colour. Finally the p layer, which took a while to design. It uses Mg impurity, deposited via a metal-organic compound called bis-cyclopentadienyl magnesium, and has to be followed by an anneal in nitrogen to activate the impurity. The resulting semiconductor slice is riddled with threading dislocations originating at the lattice mismatch with sapphire, typically about 10^{10} cm^{-2}. However, once it is metal contacted and cleaved or cut into diodes usually just under 1 mm^2 in area, the LEDs work well with quantum efficiency of 10–50%.

8.11.7   Hydride vapour phase epitaxy (HVPE) for nitride devices

The HVPE technique has attracted great interest because it can produce thick layers at high growth rates and comparatively low cost. The basic reactants are obtained from the following chemical reactions, where the bracketed (g), (l) and (s) refer to gas, liquid and solid phases, respectively.

$$2Ga(l) + 2HCl(g) = 2Ga\,Cl(g) + H_2(g)$$

and for the higher chloride

$$GaCl(l) + 2HCl(g) = GaCl_3(g) + H_2(g).$$

Further, there are two reaction pathways for GaN deposition.

$$GaCl(g) + NH_3(g) = GaN(s) + HCl(g) + H_2(g)$$

$$3GaCl(g) + 2NH_3(g) = 2GaN(s) + GaCl(g) + 3H_2(g).$$

The snag in this process is that the decomposition of ammonia is only about 3–4% even at a substrate temperature 950°C, and the excess H$_2$ and HCl are no help for the reaction and encourage GaN deposition in other parts of the apparatus. There is also the familiar problem of no GaN substrate, so sapphire (Al$_2$O$_3$) is usual, but ZnO, SiC and Si have been used frequently. As well as threading dislocations from the lattice mismatch, there are also stacking dislocations caused by uneven growth. The main results reported so far have been with layers 10–100 μm thick grown in a few hours, which have produced successful LEDs and lasers. Some layers have been grown up to 300 μm. A major interest is getting free standing GaN films by removing the substrate. One way of doing this is by focussing a laser through the layer on to the sapphire substrate, and by heating it at several points inducing it to crack off. Other substrates such as Si can be etched away with HCl. SiC can be removed by ion etching in SF$_6$. A very direct method for any substrate is mechanical abrasion with diamond impregnated cloths or a slurry.

There is a lot of work going on to improve GaN quality and one possible success route is by chemical or heat treating free standing slices for further epitaxy by one of these methods.

Exercises

8.1. Indicate the main steps (and justify the approximations) used in deriving the position of the Fermi level in intrinsic semiconductors. How near is it to the middle of the gap in GaAs at room temperature? The energy gap is 1.4 eV and the effective masses of electrons and holes are $0.067m_0$ and $0.65m_0$ respectively.

8.2. Show that the most probable electron energy in the conduction band of a semiconductor is $\frac{1}{2}kT$ above the bottom of the band (assume that the Fermi level is several kT below the conduction band). Find the average electron energy.

8.3. In a one-dimensional model of an intrinsic semiconductor the energy measured from the bottom of the valence band is

$$E = \frac{\hbar^2 k_1^2}{3m_0} + \frac{\hbar^2 (k - k_1)^2}{m_0}.$$

This is an approximate formula accurate only in the vicinity of the minimum of the conduction band, which occurs when $k = k_1 = \pi/a$, where a the lattice spacing is 0.314 nm. The Fermi energy is at 2.17 eV.

Calculate (i) the energy gap between the valence and conduction bands, and (ii) the effective mass of electrons at the bottom of the conduction band.

Assume that the Fermi level is halfway between the valence and conduction bands.

8.4. The variation of the resistivity of intrinsic germanium with temperature is given by the following table:

$T(K)$	385	455	556	714
$\rho(\Omega m)$	0.028	0.0061	0.0013	0.000274

It may be assumed, as a rough approximation, that the hole and electron mobilities both vary as $T^{-3/2}$, and that the forbidden energy gap, E_g, is independent of temperature.

(i) Determine the value of E_g.
(ii) At about what wavelength would you expect the onset of optical absorption?

8.5. What is the qualitative difference between the absorption spectra of a direct gap and that of an indirect gap semiconductor?

8.6. Considering that $N_e N_h = N_i^2$ in a given semiconductor, find the ratio N_h/N_e which yields minimum conductivity. Assume that collision times for electrons and holes are equal and that $m_e^*/m_h^* = 0.5$.

8.7. In a certain semiconductor the intrinsic carrier density is N_i. When it is doped with a donor impurity N_0, both the electron and hole densities change. Plot the relative electron and hole densities N_e/N_i and N_h/N_i as a function of N_0/N_i in the range $0 \leqslant N_0/N_i \leqslant 10$. Assume that all donor atoms are ionized.

8.8. Consider a sample of intrinsic silicon.

(i) Calculate the room temperature resistivity.
(ii) Calculate the resistivity at 350°C.
(iii) If the resistance of this sample of silicon is R find the temperature coefficient of resistance at room temperature defined as $(1/R)(dR/dT)$. How can this be used to measure temperature?

Take $E_g = 1.1\,\text{eV}$, $m_e^* = 0.26 m_0$; $m_h^* = 0.39 m_0$,
$$\mu_e = 0.15\,\text{m}^2\,\text{V}^{-1}\,\text{s}^{-1}, \quad \mu_h = 0.05\,\text{m}^2\,\text{V}^{-1}\,\text{s}^{-1}$$

8.9. A sample of gallium arsenide was doped with excess arsenic to a level calculated to produce a resistivity of 0.05 Ω m. Owing to the presence of an unknown acceptor impurity the actual resistivity was 0.06 Ω m, the sample remaining n-type. What were the concentrations of donors and acceptors present?

Take $\mu_e = 0.85\,\text{m}^2\,\text{V}^{-1}\,\text{s}^{-1}$ and assume that all impurity atoms are ionized.

8.10. Silicon is to be doped with aluminium to produce p-type silicon with resistivity 10 Ω m. By assuming that all aluminium atoms are ionized and taking the mobility of a hole in silicon to be 0.05 $\text{m}^2\,\text{V}^{-1}\,\text{s}^{-1}$ find the density of aluminium.

8.11. Estimate what proportion of the aluminium is actually ionized in exercise 8.10 at room temperature. The acceptor level for aluminium is 0.057 eV above the valence band.

8.12. A sample of silicon is doped with indium for which the electron acceptor level is 0.16 eV above the top of the valence band.

(i) What impurity density would cause the Fermi level to coincide with the impurity level at 300 K?
(ii) What fraction of the acceptor levels is then filled?
(iii) What are the majority and minority carrier concentrations?

Use data from Exercise 8.8.

8.13. A certain semiconductor is doped with acceptor type impurities of density N_A which have an impurity level at $E_A = E_g/5$. At the temperature of interest $E_g = 20kT$ and $E_F = 5kT$. The effective masses of electrons and holes are $m_e^* = 0.12 m_0$ and $m_h^* = m_0$. For $N_A = 10^{23}\,\text{m}^{-3}$ find

(i) the ionized acceptor density,
(ii) the ratio of electron density to hole density,
(iii) the hole density,
(iv) the electron density,
(v) the temperature,
(vi) the gap energy.

8.14. Show that in the low-temperature region the electron density in an n-type material varies as

$$N_e = N_c^{1/2} N_D^{1/2} \exp[-(E_g - E_D)/2kT]$$

[Hint: Assume that $N_e = N_D^+$ and that the donors are only lightly ionized in that temperature range, i.e. $E_F > E_D$.]

8.15. The Bohr radius for a hydrogen atom is given by eqn (4.24). On the basis of the model presented in Section 8.3 determine for silicon the radius of an impurity electron's orbit.

8.16. The conductivity of an n-type semiconductor is σ at an absolute temperature T_1. It turns out that at this temperature the contributions of impurity scattering and lattice scattering are equal. Assuming that in the range T_1 to $2T_1$ the electron density increases quadratically with absolute temperature, determine the ratio $\sigma(2T_1)/\sigma(T_1)$.

8.17. The rate of recombination (equal to the rate of generation) of carriers in an extrinsic semiconductor is given by eqn (8.50). If the minority carrier concentration in an n-type semiconductor is above the equilibrium value by an amount $(\delta N_h)_0$ at $t = 0$, show that this extra density will reduce to zero according to the relationship,

$$\delta N_h = (\delta N_h)_0 \exp(-t/\tau_p),$$

where

$$\tau_p = \frac{1}{\alpha N_e}.$$

[Hint: The rate of recombination is proportional to the *actual* density of carriers, while the rate of generation remains constant.]

8.18. Derive the continuity equation for minority carriers in an n-type semiconductor.
[Hint: Take account of recombination of excess holes by introducing the lifetime, τ_p.]

8.19. Figure 8.16 shows the result of a cyclotron resonance experiment with Ge. Microwaves of frequency 24 000 MHz were transmitted through a slice of Ge and the absorption was measured as a function of a steady magnetic field applied along a particular crystalline axis of the single-crystal specimen. The ordinate is a linear scale of power absorbed in the specimen. The total power absorbed is always a very small fraction of the incident power.

(i) How many distinct types of charge carriers do there appear to be in Ge from this data?
(ii) How many types of charge carriers are really there? How would you define whether they are 'real' charge carriers or not?
(iii) What are the effective masses for this particular crystal direction? Can this effective mass be directly interpreted for electrons? For holes?
(iv) If the figure were not labelled would you be able to tell which peaks referred to electrons?
(v) Estimate the collision times of the holes (Hint: Use eqn (1.61)).
(vi) Estimate the relative number density of the holes.
(vii) In Section 8.8 we talked about *three* different types of holes. Why are there only two resonance peaks for holes?

9 Principles of semiconductor devices

Les mystères partout coulent comme les sèves,
Baudelaire *Les Sept Vieillards*

This thing with knobs and a pretty light.
A. Wesker *Chips with everything*

9.1 Introduction

You have bravely endured lengthy discussions on rather abstract and occasionally nebulous concepts in the hope that something more relevant to the practice of engineering will emerge. Well, here we are; at last we are going to discuss various semiconductor devices. It is impossible to include all of them, for there are so many nowadays. But if you follow carefully (and if everything we have discussed so far is at your fingertips) you will stand a good chance of understanding the operation of all existing devices and—I would add—you should be in a very good position to understand the operation of semiconductor devices to come in the near future. This is because human ingenuity has rather narrow limits. Hardly anyone ever produces a new idea. It is always some combination of old ideas that leads to reward. Revolutions are few and far between. It is steady progress that counts.

9.2 The p–n junction in equilibrium

Not unexpectedly, when we want to produce a device, we have to put things together. This is how we get the simplest semiconductor device the p–n junction, which consists of a p- and an n-type material in contact [Fig. 9.1(a)]. Let us imagine now that we literally put the two pieces together.* What happens when they come into contact? Remember, in the n-type material there are lots of electrons, and holes abound in the p-type material. At the moment of contact the electrons will rush over into the p-type material and the holes into the n-type material. The reason is, of course, diffusion: both carriers make an attempt to occupy uniformly the space available. Some electrons, moving towards the left, collide head-on with the onrushing holes and recombine, but others will be able to penetrate farther into the p-type material. How far? Not very far; or, to put it another way, not *many* get very far because their efforts are frustrated by the appearance of an electric field. The electrons leave positively charged donor atoms behind, and similarly there are negatively charged acceptor atoms left in

* This is *not* how junctions are made.

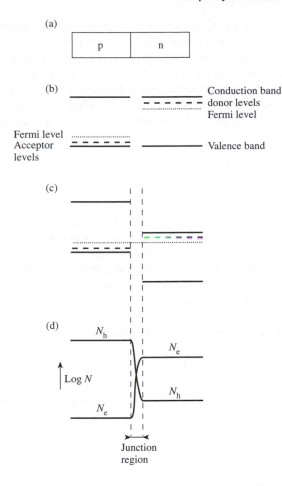

Fig. 9.1

The p–n junction. (a) A p- and an n-type material in contact, (b) the energy diagrams before contact, (c) the energy diagrams after contact, (d) electron and hole densities.

the p-type material when the holes move out. This charge imbalance will give rise to an electric field, which will increase until equilibrium is reached.

Having reached equilibrium, we can now apply a theorem mentioned before when discussing metal–metal junctions. We said that whenever two or more materials are in thermal equilibrium, their respective Fermi levels must agree.

The Fermi levels before contact are shown in Fig. 9.1(b) and after contact in Fig. 9.1(c). Here we assume that some (as yet unspecified) distance away from the junction, nothing has changed; that is, the energy diagram is unaffected, apart from a vertical shift needed to make the two Fermi levels coincide. This is not to diminish the significance of the vertical shift. It means that electrons sitting at the bottom of the conduction band on the left-hand side have higher energies than their fellow electrons sitting at the bottom of the conduction band at the right-hand side. By how much? By exactly the difference between the energies of the original Fermi levels.

You may complain that we have applied here a very profound and general theorem of statistical thermodynamics, and we have lost in the process the physical picture. This is unfortunately true, but nothing stops us returning to the physics. We agreed before that an electric field would arise in the vicinity of the metallurgical junction. Thus, the lower energy of the electrons on the right-hand side is simply due to the fact that they need to do some work against the electric field before they can reach the conduction band on the left-hand side.

What can we say about the transition region? One would expect the electron and hole densities to change gradually from high to low densities as shown in Fig. 9.1(d). But what sort of relationship will determine the density at a given point? And furthermore, what will be the profile of the conduction band in the transition region? They can all be obtained from Poisson's equation

$$\frac{d^2 U}{dx^2} = \frac{1}{\epsilon} \text{ (net charge density)}, \tag{9.1}$$

* We are in a slight difficulty here because up to now potential meant the potential energy of the electron, denoted by V. The relationship between the two quantities is $eU = V$, which means that if you confuse the two things you'll be wrong by a factor of 10^{19}.

where U is the electric potential used in the usual sense.* Since the density of mobile carriers depends on the actual variation of potential in the transition region, this is not an easy differential equation to solve. Fortunately, a simple approximation may be employed, which leads quickly to the desired result.

As may be seen in Fig. 9.1(d), the density of mobile carriers rapidly decreases in the transition region. We are, therefore, nearly right if we maintain that the transition region is completely depleted of mobile carriers. Hence we may assume the net charge densities are approximately of the form shown in Fig. 9.2(a). Charge conservation is expressed by the condition

The transition region is often called the 'depletion' region.

$$N_A x_p = N_D x_n. \tag{9.2}$$

$-x_p$ and x_n are the widths of the depletion regions in the p- and n-type materials, respectively.

Poisson's equation for the region $-x_p$ to 0 reduces now to the form

$$\frac{d^2 U}{dx^2} = \frac{e N_A}{\epsilon}. \tag{9.3}$$

Integrating once, we get

$$\mathcal{E} = -\frac{dU}{dx} = -\frac{e N_A}{\epsilon}(x + C), \tag{9.4}$$

C is an integration constant.

According to our model, the depletion region ends at $-x_p$. There is no charge imbalance to the left of $-x_p$, hence the electric field must be equal to zero at $x = -x_p$. With this boundary condition eqn (9.4) modifies to

$$\mathcal{E} = \frac{e N_A}{\epsilon}(x + x_p). \tag{9.5}$$

Similar calculation for the n-type region yields

$$\mathcal{E} = \frac{e N_D}{\epsilon}(x - x_n). \tag{9.6}$$

The electric field varies linearly in both regions, as may be seen in Fig. 9.2(b). It takes its maximum value at $x = 0$ where there is an abrupt change in its

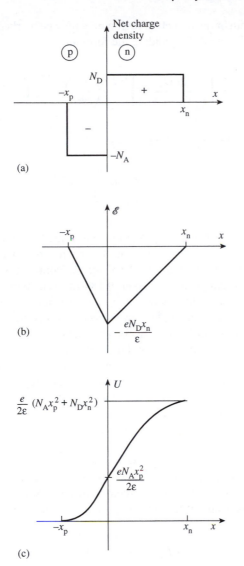

Fig. 9.2
(a) Net charge densities (b) electric field, (c) potential in the transition region of a p–n junction.

slope. The variation of voltage may then be obtained from

$$U = -\int_{-x_p}^{x} \mathcal{E}\,dx, \tag{9.7}$$

leading to the quadratic function plotted in Fig. 9.2(c). The total potential difference is

$$U_0 = U(x_n) - U(-x_p) = \frac{e(N_A x_p^2 + N_D x_n^2)}{2\epsilon}. \tag{9.8}$$

This is called the 'built-in' voltage between the p and n regions. A typical figure for it is 0.3 V.

The total width of the depletion region may now be worked out with the aid of equations (9.2) and (9.8), yielding the formula

$$w = x_p + x_n = \left\{ \frac{2\epsilon U_0}{e(N_A + N_D)} \right\}^{1/2} \left\{ \left(\frac{N_A}{N_D} \right)^{1/2} + \left(\frac{N_D}{N_A} \right)^{1/2} \right\}. \tag{9.9}$$

If, say, $N_A \gg N_D$, eqn (9.9) reduces to

$$w = \left(\frac{2\epsilon U_0}{e N_D} \right)^{1/2}, \tag{9.10}$$

which shows clearly that if the p-region is more highly doped, practically all of the potential drop is in the n-region. Taking for the donor density $N_D = 10^{21}$ m^{-3} and the typical figure of 0.3 V for the contact potential, the width of the transition region comes to about 0.18 μm. Remember this is the value for an abrupt junction. In practice, the change from acceptor impurities to donor impurities is gradual, and the transition region is therefore much wider. A typical figure is about 1 μm. Thus in a practical case we cannot very much rely on the formulae derived above, but if we have an idea how the acceptor and donor concentrations vary, similar equations can be derived.

From our simple model (assuming a depletion region) we obtained a quadratic dependence of the potential energy in the transition region. More complicated models give somewhat different dependence, but they all agree that the variation is monotonic. Our energy diagram is thus as shown in Fig. 9.3.

We can describe now the equilibrium situation in yet another way. The electrons sitting at the bottom of the conduction band at the p-side will roll down the slope because they lower their energy this way. So there will be a flow of electrons from left to right, proportional to the density of electrons in the p-type material:

$$I_{e(\text{left to right})} \sim N_{ep}. \tag{9.11}$$

The electrons in the n-type material, being the majority carriers, are very numerous. So, although most of them will be sitting at the bottom of the conduction band, there will still be a considerable number with sufficient energies to cross to the p-side. Assuming Boltzmann statistics, this number is given by

$$N_{en} \exp \left(\frac{-eU_0}{kT} \right). \tag{9.12}$$

Substituting N_{en} from eqn (8.17) we get

$$N_c \exp \left\{ \frac{-(E_g + eU_0 - E_F)}{kT} \right\}. \tag{9.13}$$

The energy difference between the bands on the p- and n-side is eU_0.

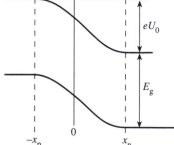

Fig. 9.3
The energy diagram for the transition region of a p–n junction.

Hence the electron current from right to left is given by

$$I_{e(\text{right to left})} \sim N_c \exp \left\{ \frac{-(E_g + eU_0 - E_F)}{kT} \right\}. \qquad (9.14)$$

In equilibrium the current flowing from left to right should equal the current flowing right to left; that is,

$$N_{ep} = N_0 \exp \left\{ \frac{-(E_g + eU_0 - E_F)}{kT} \right\}. \qquad (9.15)$$

Equation (9.15) gives nothing new. If we express the electron density in the p-type material with the aid of the Fermi level, then we could show from eqn (9.15) that eU_0 should be equal to the difference between the original Fermi levels, which we already knew. But although eqn (9.15) does not give any new information, we shall see in a moment that by describing the equilibrium in terms of currents flowing in opposite directions, the rectifying properties of the p–n junction can be easily understood.

We could go through the same argument for holes without much difficulty, provided we can imagine particles rolling uphill, because for holes that is the way to lower their energy. The equations would look much the same, and I shall not bother to derive them.

9.3   Rectification

Let us now apply a voltage as shown in Fig. 9.4. Since there are much fewer carriers in the transition region, we may assume that all the applied voltage will drop across the transition region. Then, depending only on the polarity, the potential barrier between the p and n regions will decrease or increase. If the p-side is made positive the potential barrier is *reduced*, and we talk of *forward bias*. The opposite case is known as *reverse bias*; the p-side is then negative, and the potential barrier is increased.

It is fairly obvious qualitatively that the number of electrons flowing from left to right is not affected in either case. The same number of electrons will still roll down the hill as in equilibrium. But the flow of electrons from right to left is seriously affected. For reverse bias it will be reduced and for forward bias it will significantly increase. So we can see qualitatively that the total current flowing for a voltage U_1 will differ from the current flowing at a voltage $-U_1$. This is what is meant by rectification.

It is not difficult to derive the mathematical relationships; we have practically everything ready. The current from left to right is the same; let us denote it by I_0. The current from right to left may be obtained by putting $e(U_0 - U_1)$ in place of eU_0 in eqn (9.14). [This is because we now want the number of electrons having energies in excess of $e(U_0 - U_1)$, etc.] At $U_1 = 0$, this current is equal to I_0 and increases exponentially with U_1; that is

$$I_{e(\text{right to left})} = I_0 \exp eU_1/kT. \qquad (9.16)$$

(a)

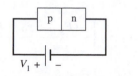

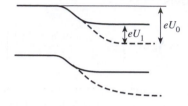

(b)

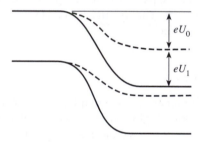

Fig. 9.4
The energy diagram of a p–n
junction for (a) forward bias and
(b) reverse bias.

Above about 1 volt bias, the final
'−1' in the rectifier equation can
be neglected.

* Adding the hole current would increase
I_0 but the form of the equation would
not change because the same exponential factor applies to the hole density in
the p-type material.

Hence the total current

$$I_e = I_{e(\text{right to left})} - I_{e(\text{left to right})} = I_0[\exp(eU_1/kT) - 1], \qquad (9.17)$$

which is known as the *rectifier equation*;* it is plotted in Fig. 9.5. For negative
values of U_1, I_e tends to I_0, and there is the exponential increase of current
with forward voltage. It is worth noting that in spite of the simple reasoning
this equation is qualitatively true for real diodes.

So, if we plot a graph of log I_e versus applied forward bias voltage, we
get a pretty good straight line for most rectifiers. However there are two
snags:

1. the slope of the line is e/mkT not e/kT, where m is a number usually lying
   between 1 and 2;
2. the current intercept, when the graph is extrapolated back to zero voltage,
   gives log I_0. But this value of I_0 is several orders of magnitude less than the
   value of I_0 obtained by measuring the reverse current (Fig. 9.5). Explaining
   this away is beyond the scope of this course; it is necessary to take into

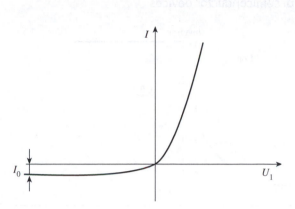

Fig. 9.5
The current as a function of applied
voltage for a p–n junction.

account recombination and generation of carriers in the depletion region.
A good account is given in the books by J.P. McKelvey and A.S. Grove
cited in the further reading list.

9.4 Injection

In thermal equilibrium the number of electrons moving towards the left is
equal to the number of electrons moving towards the right. However, when a
forward bias is applied, the number of electrons poised to move left is increased
by a factor, $\exp eU_1/kT$. This is quite large; for an applied voltage of 0.1 V
the exponential factor is about 55 at room temperature. Thus the number of
electrons appearing at the boundary of the p-region is 55 times higher than the
equilibrium concentration of electrons there.

What happens to these electrons? When they move into the p-region they
become minority carriers, rather like immigrants travelling to a new country suddenly become foreigners. But, instead of mere political friction, the
electrons' ultimate fate is annihilation. They are slain by heroic holes, who
themselves perish in the battle.* Naturally, to annihilate all the immigrants,
time and space are needed; so some of them get quite far inside foreign territ-
ory as shown in Fig. 9.6, where the density of electrons is plotted as a function
of distance in a p–n junction under forward bias. The electron density declines,
but not very rapidly. A typical distance is about 1 mm, which is about a thousand
times larger than the width of the transition region.

Let us go back now to the plight of the holes. They are there in the p-type
material to neutralize the negative charge of the acceptor atoms. But how will
space charge neutrality be ensured when electrons are injected? It can be done
in only one way; whenever the electron density is increased, the hole density
must increase as well. And this means that new holes must move in from the
contacts. Thus, as electrons move in from the right, holes must move in from
the left to ensure charge neutrality. Hence the current of electrons and holes
will be made up of six constituents, as shown in Fig. 9.7:

(i) The electron current flowing in the n-type material and providing the
     electrons to be injected into the p-type material. It is constant in the
     n-region.

* I do not think that the pacifist interpretation of the recombination of electrons and
holes (they get married, and live happily
ever after) can bear closer scrutiny. When
an electron and a hole recombine, they
disappear from the stage and that's that. I
would however, be willing to accept the
above interpretation for excitons which
are electron–hole pairs bound together by
Coulomb forces but even then one cannot
claim that they lived happily ever after
because the lifetime of excitons is less
than a picosecond, even shorter than the
expected duration of modern marriages.

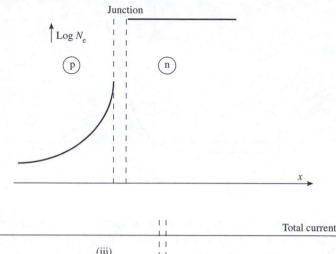

Fig. 9.6

The electron distribution in a forward biased p–n junction.

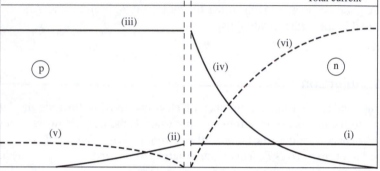

Fig. 9.7

The current distribution in a forward biased p–n junction.

(ii) A declining current of injected electrons in the p-region. The current declines because the number of electrons becomes less and less as they recombine with the holes.

(iii) The current of holes in the p-region to provide the holes to be injected into the n-region. We have not discussed this because the injection of holes is entirely analogous to the injection of electrons.

(iv) A declining current of injected holes in the n-region. The current declines because of recombination with electrons.

(v) A declining current of holes in the p-region to compensate for the holes lost by recombination.

(vi) A declining current of electrons in the n-region to compensate for the electrons lost by recombination.

Adding up the currents, you can see that the total current is constant, as it should be.

Let me emphasize again that the current in a p–n junction is quite different from the currents you have encountered before. When you apply a voltage to a piece of metal, all that happens is that the electrons, which are already there, acquire some ordered motion. When a forward bias is applied to a p–n junction minority carriers get injected into both regions. These minority carriers were *not* there originally in such a high density; they came as a consequence of the applied voltage.

The distinction between ordinary conduction and minority carrier injection is important. It is the latter which makes transistor action possible.

9.5   Junction capacity

I would like to say a few more words about the reverse biased junction. Its most interesting property (besides the high resistance) is that the presence of two layers of space charge in the depletion region makes it look like a capacitor.

We may calculate its capacitance in the following way. We first derive the relationship between the width of the depletion layer in the n-region and the voltage in the junction, which may be obtained from equations (9.2) and (9.8). We get

U_0 is the 'built-in' voltage.

$$x_n = \left\{ \frac{2\epsilon U_0 N_A}{e N_D (N_D + N_A)} \right\}^{1/2}. \tag{9.18}$$

U_1 is the applied voltage in the reverse direction.

For reverse bias the only difference is that the barrier becomes larger, that is U_0 should be replaced by $U_0 + U_1$, yielding

$$x_n = \left\{ \frac{2\epsilon (U_0 + U_1) N_A}{e N_D (N_D + N_A)} \right\}^{1/2}. \tag{9.19}$$

The total charge of the donor atoms is

$$Q = e N_D x_n = \left\{ 2\epsilon e (U_0 + U_1) \frac{N_A N_D}{N_A + N_D} \right\}^{1/2}. \tag{9.20}$$

Now a small increase in voltage will add charges at the boundary—as it happens in a real capacitance. We may, therefore, define the capacitance of the junction (per unit area) as

$$C = \frac{\partial Q}{\partial U_1} = \left\{ \frac{\epsilon e}{2(U_0 + U_1)} \frac{N_A N_D}{N_A + N_D} \right\}^{1/2}. \tag{9.21}$$

We can now assign an equivalent circuit to the junction and attribute a physical function to the three elements (Fig. 9.8). R_1 is simply the ohmic resistance of the 'normal' that is not depleted, semiconductor. $R(U_1)$ is the junction resistance. It is very small in the forward direction ($0.1 - 10\Omega$ typically) and large in the reverse direction ($10^6 - 10^8 \Omega$ typically). $C(U_1)$ is the capacitance that varies with applied voltage, given by eqn (9.21). Clearly, this equation loses validity for strong forward bias because the depletion layer is then flooded by carriers. This is actually borne out by our equivalent circuit, which shows that the capacity is shorted-out when $R(U_1)$ becomes small. With a reversed biased diode, $C(U_1)$ is easily measured, and can be fitted to an equation like (9.21) very well. The characteristic of a commercial diode is plotted in Fig. 9.9. Here C is proportional to $(U_1 + 0.8)^{-1/2}$, whence the constant 0.8 V may be identified with the 'built-in' voltage.

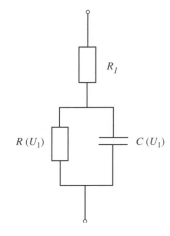

Fig. 9.8
The equivalent circuit of a p–n junction.

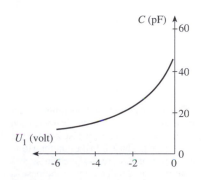

Fig. 9.9
Capacitance of a commercially available junction diode as a function of applied voltage.

9.6   The transistor

Here we are. We have arrived at last at the transistor, the most famous electronic device of the century. It was discovered at Bell Telephone Laboratories by

Bardeen, Brattain, and Shockley in 1948. Since then it conquered the world many times over. It made possible both the pocket radio and the giant computers, to mention only two applications. The number of transistors manufactured is still growing fast from year to year. The figure for 1967, when we first prepared these lectures, was a few times 10^9; that is about one transistor for every man, woman, and child living on Earth. I do not think anyone knows the number of transistors produced nowadays. We are rapidly approaching the stage when a single computer might contain as many as 10^9 transistors (or one of its close relations, namely field-effect transistors or charge-coupled devices).

I am afraid the discussion of the transistor will turn out to be an anticlimax. It will seem to be too simple. If you find it too simple, remember that we have trodden a long tortuous path to get there. It might be worthwhile to recapitulate the main hurdles we have scaled:

1. Postulation of Schrödinger's equation.
2. The solution of Schrödinger's equation for a rigid, three-dimensional potential well.
3. Postulation of Pauli's principle and the introduction of spin.
4. Formulation of free electron theory, where electrons fill up the energy levels in a potential well up to the Fermi energy.
5. Derivation of the band structure by a combination of physical insight and Schrödinger's equation.
6. Introduction of the concept of holes.
7. Demonstration of the fact that electrons at the bottom of the conduction band (and holes at the top of the valence band) may be regarded as free, provided that an effective mass is assigned to them.
8. Determination of the Fermi level and carrier densities in extrinsic semi-conductors.
9. The description of a p–n junction in terms of opposing currents.
10. Explanation of minority-carrier injection.

So if you want to go through a logical chain of reasoning and want to explain the operation of the transistor from first principles, those above are the main steps in the argument.

Now let us see the transistor itself. It consists of two junctions with one semiconductor region common to both. This is called the *base*, and the other two regions are the *emitter* and the *collector* as shown in Fig. 9.10 for a p–n–p transistor. There are also n–p–n transistors; the ensuing explanation could be made to apply to them by judicious changing of words.

Consider first the emitter–base p–n junction. It is forward biased (*positive* on p-side for those who like mnemonics). This means that large numbers of carriers flow, holes into the base, electrons into the emitter, Now the holes arriving into the base region will immediately start the process of recombination with electrons. But, as explained before, time and space are needed to annihilate the injected minority-carriers. Hence, for a narrow base region ($\ll 1$ mm), the hole current leaving the base region will be almost identical to the hole current entering from the emitter. Now what happens to the holes when they arrive to the collector region? They see a negative voltage (the base–collector junction is reverse biased) and carry on happily towards the load. Thus, practically the same current that left the emitter finds its way to the load.*

* The above argument is not quite correct because the emitter current is not carried solely by holes. There is also an electron flow from the base into the emitter. However, transistors are designed in such a way that the conductivity of the base is well below the conductivity of the emitter (a typical figure may be a factor of a hundred); thus the minority-carrier flow from the base to the emitter is usually negligible.

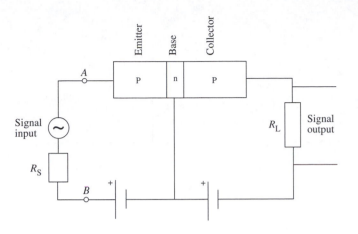

Fig. 9.10
The p–n–p transistor as an amplifier.

So far there is nothing spectacular; the current gain of the device is somewhat below unity. Why is this an amplifier? It is an amplifier because the voltage gets amplified by a large factor. This is because the input circuit is a low-impedance circuit; a low voltage is thus sufficient to cause a certain current. This current reappears in the high impedance output circuit and is made to flow across a large load resistance, resulting in a high output voltage. Hence the transistor in the common base circuit is a voltage amplifier.

We should, however, know a little more about this amplifier. Can we express its properties in terms of the usual circuit parameters: impedances, current sources, and voltage sources? How should we attempt the solution of such a problem? Everything is determined in principle. If the bias voltages are fixed and an a.c. voltage is applied to the input of the transistor in Fig. 9.10, then the output current is calculable. Is this enough? Not quite. We have to express the frequency dependence in the form of rational fractions (this is because impedances are either proportional or inversely proportional to frequency) and then an equivalent circuit can be defined. It is a formidable job; it can be done and it has been done, but, of course, the calculation is far too lengthy to include here. Although we cannot solve the complete problem, it is quite easy to suggest an approximate equivalent circuit on the basis of our present knowledge.

Looking in at the terminals A and B of Fig. 9.10, what is the impedance we see? It comprises three components: the resistance of the emitter, the resistance of the junction, and the resistance of the base. The emitter is highly doped in a practical case, and we may therefore neglect its resistance, but the base region is narrow and of lower conductivity and so we must consider its resistance. Hence we are left with r_e (called misleadingly the *emitter resistance*) and r_b (base resistance), forming the input circuit shown in Fig. 9.11(a).

r_e is in fact the resistance of the junction.

What is the resistance of the output circuit? We must be careful here. The question is how will the a.c. collector current vary as a function of the a.c. collector voltage? According to our model, the collector current is quite independent of the collector voltage. It is equal to αi_e, where i_e is the emitter current, and α is a factor very close to unity. Hence, our first equivalent output circuit must simply consist of the current generator shown in Fig. 9.11(b). In practice the impedance turns out to be less than infinite (a few hundred thousand ohms

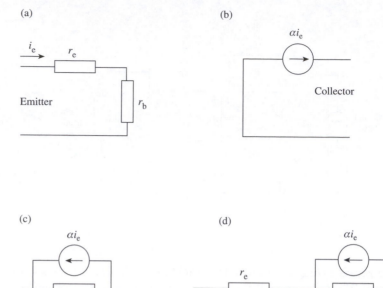

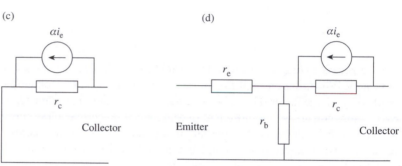

Fig. 9.11

The construction of an equivalent circuit of a transistor. (a) The emitter–base junction. (b) In first approximation the collector current depends only on the emitter current. (c) In a more accurate representation there is a collector resistance r_c as well as the collector circuit. (d) The complete low-frequency equivalent circuit.

---

* This exceedingly simple construction cannot be done in general but is permissible in the present case when $r_c \gg r_b$.

is a typical figure); so we should modify the equivalent circuit as shown in Fig. 9.11(c).

Having got the input and output circuits, we can join them together to get the equivalent circuit of the common base transistor* [Fig. 9.11(d)].

We have not included any reactances. Can we say anything about them? Yes, we can. We have already worked out the junction capacity of a reverse biased junction. That capacity should certainly appear in the output circuit in parallel with r_c. There are also some other reactances as a consequence of the detailed mechanism of current flow across the transistor. We can get the numerical values of these reactances if we have the complete solution. But luckily the most important of these reactances, the so-called *diffusion reactance*, can be explained qualitatively without recourse to any mathematics.

Let us look again at the p–n junction of the p–n–p transistor. When a step voltage is applied in the forward direction, the number of holes able to cross into the n-region suddenly increases. Thus, in the first moment, when the injected holes appear just inside the n-region, there is an infinite gradient of hole density, leading to an infinitely large diffusion current. As the holes diffuse into the n-region, the gradient decreases, and finally the current settles down to its new stationary value as shown in Fig. 9.12. But this is exactly the behaviour one would expect from a capacitance in parallel with a resistance. Thus, when we wish to represent the variation of emitter current as a function of emitter voltage, we are entitled to put a capacitance there. This is not a real honest-to-god, capacitance; it just looks as if it were a capacitance, but that is all that matters. When drawing the equivalent circuit, we are interested in appearance only!

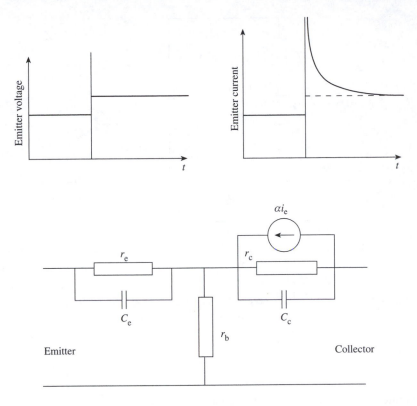

Fig. 9.12
The emitter current as a function of time when the emitter voltage is suddenly increased. It looks like the current response of a parallel *RC* circuit.

Fig. 9.13
A good approximation to the high-frequency equivalent circuit of a transistor.

Including now both capacitances, we get the equivalent circuit of Fig. 9.13. We are nearly there. There is one more important effect to consider: the frequency-dependence of α. It is clear that the collector current is in phase with the emitter current when the transit time of the carriers across the base region is negligible, but α becomes complex (and its absolute value decreases) when this transit time is comparable with the period of the a.c. signal. We cannot go into the derivation here, but α may be given by the simple formula*

$$\alpha = \frac{\alpha_0}{1 + j(\omega/\omega_\alpha)}, \tag{9.22}$$

* Not to depart from the usual notations, we are using j here as honest engineers do, but had we done the analysis with our chosen $\exp(-i\omega t)$ time dependence, we would have come up with $-i$ instead of j.

where ω_α is called the *alpha cut-off frequency*. The corresponding equivalent circuit is obtained by replacing α in Fig. 9.13 by that given in eqn (9.22). And that is the end as far as we are concerned. Our final equivalent circuit represents fairly well the frequency-dependence of a commercially available transistor.

We have seen that the operation of the transistor can be easily understood by considering the current flow through it. The frequency dependence is more complicated, but still we have been able to point out how the various reactances arise.

It has been convenient to describe the common base transistor configuration, but of course the most commonly used arrangement is the common emitter, shown in Fig. 9.14(a). Again, most of the current i_e from the forward-biased

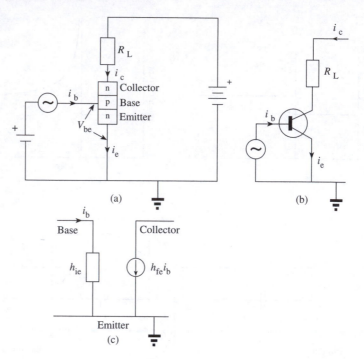

Fig. 9.14

The common emitter transistor.
(a) General circuit arrangement,
showing current and voltage
nomenclature. (b) Circuit diagram.
(c) Equivalent circuit of the transistor.

[†] The full expression for i_e should contain a term dependent on the emitter-to-collector voltage. This is usually small. Look it up in a circuitry book if you are interested in the finer details.

emitter–base junction gets to the collector, so we can write[†]

$$i_c = \alpha i_e, \tag{9.23}$$

as before, and

$$i_c = i_e - i_b = \frac{\alpha i_b}{1 - \alpha} = h_{fe} i_b, \tag{9.24}$$

where we have introduced a current gain parameter, h_{fe}, which is usually much greater than unity. This fixes the right-hand side of the equivalent circuit of Fig. 9.14(c) as a current generator h_{fe} times greater than the input current. The input side is a resistance, h_{ie}, which again includes the series resistance of the base and emitter contact regions.

Note that the major part of transistor amplifier design is based on the simple equivalent circuit of Fig. 9.14(c). At high frequencies, of course, the capacitances discussed have to be added.

I have so far talked about the applications of transistors as amplifiers, that is, analogue devices. Historically, these applications came first because at the time of the invention of the transistor there was already a mass market in existence eager to snap up transistor amplifiers—particularly for portable devices. The real impact of the transistor came, however, not in the entertainment business but in computers. Admittedly, computers did exist before the advent of the transistor, but they were bulky, clumsy, and slow. The computers you know and respect, from giant ones down to pocket calculators, depend on the good services of transistors. One could easily write a thousand pages about the circuits used in various computers—the trouble is that by the time the thousandth page is jotted down, the first one is out of date. The rate of technical change in this field is simply breathtaking, much higher than ever before in any branch of technology. Fortunately, the principles are not difficult. For building a logic circuit all we need is a device with two stable states, and that can be easily provided by a transistor, for example in a form (Fig. 9.15) quite similar to its use as an amplifier. When the base current, $I_B = 0$ (we use capital letters to describe the d.c. current), no collector current flows, $I_c = 0$, and consequently $U_{CE} = E$. If a base current is impressed upon the circuit, then a collector current flows, and U_{CE} is close to zero. Hence, we have a 'high' and a 'low' output voltage which may be identified with a logical '1' or '0' (or the other way round). I shall not go into any more details, but I would just like to mention some of the acronyms in present-day use for which transistors are responsible. They include TTL (transistor–transistor logic), ECL (emitter coupled logic) and I^{2}L (integrated injection logic).

9.7 Metal–semiconductor junctions

Junctions between metals and semiconductors had been used in radio engineering for many years before the distinction between p- and n-type semiconductors was appreciated. Your great grandfathers probably played about with 'cat's whiskers' in their early 'crystal sets', as radios were then called, stressing the importance of the piece of coal or whatever was used as the semiconductor detector.

The behaviour of metal–semiconductor junctions is more varied to describe than that of p–n junctions. We find that there is different behaviour on the one hand, with p- and n-type semiconductors, than on the other, when the metal work function is greater or less than that of the semiconductor.

We shall first consider the case of an n-type semiconductor in contact with a metal, whose work function is greater than that of the semiconductor. The semiconductor work function (ϕ_S in Fig. 9.16) is defined as the energy difference between an electron at the Fermi energy and the vacuum level. The fact that

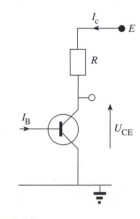

Fig. 9.15
A transistor as a logic element.

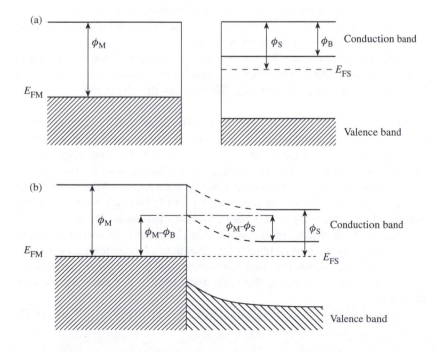

Fig. 9.16
Energy diagrams for a junction between a metal and an n-type semiconductor ($\phi_M > \phi_S$), (a) before contact, (b) after contact the Fermi levels agree ($E_{FM} = E_{FS}$).

there are usually no electrons at the Fermi energy need not bother us—we do not have to explain definitions. Another measure often used is the electron affinity, ϕ_B.

The band structure of the two substances is shown in Fig. 9.16(a). When they are joined together, we may apply again our general theorem and make the Fermi levels equal. Thus, we may start the construction of Fig. 9.16(b) by drawing a horizontal line for the Fermi energy, and a vertical one for the junction. We leave the metal side unchanged because we shall assume that 'band-bending' cannot occur in a metal.* We are really saying here that all the potential drop will take place in the semiconductor, which, in view of the much smaller number of carriers there, is a reasonable assumption.† Away from the junction we draw the valence band edge, the conduction band edge, and the vacuum level in the same position (relative to the Fermi level) as for the bulk material, in Fig. 9.16(a). These are shown as solid lines. Now with an infinitely small gap the vacuum levels are equal; thus, we may join them with a dotted line in Fig. 9.16(b); the conduction and valence band edge must also be continued parallel to the vacuum level.

What can we say about the charges? We may argue in the same fashion as for a metal–metal junction. In the first instance, when the metal and the semiconductor are brought together, the electrons from the conduction band cross over into the metal in search of lower energy. Hence a certain region in the vicinity of the junction will be practically depleted of mobile carriers. So we may talk again about a depletion region and about the accompanying potential variation, which is incidentally, the same thing as the 'band-bending' obtained from the band picture.

So the two pictures are complementary to a certain extent. In the first one the 'band-bending' is a consequence of the matching of the Fermi levels and vacuum levels, and the charge imbalance follows from there. In the second picture electrons leave the semiconductor, causing a charge imbalance and hence a variation in the potential energy. Whichever way we look at it, the outcome is a potential barrier between the metal and the semiconductor. Note that the barrier is higher from the metal side.

In dynamic equilibrium the number of electrons crossing over the barrier from the metal to the semiconductor is equal to the number crossing over the barrier from the semiconductor side. We may say that the current I_0 flows in both directions.

Let us apply now a voltage; according to the polarity, the electrons' potential energy on the semiconductor side will go up or down. For a forward bias it goes up, which means that we have to draw the band edges higher up. But the vacuum level at the junction stays where it was. So the effect of the higher band edges is smaller curvature in the vicinity of the junction and a reduced potential barrier, as shown in Fig. 9.17. Now all electrons having energies above $\phi_M - \phi_S - eU_1$ may cross into the metal. By analogy with the case of the p–n junction it follows that the number of carriers (capable of crossing from the semiconductor into the metal) has increased by a factor $\exp eU_1/kT$, and hence the current has increased by the same factor. Since the current from the metal to the semiconductor has not changed, the total current is

$$I = I_0[\exp(eU_1/kT) - 1];\qquad(9.25)$$

*In a metal the charge inequality is confined to the surface.

†We met a very similar case before when discussing p–n junctions. If one of the materials is highly doped, all the potential drop takes place in the other material.

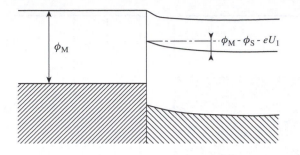

Fig. 9.17
The junction of Fig. 9.16 under
forward bias. The potential barrier
for electrons on the semiconductor
side is reduced by eU_1.

that is, a junction of this type is a rectifier.

There is one point I want to make concerning the potential barrier in this junction. One of the electrodes is a metal, and there are charged carriers in the vicinity of the metal surface. Does this remind you of any physical configuration we have studied before? Where have we met a potential barrier and charged carriers giving rise to image charges? In the study of electron emission in Chapter 6 we came to the conclusion that the image charges lead to a lowering of the potential barrier, and you may remember that it was called the Schottky effect. According to the formula we derived there, the reduction was proportional to $(\mathcal{E}/\epsilon_0)^{1/2}$. Well, the same thing applies here with the difference that ϵ_0 should be replaced by $\epsilon_r \epsilon_0$, where ϵ_r is the relative dielectric constant of the semiconductor. For silicon, for example, $\epsilon_r = 12$, the effect is therefore smaller. So the Schottky effect is not very large, but it happened to give its name to these particular junctions. They are usually referred to as *Schottky diodes* or *Schottky barrier diodes*.

Let us now investigate the case when the work function of the metal is smaller than that of the n-type semiconductor. The situation before and after contact is illustrated in Fig. 9.18. Now, to achieve equilibrium, electrons had to move from the metal to the semiconductor, establishing there an accumulation region. There is no potential barrier now from whichever side we look at the junction. As a consequence the current flow does not appreciably depend on the polarity of the voltage. This junction is *not* a rectifier.

9.8   The role of surface states; real metal–semiconductor junctions

The theory of metal–semiconductor junctions as presented above is a nice, logical, consistent theory that follows from the physical picture we have developed so far. It has, however, one major disadvantage; it is not in agreement with experimental results, which seem to suggest that all metal–semiconductor junctions are rectifiers independently of the relative magnitudes of the work functions. This does not necessarily mean that the theory is wrong. The discrepancy may be caused by the physical realization of the junction. Instead

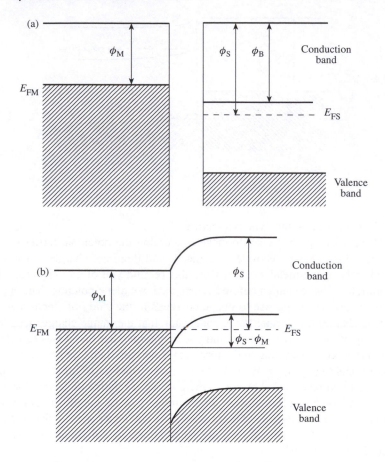

Fig. 9.18

Energy diagrams for a junction between a metal and an n-type semiconductor ($\phi_M < \phi_S$), (a) before contact, (b) after contact the Fermi levels agree ($E_{FM} = E_{FS}$).

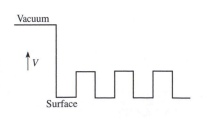

Fig. 9.19

The potential energy distribution near the surface of a crystal.

of two clean surfaces lining up, there might in practice be some oxide layers, and the crystal structure might be imperfect. This may be one of the reasons why 'real' junctions behave differently from 'theoretical' junctions. The other reason could be that the theory, as it stands, is inadequate, and to get better agreement with experiments, we must take into account some hitherto neglected circumstance.

Have we taken into account anywhere that our solids are of finite dimensions? Yes, we have; we determine the number of allowed states from the boundary conditions. True, but that is not the only place where the finiteness of the sample comes in. Remember, in all our models leading to the band picture we have taken the crystal as perfectly periodic, and we have taken the *potential* as perfectly periodic. This is surely violated at the surface. The last step in the potential curve should be different from the others, that is, the potential profile in the solid should rather be chosen in the form displayed in Fig. 9.19. It was shown some years ago that the assumption of such a surface barrier would lead to the appearance of some additional discrete energy levels in the forbidden gap, which are generally referred to as surface states.

If we assume a semiconductor is n-type, some of these surface states may be occupied by electrons that would otherwise be free to roam around. Some of the donor atoms will therefore have uncompensated positive charges leading to 'band-bending' as shown in Fig. 9.20. Thus, the potential barrier is already there before we even think of making a metal contact.

What happens when we do make contact between the semiconductor and the metal? Let us choose the case when the metal has the lower work function, when according to our previous theory the junction is not rectifying. Then, as we have agreed before (and it is still valid) electrons must flow from the metal to the semiconductor until equilibrium is established. But if there is a sufficient number of empty surface states still available, then the electrons will occupy those without much effecting the height of the potential barrier. So the potential barrier stays, and the junction is rectifying.

It would be difficult in a practical case to ascertain the share of these 'theoretical' surface states (called also Tamm states) in determining the behaviour of the junction because surface imperfections are also there, and those can trap electrons equally well. It seems, however, quite certain that it is the surface effects that make all real metal–semiconductor junctions behave in a similar manner.

Finally, I would like to mention ohmic contacts, that is, contacts that do not care which way the voltage is applied. To make such a contact is not easy; it is more an art than a science. It is an important art though, since all semiconductor devices have to be connected to the outside world.

The two most often used recipes are: (i) to make the contact with alloys containing metals (e.g. In, Au, Sn) that diffuse into the surface forming a gradual junction; or (ii) to make a heavily doped semiconductor region (usually called n^+ or p^+) with about 10^{24} carriers per cubic metre in between the metal and semiconductor to be connected.

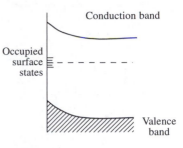

Fig. 9.20

In a real semiconductor electrons may occupy surface states. The donor atoms left behind have a positive charge which leads to the 'bending' of the band.

9.9 Metal–insulator–semiconductor junctions

Let us now make life a little more complicated by adding one more component and look at metal–insulator–semiconductor junctions. What happens as we join the three materials together? Nothing. If the insulator is thick enough to prevent tunnelling (the situation that occurs in all practical devices of interest), the metal and the semiconductor are just unaware of each other's existence.

What does the energy diagram look like? For simplicity we shall assume that the Fermi levels of all three materials coincide before we join them together. The energy diagram then takes the form shown in Fig. 9.21, where the semiconductor is taken as n-type.

Are there any surface states at the semiconductor–insulator interface? In practice there are, but their influence is less important than for metal–semiconductor junctions, so we shall disregard them for the time being.

Let us now apply a positive voltage to the metal as shown in Fig. 9.22(a). Will a current flow? No, there can be no current through the insulator. The electrons will nevertheless respond to the arising electric field by moving towards the insulator. That is as far as they can go, so they will accumulate in front of the insulator. Their distribution will be something like that shown in Fig. 9.22(b), where N_{en} is the equilibrium concentration in the bulk semiconductor, and x is the distance away from the insulator. The shape of the curve may be obtained from the same considerations as in a p–n junction. The diffusion current (due to the gradient of the electron distribution) flowing to the right must be equal

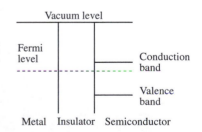

Fig. 9.21

Energy diagram for a metal–insulator–n-type semiconductor junction at thermal equilibrium.

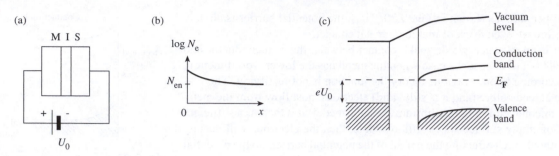

Fig. 9.22

A metal–insulator–n-type semiconductor junction under forward bias. (a) Schematic representation. (b) Variation of electron density in the semiconductor as a function of distance. (c) Energy diagram.

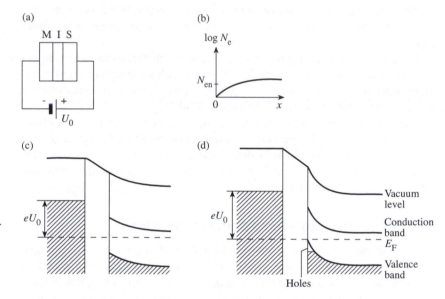

Fig. 9.23

A metal–insulator–n-type semiconductor junction under reverse bias. (a) Schematic representation. (b) Variation of electron density in the semiconductor as a function of distance. (c) Energy diagram at moderate voltage. (d) Energy diagram at a voltage high enough for producing holes.

to the conduction current (due to the applied field) flowing to the left. The corresponding energy diagram is shown in Fig. 9.22(c), where the Fermi level in the semiconductor is taken as the reference level. Looking at the energy diagram, we may now argue backwards and say that eqn (8.17) must still be roughly valid so the electron density is approximately an exponential function of the distance of the Fermi level from the bottom of the conduction band. Hence the electron density is increasing towards the insulator.

Next, let us apply a negative voltage to the metal [Fig. 9.23(a)]. The electrons will be repelled, creating a depletion region, as in a reverse biased p–n junction. In fact, we could determine the width of the depletion region (see Example 9.4) by a method entirely analogous to that developed in Section 9.2. Alternatively, we can argue that the electron distribution will be of the shape shown in Fig. 9.23(b), and we may talk again about the balance of diffusion and conduction currents. Finally the band bending picture is shown in Fig. 9.23(c), from which we can also conclude that the electron density is decreasing towards the insulator. What will happen as we apply higher and higher reverse bias? The obvious answer is that the depletion region will widen. What else could one expect? It is difficult to believe at first hearing, but the fact (fortunate as it happens) is that holes will appear. Can we explain this phenomenon by any of our models? If we consider only ionized donor atoms and mobile electrons, as in the model developed in Section 9.2, we have not got the slightest chance of creating holes. On the other hand if we adopt the notion that the density of a carrier at any point is determined by the distance in energy from the Fermi level to the edge of the particular band, then holes have acquired the right to appear. All we need to do is to apply a sufficiently large reverse bias [Fig. 9.23(d)] which will bring the Fermi level right down, close to the top of the valence band. Thus, according to this model, holes may become the majority carriers near to the surface of an n-type semiconductor. Odd, is it not? The problem still remains, though, that the holes must come from somewhere. The only process known to produce holes in an n-type semiconductor is thermal generation of electron–hole pairs. But are not the rates of generation and recombination equal? Would not the holes generated thermally immediately disappear by recombination? This is true indeed under thermal equilibrium conditions, but our junction is not necessarily in thermal equilibrium.

Let us look again at the whole process, considering time relationships as well. At $t = 0$ we apply a negative voltage to the junction. Most of the electrons clear out by t_1, leaving a depletion region of the order of 1 μm behind. What happens now to thermally generated electron–hole pairs? The electrons move away from the insulator, and the holes move towards the insulator. Not much recombination will occur because both the electron and hole densities are small, and they are separated in space. What will happen to the holes? They will congregate in the vicinity of the insulator, where they can find a nice comfortable potential minimum.

The conditions of equilibrium are rather complicated. At the end, say by t_2, the hole diffusion current away from the insulator must be equal to the hole conduction current towards the insulator, and the rates of generation and recombination must balance each other. It is then quite reasonable to conclude that if the applied negative voltage is large enough, that is the potential minimum at the insulator surface is deep enough, then a sufficient number of holes can congregate, and the part of the n-type semiconductor adjacent to the insulator will behave as if it were p-type. This is called inversion.

Our conclusion so far is that under equilibrium conditions inversion may occur. Whether equilibrium is reached or not depends on the time constants t_1 and t_2. How long is t_1? As far as we know no one has measured it, but it can not take long for electrons to clear out of a 1 μm part of the material. If we take a snail moving with a velocity of 1 m h^{-1} it will need about 3 ms to cover 1 μm. Thus electrons, which may be reasonably expected to move faster than snails, would need very little time indeed to rearrange themselves and create a depletion region. On the other hand, in sufficiently pure materials the thermal generation time constant might be as long as a few seconds. Thus, if all the operations we perform in a metal–insulator–semiconductor junction are short in comparison with the generation time of electron–hole pairs, then the minority carriers will not have the time to appear, a mode of operation called the *deep depletion mode*.

The phenomenon of inversion is not restricted of course to n-type semiconductors. Similar inversion occurs in a metal–insulator–p-type semiconductor junction.

Inversion and deep depletion are some further representations of the multifarious phenomena of semiconductor physics. They are certainly interesting, but are they useful? Can a device through which no current flows be of any use at all in electronics? The secret is that current can flow *along* the insulator surface. The emerging devices are very important indeed. Under acronyms like MOSFETs and CCDs they are the flagbearers of the microelectronics revolution. I shall talk about them a little later.

9.10 The tunnel diode

So far, we have considered impurity semiconductors with very low impurity contents, typically less than one part per million. We have characterized the impurity type and density by working out where the Fermi level is, and have found that in all cases it is well within the energy gap. This has meant, among other things, that the sums are much simpler, for we are able to approximate to the Fermi function. However, when the impurity level becomes very high (typically about 10^{24}m^{-3} or about 0.01%) the Fermi level moves right up into the conduction band (or down into the valence band for a p-type impurity). The semiconductor is then said to be 'degenerate'. The tunnel diode is the only device we shall mention in this chapter in which degenerate semiconductors are used on both the p- and n-sides of the junction. We shall, however, return to degenerate junctions in Section 12.7, where laser diodes are discussed.

The energy diagram at thermal equilibrium is given in Fig. 9.24(a), where, for simplicity, we take the difference between the Fermi level and the band edge as the same on both sides. It is interesting to see that the 'built-in' potential is larger than the energy gap; thus the number of electrons crossing over the potential barrier at thermal equilibrium must be small. Hence, I_0 in the rectifier

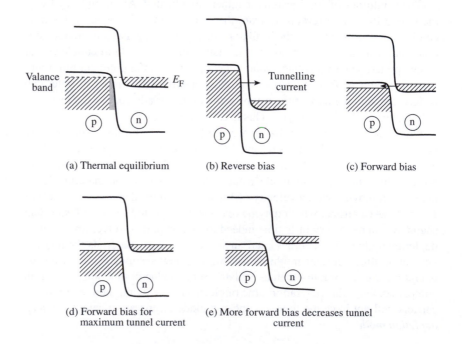

(a) Thermal equilibrium   (b) Reverse bias   (c) Forward bias

(d) Forward bias for maximum tunnel current   (e) More forward bias decreases tunnel current

Fig. 9.24

Energy diagrams for the tunnel diode.

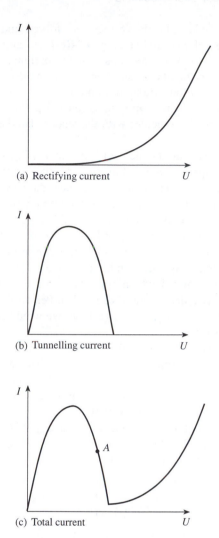

(a) Rectifying current U

(b) Tunnelling current U

(c) Total current U

Fig. 9.25
The current in a tunnel diode is the sum of the tunnelling current and of the usual rectifying current.

equation is small, and the rectifying characteristic is rather elongated, as shown in Fig. 9.25(a).

Looking carefully at the diagram you may realize that another mechanism of electron flow may also be effective. Remember, tunnel diodes are highly doped, and high doping means a narrow transition region. Thus, electrons as well as moving *over* the potential barrier may also tunnel *through* the potential barrier, and if one puts in the figures, it turns out that the tunnelling current is the larger of the two. Hence, we may imagine thermal equilibrium as the state in which the tunnelling currents are equal and in opposite directions.

What happens now if we apply a reverse bias? As may be seen in Fig. 9.24(b), the number of electrons tunnelling from left to right is increased because the electrons on the p-side face a large number of empty states on the n-side. We could work out this current by considering a rectangular potential barrier (the one we so skilfully solved when first confronted with Schrödinger's equation) and taking account of the occupancy of states on both sides. It is one of those fairly lengthy and tedious calculations that are usually left as an exercise for the student. So I would rather leave the work to you and go on to discuss qualitatively the case of forward bias [Fig. 9.24(c)]. The situation is essentially the same as before, but now the electrons tunnel from right to left. If the applied voltage is increased, the number of states available on the p-side increases, and so the current increases too. Maximum current flows when electrons on the n-side have access to all the empty states on the p-side, that is, when the Fermi level on the n-side coincides with the valence band edge on the p-side [Fig. 9.24(d)].

If the bias is increased further, there will be an increasing number of electrons finding themselves opposite the forbidden gap. They cannot tunnel because they have no energy levels to tunnel into. Hence, the current must decrease, reaching zero when the top of the valence band on the p-side coincides with the bottom of the conduction band on the n-side [Fig. 9.24(e)]. Therefore, the plot of current against voltage must look like that shown in Fig. 9.25(b). But this is not the total current; it is the current due to tunnelling alone. We can get the total current by simply taking the algebraic sum of the currents plotted in Fig. 9.25(a) and (b), which is a permissible procedure, since the two mechanisms are fairly independent of each other. Performing the addition, we get the $I - U$ characteristics [Fig. 9.25(c)] that we would be able to measure on a real tunnel diode.*

You know now everything about the tunnel diode with the exception of the reason why it can perform some useful function. The answer follows from the $I - U$ characteristics. There is a region where the slope is negative that is usually referred to as a negative resistance. In case you have not heard this curious phrase before we shall briefly explain it.

Consider an ordinary tuned circuit as shown in Fig. 9.26. If we start it oscillating in some way, and then leave it, the oscillations will decay exponentially, their amplitude falling with time according to

$$\exp\left(-\frac{R}{2L}t\right). \tag{9.26}$$

Physically, the resistance R absorbs the oscillating energy and gets a little hotter. If we now put a negative resistance in series with R, odd things happen. In the particular case when the negative resistance $(-R_1)$ is equal in magnitude to R, the total resistance becomes

$$R - R_1 = 0, \tag{9.27}$$

and the exponential becomes unity. This means that an oscillation, once started in the circuit, will continue with no decay, the negative resistance replenishing all the energy dissipated as heat in the real resistance.

If we could get energy like this out of a simple circuit isolated from the rest of the world except for the R, L, and C we have drawn, it would contravene the second law of thermodynamics and make perpetual motion fairly straightforward. As this does not happen, we can conclude that a 'negative resistance' has to be an *active* circuit device that is connected to a power supply other than the

* In fact, there is some deviation from this characteristic owing to the inevitable presence of some energy levels in the forbidden gap. There is thus some additional tunnelling, which becomes noticeable in the vicinity of the current minimum. But even including this effect the ratio of current maximum to current minimum may be as high as 15 in a practical case.

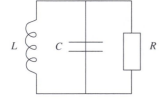

Fig. 9.26
A tuned circuit.

oscillating signal with which it interacts. This is very true of the tunnel diode, since to act as a negative resistance, it has to be biased with a battery to the point A in Fig. 9.25(c). The power to overcome the circuit losses comes from this battery.

If the magnitude of the negative resistance in Fig. 9.25(c) is greater than the loss resistance R, the initiatory signal will not only persist; it will grow. Its magnitude will, of course, be limited by the fact that the tunnel diode can be a negative resistance for only a finite voltage swing (about 0.2 V). Thus, given a negative resistance circuit engineers can make oscillators and amplifiers. The particular advantage of tunnel diodes is that, as the junctions are thin, the carrier transit times are shorter than in a transistor, and very high-frequency operation (up to about 10^{11} Hz) is possible. Their limitation is that with their inherently low voltage operation, they are very low-power devices.

9.11   The backward diode

This is essentially the same thing as the tunnel diode, only the doping is a little lighter. It is called a *backward diode* because everything is the other way round. It has low impedance in the reverse direction and high impedance in the forward direction, as shown in Fig. 9.27.

The secret of the device is that the doping is just that much lighter (than that of the tunnel diode) as to line up the band edges (the top of the valence band on the p-side to coincide with the bottom of the conduction band on the n-side) at zero bias. Hence, for a forward bias there is no tunnelling, just the 'normal' flow, which is very small. In the reverse direction, however, a large tunnelling current may flow.

The backward diode is a very efficient rectifier (of the order of one to a thousand) for low voltages. For higher voltages, of course, the 'forward' current may become significant.

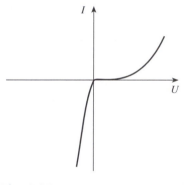

Fig. 9.27
The current voltage characteristics of a backward diode.

9.12   The Zener diode and the avalanche diode

You should not dwell too heavily on the memory of the backward diode; it is rather exceptional. I am pleased to say that from now on *forward* means forward and *reverse* means reverse.

We shall now consider what happens at higher voltages. In the forward direction the current goes on increasing, and eventually the diode will be destroyed when more energy is put in than can be conducted away. This is a fascinating topic for those engineers whose job is to make high-power rectifiers, but it is of limited scientific interest for the rest of us.

There is considerably more interest in the reverse direction. It is an experimental fact that breakdown occurs very sharply at a certain reverse voltage as shown in Fig. 9.28. Since the 'knee' of this breakdown curve is much sharper than the current rise in the forward direction, and since the knee voltage can be controlled by the impurity levels, this effect has applications whenever a sudden increase in current is required at a certain voltage. The diode can therefore be used as a voltage stabilizer or a switch. In the latter application it has

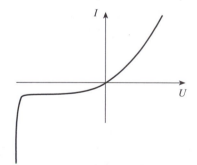

Fig. 9.28
The current voltage characteristics of a p–n junction showing the sudden increase in current at a specific value of reverse voltage.

the further advantage that the breakdown is not only sharp but occurs very fast as well.

9.12.1   Zener breakdown

The breakdown may occur by two distinct mechanisms: (1) Zener breakdown, (2) avalanche breakdown.

The mechanism suggested by Zener (1934) may be explained as follows.*At low reverse bias there is only the flow of minority electrons from the p-side to the n-side. As the reverse bias is increased, at a certain voltage the bands begin to overlap, and tunnelling current may appear. The tunnelling current does appear if the doping is large enough and the junction is narrow enough. But Zener diodes (in contrast to tunnel diodes and backward diodes) are designed in such a way that practically no tunnelling occurs when the bands just overlap; the potential barrier is too wide [Fig. 9.29(a)]. However, as the reverse bias is increased, the width of the barrier decreases [Fig. 9.29(b)] leading—above a certain voltage—to a very rapid rise in current.

9.12.2   Avalanche breakdown

Avalanche diodes differ from Zener diodes by having somewhat smaller impurity density. The depletion layer is then wider, and the Zener breakdown would occur at a considerably higher voltage. However, before the tunnelling current has a chance to become appreciable another mechanism takes over, which—very aptly—is designated by the word *avalanche*.

You know that electrons in a solid are accelerated by the applied electric field. They give up the kinetic energy acquired when they collide with lattice atoms. At a sufficiently high electric field an electron may take up enough energy to ionize a lattice atom, that is to create an electron–hole pair. The newly created electrons and holes may in turn liberate further electron–hole pairs, initiating an avalanche.

Note that the two mechanisms are quite different, as may be clearly seen in Fig. 9.30. In both cases the electron moves from the valence band of the p-type material into the conduction band of the n-type material, but for Zener breakdown it moves horizontally, whereas for avalanche breakdown it must move vertically. But although the mechanisms are different, nevertheless, in a practical case it is difficult to distinguish between them. The diode breaks down, and that is the only experimental result we have. One may attempt to draw the distinction on the basis of the temperature-dependence of the two breakdown mechanisms but, in general, it is not worth the effort. For a practical application all that matters is the rapid increase in current, whatever its cause.

Avalanche diodes may also be used for generating microwaves. The principles of operation (as for most microwave oscillators) are fairly complicated. The essential thing is that both during the avalanche process and the subsequent drift of the created carriers, the current and the electric field are not in phase with each other. By judicious choice of the geometry one may get 180° phase difference between voltage and current, at least for a certain frequency range. But this is nothing more than a frequency-dependent negative resistance. Putting the diode in a cavity, the oscillator is ready.

* There were, of course, no p–n junctions at the time. The mechanism was suggested for bulk breakdown to which, incidentally, it does not apply. The explanation turned out to be applicable to breakdown in p–n junctions.

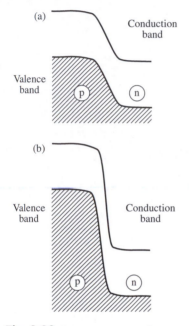

Fig. 9.29
A heavily doped p–n junction (a) in thermal equilibrium. (b) at reverse bias. The width of the potential barrier decreases as the bias is increased.

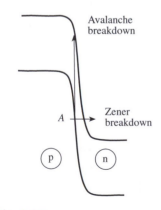

Fig. 9.30
The mechanism of Zener and avalanche breakdown.

9.13   Varactor diodes

As I have mentioned above, and have shown mathematically in eqn (9.21), the capacitance of a reverse-biased p–n junction is voltage-dependent. In other words the capacitance is *variable*, and that is what the name 'varactor' seems to stand for.

Varactor diodes are p–n junctions designed for variable-capacitance operation. Is a variable capacitance good for anything? Yes, it is the basis of the so-called 'parametric amplifier'. How does a parametric amplifier work? This is really a circuit problem, but I had better explain its operation briefly.

Imagine just an ordinary resonant circuit oscillating at a certain frequency. The charge on the capacitor then varies sinusoidally as shown in Fig. 9.31(a).

Suppose the plates of the capacitor are pulled apart when Q reaches its maximum and are pushed back to the initial separation when Q is zero. This is shown in Fig. 9.31(b), where d is the distance between the plates. When Q is finite and the plates are pulled apart, one is doing work against coulombic attraction. Thus, energy is pumped into the resonant circuit at the times t_1, t_3, t_5, etc. When Q is zero, no energy need be expended to push the plates back. The energy of the resonant circuit is therefore monotonically increasing.

To see more clearly what happens, let us try to plot the voltage against time. From t_0 to t_1 it varies sinusoidally. At t_1 the separation between the plates is suddenly increased, that is, the capacitance decreased. The charge on the plates could not change instantaneously; so the reduced capacitance must lead to increased voltage ($Q = CU$ must stay constant). The voltage across the capacitor therefore jumps abruptly at t_1, t_3, t_5, etc., and it is unaffected at t_2, t_4, t_6, etc., as shown in Fig. 9.31(c).

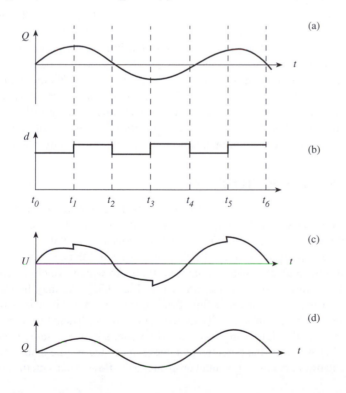

Fig. 9.31
Illustration of the basic principles of parametric amplification.

We may argue in a similar manner that the charge must also increase. When the plate is pushed back at t_2, the voltage is not affected, but the capacitance goes back to its original value. Hence, when the capacitor is charged again, Q will come to a higher peak value, as shown in Fig. 9.31(d).

Note that the important thing is to vary the capacitance in the resonant circuit at twice the resonant frequency. Amplification is achieved then at the expense of the energy available to vary the capacitance. It may be shown (both theoretically and experimentally) that the variation of the capacitance need not be abrupt. Any reasonable variation of the capacitance at twice the rate of the signal frequency would do.*

It is interesting to note that the possibility of parametric amplification had been known for over fifty years, but it has become practical only after the advent of the p–n junction.

Now with what sort of properties would we like to endow our p–n junction to make it suitable for this particular application? Well, it will be the integral part of some sort of tuned circuit, where losses are generally unwelcome. Hence, we shall use heavy doping to reduce the resistance. We should not dope too much, however, because that would lead to narrow depletion regions and low Zener breakdown. Since the varactor diode must operate under reverse bias (to get the capacitance) its useful range is between $U = 0$ and $U = U_B$; a low-breakdown voltage is obviously undesirable.

In practice the p-side of the junction is usually doped very heavily, so that it does not contribute at all to the total series resistance. All the depletion layer is then in the n-type material, whose length is limited to the possible minimum. It is equal to the length of the depletion region just below breakdown (when the depletion region is the longest).

That is roughly how parametric amplifiers work. But is it worth making a complicated amplifier which needs a high-frequency pump oscillator when a 'simple' transistor will amplify just as well? The limitation of a transistor is that it will be a source of noise as well as gain. All amplifiers introduce additional noise[†] due to the random part of their electronic motion, so the emitter and collector currents in a transistor are fairly copious noise sources. As there is almost no standing current in a varactor, it introduces very little noise; so parametric amplifiers are worth their complications in very sensitive receivers, for example for satellite communication links, radio astronomy, and radar.

9.14   Field-effect transistors

We have seen a number of two-terminal devices; let us now discuss a device that has a control electrode as well, like the vacuum triode or the transistor. It was originally proposed in the 1920s but realized only in the 1950s, receiving the rather inappropriate name of field-effect transistor (FET). It consists simply of a piece of semiconductor—let us suppose n-type—to which two ohmic contacts, called the *source* and the *drain*, are made (Fig. 9.32). As may be seen, the drain is positive: thus electrons flow from source to drain. There is also a gate electrode consisting of a heavily doped p-type region (denoted by p^+). Let us assume for the time being that U_{SG}, the voltage between source and gate, is zero. What will be the potential at some point in the n-type material? Since there is an ordinary ohmic potential drop due to the flow of current, the potential

---

* The magic factor 2 in frequency is not necessary either. One may apply a 'pump' at a frequency ω_3, amplify the 'signal' at ω_1, while the so-called 'idler' has a frequency of $\omega_2 = \omega_3 - \omega_1$. We shall come across a similar relationship in the optical parametric oscillator discussed in Section 12.9.

There must be a compromise between reducing the resistivity and ensuring a high breakdown voltage.

† Noise due to electric currents was mentioned briefly in a previous footnote (p. 4). A fuller discussion is beyond our present scope, but if you wish for further reading in this interesting topic see F.N.H. Robinson, *Noise and fluctuations in electronic devices and circuits*, O.U.P., 1974.

grows from zero at the earthed source terminal to U_{DS} at the drain. Hence, the
p$^+$n junction is always reverse biased with the reverse bias increasing towards
the drain. As a consequence, the depletion region has an asymmetrical shape
as shown in Fig. 9.32. The drain current must flow in the channel between the
depletion regions.

If we make the gate negative, then the reverse bias, and with it the depletion
region, increases, forcing the current to flow through a narrower region, that is
through a higher resistance. Consequently, the current decreases. Making the
gate more and more negative with respect to the source, there will obviously
be a voltage at which the depletion regions join and the drain current decreases
to practically zero as shown in Fig. 9.33(a). This I_D versus U_{GS} characteristic
is strongly reminiscent of that of anode current against grid voltage in a good
triode, the product of a bygone age when the subject of electronics was nice
and simple.

The physical picture yielding the I_D versus U_{DS} characteristics is a little
more complicated. As U_{DS} increases at constant gate voltage, there are two
effects occurring simultaneously: (i) the drain current increases because U_{DS} has
increased, a simple consequence of Ohm's law; (ii) the drain current decreases
because increased drain voltage means increased reverse bias and thus a smaller
channel for the current to flow. Now will the current increase or decrease?
You might be able to convince yourself that when the channel is wide, and
the increase in U_{DS} means only a relatively small decrease in the width of the

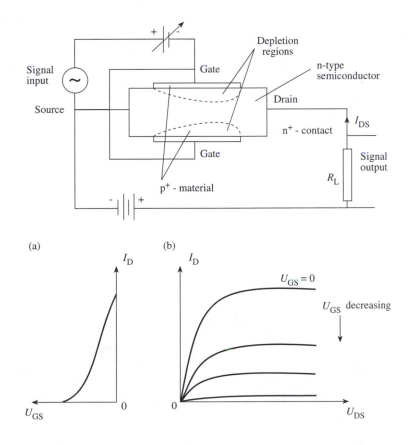

Fig. 9.32
Schematic representation of a field
effect transistor (FET). The current
between *source* and *drain* is
controlled by the voltage on the *gate*
electrodes.

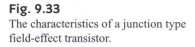

Fig. 9.33
The characteristics of a junction type
field-effect transistor.

channel, the second effect is small, and the current increases. However, as the channel becomes narrower the second effect gradually gains importance, and the increase of I_D with U_{DS} slows down, as shown in Fig. 9.33(b). At the so-called pinch-off voltage, the two effects cancel each other, and they keep their balance for voltages beyond that. The current stays constant; it has reached saturation. The actual value of the saturation current would naturally depend on the gate voltage. At lower gate voltages the saturation current is smaller.

The physical mechanism of current flow in an FET is entirely different from that in a vacuum tube, but the characteristics are similar, and so is the equivalent circuit. A small change in gate voltage, u_{gs}, results in a large change in drain current. Denoting the proportionality factor by g_m, called the mutual conductance, a drain current equal to $g_m u_{gs}$ appears. Furthermore, one needs to take into account that the drain current varies with drain voltage as well. Denoting the proportionality constant by r_d (called the drain resistance), we may now construct the equivalent circuit of Fig. 9.34, where i_d, u_{gs}, and u_{ds} are the small a.c. components of drain current, gate voltage, and drain voltage, respectively.

A modern and more practical variant of this device is the metal–oxide–semiconductor transistor or MOST, also known as metal–oxide–semiconductor field-effect transistor or MOSFET. It is essentially a metal–insulator–semiconductor junction provided with a source and a drain as shown in Fig. 9.35(a). To be consistent with our discussion in Section 9.9, we shall assume that the substrate is an n-type semiconductor, and the source and drain are made of p$^+$ material. At zero gate bias, no drain current flows because one of the junctions is bound to be reverse biased. What happens as we make the gate negative? Remembering the physical phenomena described in Section 9.9, we may claim that at sufficiently large negative gate voltage inversion will occur, that is the material in the vicinity of the insulator will turn into a p-type semi-conductor. Holes may then flow unimpeded from source to drain. The rest of the story is the same as for an ordinary FET (note that since the advent of the MOS-FET, ordinary FETs got dissatisfied with their simple sounding acronym and

Fig. 9.34

Equivalent circuit of a field-effect transistor.

Fig. 9.35

Schematic representation of a MOSFET. (a) Zero gate bias. (b) Forward bias inducing a p-channel.

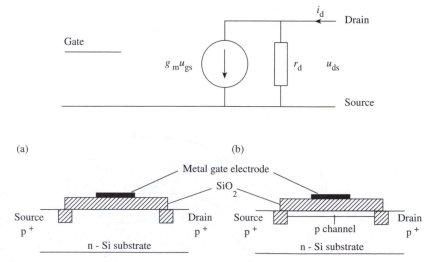

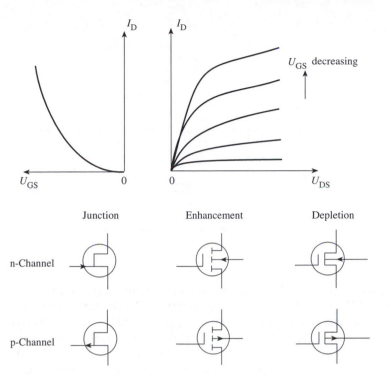

Fig. 9.36
The characteristics of a MOSFET.

Junction   Enhancement   Depletion

n-Channel

p-Channel

Fig. 9.37
Symbols for FETs.

opted for the more sonorous title of JUGFET or junction-gate field-effect transistor) and the characteristics are fairly similar, as shown in Fig. 9.36, though in the present case there is no proper current saturation, only a knee in the I_D versus U_{DS} characteristics.

The devices in which conduction occurs by inversion are said to operate in the enhancement mode. There is also a *depletion* mode device in which one starts with a p-channel [Fig. 9.35(b)] and depletes the holes by applying a positive bias to the gate. This is more similar to the traditional FETs.

Naturally, both the enhancement and depletion devices described have their counterparts with n^+ drains and sources and p-type substrate. In principle there is no difference between them. In practice there is some difference, because the surface potential at the $Si - SiO_2$ interface tends to be positive, thus it is easier to achieve inversion in an n-type material.

Having so many different types of FETs has tested the ingenuity of those whose job is to think up symbols for new devices. The solutions they came up with are shown in Fig. 9.37.

Applications for performing logic functions are obvious. Depending on the gate voltage the FET of Fig. 9.35 is either ON (U_{DS} low) or OFF (U_{DS} high). There are naturally many varieties on the basic theme; I want to mention only one of them known as CMOS (complementary MOS) which rose to fame owing to its low power consumption. The simplest CMOS circuit is the inverter shown in Fig. 9.38. The upper device M1 is a p-channel MOS, whereas the lower device M2 is of the n-channel variety. The drains and gates are connected, and the source is connected to the substrate in each device. Let us see what happens when the input voltage U_{in} is approximately zero. For M2, a positive U_{GS} is

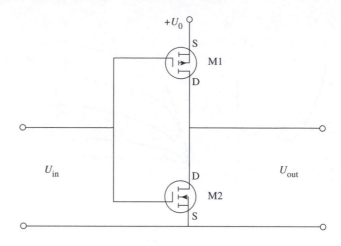

Fig. 9.38
A CMOS inverter.

required to turn it ON. In the absence of a positive U_{GS}, M2 is OFF. What about M1? It needs a negative U_{GS} to turn it ON. But that is exactly what it has. The source is at a high potential U_0 (say 5 V) and the gate is at about zero potential hence M1 is ON, the potential drop across the device is small, and U_{out} is approximately equal to U_0. Thus when U_{in} is low, U_{out} is high. This is what an inverter is supposed to do. You may easily convince yourself that when U_{in} is equal to U_0, M1 is OFF and M2 is ON leading to an output voltage of about zero.

Why is this inverter better than other inverters? On account of the low currents flowing. But surely when a device is ON, there will be a lot of current flowing. Is it not the best analogy a floodgate? When the gates are closed there is only a trickle, but when they are open, there is a torrent of water. This analogy does indeed apply to many electronic devices (e.g. to the transistor of Fig. 9.15 or to the FET of Fig. 9.35) but not to the CMOS inverter. The current through M1 cannot be large when it is ON because there is nowhere the current can flow. In fact, the current must be equal to that flowing in M2 when it is OFF, and that is equal to the current through a reverse biased p–n junction, say 50 nA. And of course the situation is similar when M1 is OFF and M2 is ON. With U_0 equal to 5 V, the power dissipation is a quarter of a microwatt. There is of course a small gate current, as well, but it will still leave power dissipation well below a microwatt*, many orders of magnitude smaller than that of competitive semiconductor devices. Thus CMOS circuits are natural candidates for all battery powered devices. In particular, their advent made possible the birth of the digital watch.

* It must be admitted that there is some more appreciable power dissipation during switching due to capacitors being charged up, and there is also a brief interval when both devices are ON.

9.15 Heterostructures

All the devices mentioned so far have been made using the same material. Some properties of the material were tampered with, for example doping made one part of the material n-type and another part p-type, but the energy gap always remained the same. We may call such structures *homostructures* in contrast to *heterostructures*, which consist of materials of different energy gaps. Is it in

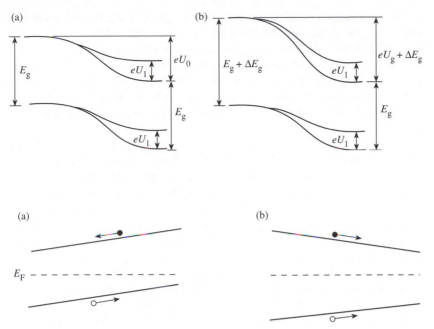

Fig. 9.39
The energy diagram of (a) a homojunction, (b) a heterojunction.

Fig. 9.40
The energy diagram of semiconductor crystals. (a) Variable doping, constant energy gap. (b) Undoped, variable energy gap.

any way desirable to change the energy gap? The simple answer is, yes, it gives us a new degree of freedom.

As an example let us look at the energy diagram of a forward biased p–n junction [Fig. 9.39(a)] in which both parts of the junction have the same energy gap, E_g. The built-in voltage is U_0, and there is a forward bias U_1.

For the heterojunction the energy gap of the n-type material is still E_g, but the energy gap of the p-type material is now larger, $E_g + \Delta E_g$. The additional gap energy is assumed here to appear entirely in the conduction band. Then, for the holes, there is no difference between the homojunction and the heterojunction; the amount of hole injection is the same in both cases. But the energy gap of the p-type material being wider in Fig. 9.39(b) means that the electrons see a higher barrier against them and, consequently, electron injection is much smaller in the heterojunction than in the homojunction. Is this good? If we imagine that this is the emitter–base junction of a p–n–p transistor then this is certainly something desirable. The hole current from emitter to base is the current upon which the operation of the transistor is based. We need that. The electron current from base to emitter does, however, no good. We are better off without it. The conclusion is that by using a heterojunction we can produce a transistor with properties superior to that of a homojunction. The transistor thus obtained has even got a name. It is known as HBT or Heterostructure Bipolar Transistor.

The essential thing is that when we turn to heterojunctions, the fates of electrons and holes are no longer tied together. Perhaps an even better illustration of their independence is provided by Fig. 9.40. Figure 9.40(a) shows the energy diagram of a homogeneous bulk semiconductor doped so that the acceptor concentration increases from left to right. Clearly, the electrons will slide down the slope, and the holes have no other option but to slide up the slope. The flow of electrons and holes is in the opposite direction.

Take now a piece of undoped semiconductor crystal, grown so that the energy gap gradually shrinks from left to right [Fig. 9.40(b)]. The slopes of the potential energy diagram are now such that electrons and holes move in the same direction. So, there is no doubt, heterojunctions offer more freedom in designing devices.

Are the advantages of heterojunctions limited to bipolar devices? One might say, yes, remembering that the separate control of electrons and holes is of no benefit to FETs. In fact, FETs also draw advantages from the availability of heterostructures but in a quite different form. What is it that we are aiming at? We want to have high mobility. But surely that depends on the choice of the material, on the temperature, and on the impurities. Once they are chosen, we have no longer any freedom. Surprisingly, it turns out that we still have some freedom. We can have our cake and eat it. More precisely, we can produce our carriers from impurities without the disadvantage of impurity scattering and the corresponding reduction in mobility. The FET which can incorporate these features is called appropriately the High Electron Mobility Transistor or HEMT (some people call it MODFET an acronym for Modulation-Doped Field Effect Transistor or TEGFET standing for Two-dimensional Electron Gas Field Effect Transistor).

What kind of properties should materials A and B, the two semiconductors to be joined in holy matrimony, possess? One should provide the electrons, so it should be doped, and the other one should provide an impurity-free environment, so it should not be doped. The problem is then to persuade the electrons that, once created in the doped material, they should move over into the undoped material. This can be achieved by a junction between a doped high-bandgap material (A) and an undoped lower-bandgap material (B). The energy diagrams of the two materials before they are joined are shown in Fig. 9.41(a). The Fermi energy of material A is quite close to the conduction band. The Fermi energy of material B may be seen to be somewhat above the middle of the gap on account of the electron effective mass being smaller than the hole effective mass [see eqn (8.24)].

Note that there are now band offsets, both at the conduction and valence bands, denoted by ΔE_c and ΔE_v, respectively.

Next, we need to match the Fermi energies. That can be easily done as shown in Fig. 9.41(b), where we have now dispensed with the vacuum levels. However, the step that follows now is far from being trivial. It is *not* simply a question of joining together the two band edges. We have to do the construction separately, left and right of the metallurgical junction. First, let us figure out how the bands bend on the left-hand side. When the two materials are joined, the conduction band of material *A* is at a higher energy than that of B. We may therefore expect electrons to move from A to B. Hence, there is a depletion region in material *A* and, consequently, the band edges curve upwards reaching the metallurgical junction at points c_A and v_A as shown in Fig. 9.41(c). At this point the bandgap will suddenly change. So we need to move down from c_A an amount of ΔE_c, and move up from v_A by an amount of ΔE_v to reach the points c_B and v_B, respectively. Now we can do the construction on the right-hand side. We join c_B to the conduction band edge and v_B to the valence band edge of material B.

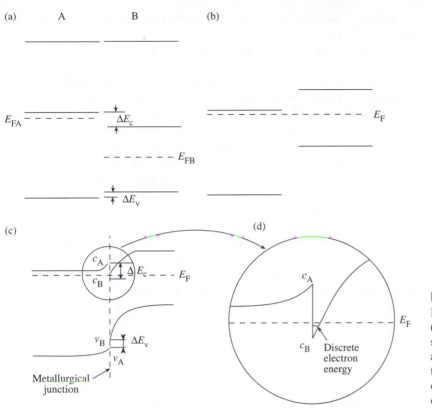

Fig. 9.41

Energy diagram for a hetero-junction. (a) Energy levels before joining the semiconductors. (b) The Fermi levels agree. (c) Energy levels after joining the semiconductors. (d) Details of the conduction band edge in the vicinity of the metallurgical junction.

The energy diagram of Fig. 9.41(c) looks quite different from anything we have seen so far. Not surprisingly, this junction has quite striking properties as may be seen in Fig. 9.41(d), where the central part of the diagram is magnified. The remarkable thing is that c_B, the deepest point in the potential well, is below the Fermi energy. Thus according to the rules of the game, the electron density there is much higher than in any other parts of the two materials. The second striking property is that the well is very narrow. How narrow? For the materials usually used, the width of the well comes out of the calculations (they are pretty complicated, one needs to solve simultaneously Poisson's equation in combination with Schrödinger's equation) as about 8 nm, comparable with atomic dimensions. So the electron is confined in one dimension. We have, in fact, a two-dimensional electron gas, which has discrete energy levels.* In the present example we are showing just one such level.

Let us return now to our original aim. We wanted to produce a Field Effect Transistor which works faster than those using homojunctions on account of the higher mobility. So what kind of materials are we going to use? For the undoped material we shall choose one which has a high mobility. GaAs with a bandgap of 1.43 eV and a mobility of $0.85 \, \text{m}^2 \, \text{V}^{-1} \, \text{s}^{-1}$ at room temperature (up to about $7.5 \, \text{m}^2 \, \text{V}^{-1} \, \text{s}^{-1}$ at 77 K) is clearly suitable for the purpose. The doped material should have a considerably higher bandgap and must be suitable for growing on GaAs. All these requirements are satisfied by $Al_x Ga_{1-x} As$.[†] With

* The discrete character of the energy levels does not much affect the argument here. They are, however, of great importance for the Resonant Tunnelling Diode to be discussed in Section 9.25 and for the latest versions of semiconductor lasers which will come up in Section 12.7.

[†] Remember our discussion in Section 8.6: Al is higher than both Ga and As in the periodic table, hence adding Al to GaAs leads to a material of higher energy gap.

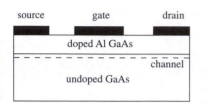

Fig. 9.42
A schematic drawing of a HEMT.

$x = 0.3$, the bandgap is 1.85 eV, ΔE_c is 0.28 eV, and ΔE_v is 0.15 eV. The dopant is silicon.

A schematic drawing of the device may be seen in Fig. 9.42. There are very few electrons in the AlGaAs, therefore the current from source to drain flows along the channel, where the electrons congregate in the undoped GaAs. Thus, at zero gate voltage we can have a flow of electrons. But can we control the current by changing the gate voltage? Yes, we can, by applying a reverse bias, which will lift the whole energy diagram. Not uniformly of course, there will be greater lift where there are fewer carriers, but the essential thing is that with a reasonable voltage (say, 0.5 V), the bottom of the potential well could be lifted above the Fermi energy, which would reduce the current very close to zero. So this device could work as a fast switch or as an amplifier in analogue circuits. Due to the high speed of the electrons, these amplifiers may work up to the mm wavelength region.

So much about n-channel devices. Can we have p-channel devices as well? We can. There is no difficulty in doping AlGaAs so as to produce a p-type material. But there is a snag. AlGaAs, like most semiconductors, has both heavy and light holes in the valence band. Since the speed of the device will be determined by the sluggishness of the heavy holes, there seems to be no point in producing p-channel HEMTs. This is indeed true for most purposes, but let us remember that complementary logic needs both n-type and p-type devices. So what can be done?

By applying a pressure, we can shift the band edge in the energy diagram. In particular, we can effectively suppress the heavy holes in favour of light holes. Further, there is no need actually to apply a pressure, as it may appear due to the inherent properties of the junction. This is what actually happens when InGaAs is grown on GaAs. Since the two materials have different lattice constants, a strain will appear in the InGaAs layer, which will suppress the hole states at the top of the valence band. By good luck the bandgap of InGaAs happens to be *smaller* than that of GaAs. Hence a fast p-channel HEMT may be constructed by doping an epitaxially grown GaAs layer with acceptors and then growing a layer of undoped InGaAs on top. The p-channel will be made up of lighter holes just inside the InGaAs layer.

What can we say about the future? As the new methods of production (MBE and MOCVD) will be more widely used, heterojunctions will hardly cost more than homojunctions. The point is that for economic reasons one wishes to avoid multiple growth (remove the wafer, do some intermediate processing and then resume the growth) but it does not matter how complicated an individual run is. So heterojunctions are here to stay.

9.16   Charge-coupled devices

Charge-coupled devices, abbreviated as CCDs, look very similar to MOSFETs, and in today's world looking similar is half the battle. Since companies are rather reluctant to invest into new types of manufacturing processes, a new device that can be made by a known process and is compatible with existing devices is an attractive proposition.

A charge-coupled device is essentially a metal–insulator–semiconductor junction working in the deep depletion mode. As mentioned in Section 9.9, the carriers are not in thermal equilibrium. There is a potential well for holes as was shown in Fig. 9.23(d), but owing to the long generation–recombination time in a pure material, it is not occupied by holes. The secret of the device is, first that the holes are introduced externally and, second, that the charge can be transferred along the insulator surface by applying judiciously chosen voltages to a set of strategically placed electrodes.

Let us look first at three electrodes only, as shown in Fig. 9.43(a). There is again an n-type semiconductor upon which an oxide layer is grown, and the metal electrodes are on the top, insulated from each other. We can look at each electrode as part of a metal–insulator–semiconductor junction which can be independently biased. In Fig. 9.43(b) there are some holes under electrode 1. They had to get there somehow, for example they could have got there by injection from a forward biased p–n junction. The essential thing is that they got there, and the question is how that positive charge can be transferred from one electrode to the next one.

At $t = t_1$ (see Fig. 9.44) the three electrodes are biased to voltages $-A, 0, 0$ respectively. The corresponding surface potential distribution is shown in Fig. 9.43(b). The holes sit in the potential well. At $t = t_2$ we apply a voltage $-A$ to electrode 2, leading to the surface potential distribution of Fig. 9.43(c). The holes are still sitting under electrode 1 but suddenly the potential well has become twice as large. Since the holes wish to fill uniformly the space available, some of them will diffuse to electrode 2. At the same time, just to give the holes a gentle nudge, U_1 is slowly returning to zero, so that by t_3 the potential well is entirely under electrode 2. Thus the transfer of charge from electrode 1 to electrode 2 has been completed [Fig. 9.43(d)].

Let me reiterate the aim. It is to transfer various sizes of charge packet along the insulator. Thus, when we have managed to transfer the charge from electrode 1 to electrode 2, the space under electrode 1 is again available for receiving a new charge packet. How could we create favourable conditions

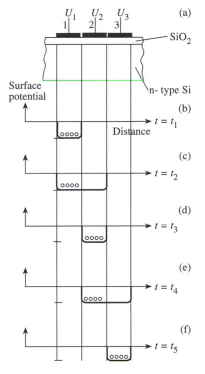

Fig. 9.43
A section of a CCD illustrating the basic principles of a charge transfer.

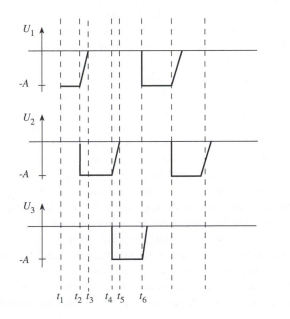

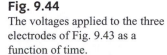

Fig. 9.44
The voltages applied to the three electrodes of Fig. 9.43 as a function of time.

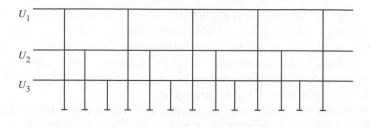

Fig. 9.45

The array of electrodes in a CCD.

for a new charge packet to reside under electrode 1? We should lower U_1. But if we lower U_1 to $-A$, what will prevent the charge under electrode 2 from rolling back? Nothing. Thus, we cannot as yet introduce a new charge packet. First we should move our original packet of holes further away from electrode 1. Therefore, our next move, at $t = t_4$, is to apply $-A$ to U_3 and increase U_2 to zero between t_4 and t_5. The surface potential distributions at t_4 and t_5 are shown in Figs 9.43(e) and (f) respectively. The period ends at t_6. We can now safely lower U_1 and receive a new packet of charge under electrode 1.

In practice, of course, there is an array of electrodes with each third one joined together as shown in Fig. 9.45. When U_1 is lowered at t_6, our original charge packet will start moving to the next electrode, simultaneously with the new charge packet entering the first electrode. With 3000 electrodes in a line, we can have 1000 charge packets stored in the device.

How many elements can be in series? It depends on the amount of charge lost at each transfer. And that is actually the limiting factor in speed as well. If we try to transfer the charge too quickly, some of it will get stuck and the information will be gradually corrupted. The troublemakers are the surface states again. They trap and release charge carriers randomly, thereby interfering with the stored information. Thus, the best thing is to keep the charge away from the surface. This can be done by inserting an additional p-channel into the junction in much the same way as in Fig. 9.35(b). The potential minimum is then in the p-channel, which under reverse bias conditions is entirely depleted of its 'own' carriers and is ready to accept charge packets from the outside. These are called buried-channel devices.

What about other limitations? Well, there is a maximum amount of charge storable above which the potential minimum disappears. There is a minimum frequency, with which the charge can be transferred, below which the information is corrupted by the thermally generated carriers. There is also a minimum size for each cell determined by tunnelling effects (if the cells are too close to each other) and dielectric breakdown (if the insulator is too thin).

What can CCDs be used for? The most important application is for optical imaging for which we do of course need a two-dimensional array. If a picture is focused upon the surface of the device (which in this case has transparent electrodes) the incident light creates electron–hole pairs proportional to its intensity. The process now has two steps: the 'integrate' period, during which U_1 is set to a negative voltage and the holes (in practical devices electron packets are used and everything is the other way round but the principles are the same) are collected in the potential minima, and the 'readout' period, during which the information, is read out. Light may still be incident upon the device during readout, but if the readout period is much shorter than the integration period, the resulting distortions of the video signal are negligible.

How many elements can we have? Arrays of about 3000 by 4000 pixels are now commercially available. Does it mean the end of the film industry? Will all conventional cameras disappear? They might hold out for a few more years but not much longer. It is interesting to note that in spite of their undoubted success story the CCDs lead is being eroded, at least at the lower end of the market, by arrays of two-dimensional CMOS devices. These newcomers to the scene have two great advantages: (i) they are much cheaper to produce and (ii) they consume much less power. They have, though, the handicap of lower light sensitivity and higher noise and that disqualifies them from attacking the top of the market for the time being.

It may be worth mentioning here that CCDs can also be used as detectors for high energy particles, which might simultaneously knock out thousands of electron-hole pairs.

9.17   Silicon controlled rectifier

This has four semiconductor layers, as shown in Fig. 9.46. Apart from the ohmic contacts at the end, there are three junctions. Suppose that junctions (1) and (3) are forward biased by the external supply, so that (2) must be reverse biased. As the supply voltage increases, the current will be limited by junction (2) to a low value, until it gets to the reverse avalanche breakdown point. Then its resistance falls very rapidly, and the current through the whole device 'switches' to follow a curve approximating to the forward bias junction characteristic, starting at this breakdown point, U_s (Fig. 9.47). So far we have described a self-switching arrangement: at a certain applied voltage the device resistance might fall from several megohms to a few ohms. The switch is made even more useful by the additional contact shown in Fig. 9.46, which injects holes into the n-region between junctions (1) and (2) by means of an external positive *control* bias. By injecting extra minority carriers into the junction (2) region the current I_c controls the value of V_s at which the device switches to the *on* position.

This device is used as a switch and in variable power supplies: broadly speaking it is the solid-state version of the gas-filled triode or thyratron. By analogy the name *thyristor* is also used to describe it.

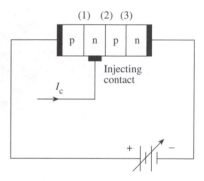

Fig. 9.46
The silicon controlled rectifier (SCR).

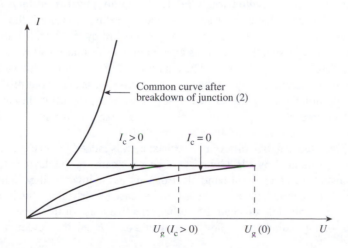

Fig. 9.47
The current voltage characteristics of a SCR. The switching voltage U_s may be controlled by the injected current I_c.

9.18   The Gunn effect

In this and the next two sections I am going to discuss devices that for a change, do not depend on a p–n junction but rather on the bulk properties of semiconductors.

We have shown how desirable it is to have negative resistance and how it can be achieved with a tunnel junction. But an inherent snag with any p–n junction is that it must behave as if there were a capacitor in parallel with the device—we worked out its value in Section 9.5. So at high frequencies this capacitor lowers the impedance and causes a falling-off of efficiency. Can we get round this problem by having a negative resistance characteristic, like that of a tunnel diode, in a bulk semiconductor? This is a long-established El Dorado of semiconductor device engineers. Nearly all semiconductors *should* behave like this.

Look again at an E–k curve that we drew earlier [Fig. 7.12(a)]. If this represents the conduction band, the electrons will be clustered about the lowest energy state: $E = 0$, $k = 0$. Now apply a field in the x-direction which accelerates the electrons, so their momentum (which, as we have mentioned before, is proportional to k) will increase as well. This means that our electrons are climbing up the E–k curve. At a certain point the effective mass changes sign as shown in Fig. 7.12(c). Now the effective mass is just a concept we introduce to say how electrons are accelerated by a field; so this change of sign means that the electrons go the other way. Current opposing voltage is a negative resistance situation. It seems that there should be a good chance of *any* semiconductor behaving like this, but in fact so far this effect has not been discovered. The reason must be that the electrons move for only a short time without collisions. So to get within this time into the negative mass region, very high fields are necessary, which cause some other trouble, for example breakdown or thermal disintegration.

As a matter of fact, we do not really need to send our electron into the negative mass region to have a negative differential resistance. If the effective mass of the electron increases fast enough as a function of the electric field, then the reduced mobility (and conductivity) may lead to a reduction of current—and that is a negative *differential* resistance, so there seems no reason why our device could not work in the region where m^* tends rapidly towards infinity. It is a possibility, but experiments have so far stubbornly refused to display the effect.

An improvement on the latter idea was put forward by Watkins, Ridley, and Hilsum, who suggested that electrons excited into a subsidiary valley of GaAs (see Fig. 8.10) might do the trick. The curvature at the bottom of this valley is smaller; so the electrons acquire the higher effective mass that is our professed aim. In addition there is a higher density of states (it is proportional to $m^{*3/2}$); and furthermore, it looks quite plausible that, once an electron is excited into this valley, it would stay there for a reasonable time.

The predicted negative differential resistance was indeed found experimentally a few years later by J.B. Gunn, who gave his name to the device. At low fields most of the conduction-band electrons are in the lower valley. When an electric field is applied, the current starts to increase linearly along the line OA in Fig. 9.48. If all electrons had the higher effective mass of the upper valley, then the corresponding Ohm's law curve would be OB. As the field increases,

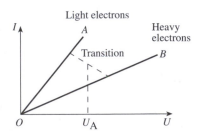

Fig. 9.48
Linear current voltage characteristics for GaAs assuming that only light electrons (OA) or only heavy electrons (OB) are present! The actual characteristics follow the OA line for low voltages and the OB line for high voltages. The transition is shown with dotted lines.

some electrons (as we mentioned before) gain enough energy (0.36 eV) to get into the higher valley, and eventually most of them end up there. So the actual I–U curve will change from something like OA at low fields to something like OB at high fields. This transition from one to the other can (and in GaAs does) give a negative differential resistance.

Having got the negative resistance, all we should have to do is to plug it into a resonant circuit (usually a cavity resonator at high frequencies) and it will oscillate. Unfortunately it is not as simple as that. A bulk negative resistance in a semiconductor is unstable, and is unstable in the sense that a slight perturbation of the existing conditions will grow.

Let us apply a voltage U_A in the negative-resistance region (Fig. 9.48). The expected electric field $\mathcal{E}_A = U_A/d$ (d is the length of the sample), and the expected potential variation, $U = \mathcal{E}_A x$, are shown in Fig. 9.49 by curves (i). It turns out that the expectations are wrong because a negative resistance in a bulk material nearly always leads to an instability. In the present case it may be shown that the instability appears in the form of the heavy electrons accumulating in a high field domain, which travels from the cathode to the anode. The potential and field distributions at a particular moment in time, when the high field domain is in transit, are shown in Fig. 9.49 by curve (ii).

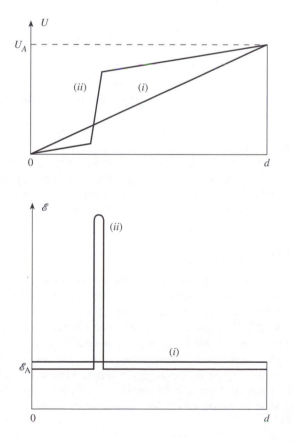

Fig. 9.49
The high field domain fully formed.

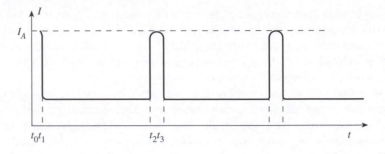

Fig. 9.50
The current as a function of time in GaAs when high field domains move across the material.

So why is this device an oscillator? Because it provides a periodically varying current. How? When the voltage U_A is switched on at t_0, the current is I_A, as shown in Fig. 9.50. Between t_0 and t_1 the high field domain is formed at the cathode. This is equivalent to the insertion of a high resistance material, hence the current must suddenly decline. It remains constant while the high-field domain moves along the material. At $t = t_2$ (where $t_2 - t_1 = d/v_{domain}$, and the velocity of the domain is roughly the same as the drift velocity of the carriers) the domain reaches the anode. The high-resistance region disappears, and the current climbs back to I_A. By the time, t_3, the domain is newly formed at the cathode, and everything repeats itself. We have obtained a periodic current waveform rich in harmonics with a fundamental frequency,

$$f = 1/(t_3 - t_1) \cong v_{domain}/d. \tag{9.28}$$

Thus, the Gunn diode has an oscillation frequency governed by the domain transit time. The velocity of the domain is more or less determined by the voltage producing the effect; so in practice the frequency is selected by the length of the device.

A typical Gunn diode is made by growing an epitaxial layer of n-type GaAs, with an electron concentration of 10^{21}–10^{22} m^{-3} on to an n$^+$-substrate (concentration about 10^{24} m^{-3}). The current flow in the device (Fig. 9.51) is through the thickness of the epitaxial layer. For good quality GaAs the domain velocity is about 10^5 m s^{-1}; a 10 μm layer will therefore make an oscillator in the 10^{10} Hz frequency band (the so called X-band of radar).

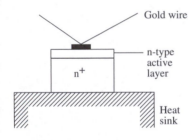

Fig. 9.51
Sketch of a typical Gunn diode.

Note that the transit time mode is not the only mode of operation for this GaAs oscillator. By preventing the formation of domains the bulk negative resistance can be directly utilized.

9.19   Strain gauges

We have noticed before (in the case of thermal expansion) that a change in lattice dimension causes a change in the energy gap as well as in the value of k at the band edge. These changes will also occur if the expansion or contraction is caused by applied stresses. The changes are slight and with intrinsic semiconductors would cause only a small change in resistance. If, however, we have a p-type semiconductor with impurities only partially ionized, a very small change in the energy bands can cause a large percentage change in the energy difference between the impurity levels and the band edge. Thus, the change in resistance of the material with stress (or strain) is large (Fig. 9.52).

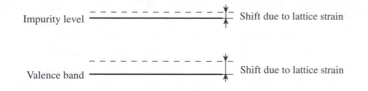

Fig. 9.52
The shift of the energy diagram with strain; this makes semiconductors suitable materials for strain gauges.

Semiconductor strain gauges are pieces of semiconductor with two ohmic contacts that are of a suitable shape to glue on to the component under test. In general, the resistance R can be written

$$R = K(\rho L/A). \qquad (9.29)$$

Thus,

$$\frac{dR}{R} = \frac{dL}{L} - \frac{dA}{A} + \frac{d\rho}{\rho}, \qquad (9.30)$$

which can be rearranged as

$$\frac{\text{Fractional resistance change}}{\text{Strain}} = \frac{dR/R}{dL/L} = 1 + 2p + \frac{d\rho/\rho}{dL/L}. \qquad (9.31)$$

The last term on the right-hand side is called the *gauge factor*,

$$G = \frac{d\rho/\rho}{dL/L}. \qquad (9.32)$$

K is a geometrical constant, L is the length, A the cross-sectional area, and ρ the resistivity of the semiconductor.

p is Poisson's ratio.

For p-type silicon this factor can be between 100 and 200. Of course, for a metal $G \sim 0$ and, since the other two terms on the right-hand side of eqn (9.31) are of order of unity, the gauge factor gives a measure of the increased sensitivity of strain gauges since semiconductor strain gauges became generally available.

9.20 Measurement of magnetic field by the Hall effect

We can rewrite the Hall-effect equation (1.20) in terms of the mobility and of $\mathcal{E}_l$, the applied longitudinal electric field, as

$$\mathcal{E}_H = B\mathcal{E}_l\mu. \qquad (9.33)$$

Hence, B may be obtained by measuring the transverse electric field, the sensitivity of the measurement being proportional to mobility. One semiconductor is quite outstanding in this respect, n-type indium antimonide. It has an electron mobility of about $8\,m^2\,V^{-1}\,s^{-1}$, an order of magnitude greater than GaAs and about fifty times greater than Si. In general, this is a simpler and more sensitive method of measuring a magnetic field than a magnetic coil fluxmeter, and the method is particularly useful for examining the variation of magnetic field over short distances, because the semiconductor probe can be made exceedingly small. The disadvantages are that the measurement is not absolute, and that it is sensitive to changes in temperature.

9.21   Gas sensors

A quite sophisticated effect is that traces of particular oxidizing, reducing or other reactive gases will modify device performance, for example by changing the conductivity of a semiconductor, such as doped tin oxide. This is a very versatile 'Varistor', which can be doped to sense various gases. For example, in its n-type form it has a lattice deficient in O. Oxygen is chemisorbed and removes conduction band electrons by trapping them on the surface. A reducing gas has the opposite effect. This can be used as a simple sensor which enables a gas company service engineer to read off the fraction of CO in flue gases with a simple instrument in a few seconds—a task that used to require a team of analytical chemists.

Forms of the varistor can be targetted on the gases associated with explosives or drugs. Hence, these 'electronic noses' are replacing the 'sniffer dogs' often featured in news items. It is a pity that dogs as well as people are being made redundant by semiconductors, but at least there is now less chance that a cat crossing the road will cause a dog misbehave and misdiagnose.

9.22   Microelectronic circuits

We shall conclude the discussion of semiconductor devices by saying a few words about the latest techniques for producing them. Since the techniques are suitable for producing very small electronic circuits, they are called *micro-electronic* circuits; and because these circuits can be interconnected, they are often referred to as *integrated circuits*. The material most often used is silicon; the small piece of material upon which one unit of the manufactured device is presented is known as a *chip*. Once the exclusive preserve of a few engineers, today even sociologists and politicians know about the silicon chip. They learned to love it or hate it; indifference is no longer possible.

The crucial property of silicon that made this technology possible is its ability to acquire a tenacious 'masking' layer of silicon dioxide. SiO_2 is familiar in an impure state as sand on the beach; it has a ceramic form used for furnace tubes and a crystalline form (quartz) that has good acoustic and optical properties. It is very hard, chemically resistant, an insulator, and has a high melting point ($\approx 1700°C$). An oxide layer can be grown by heating the silicon to 1200°C in an oxygen atmosphere. The growth rate is very slow, about 1 μm per hour, and the thickness is thus easily controlled.

What else needs to be done besides growing an oxide layer? Quite a lot, there are a number of other operations to be performed. Instead of discussing them separately, we shall give here a brief description of the production of a single n–p–n transistor, and shall introduce the various techniques as we go along.

We shall start with a p-type substrate which has an n-type epitaxial layer of about 4–8 μm on the top as shown in Fig. 9.53(a). Next, we grow an oxide layer, then cover it by a thin film of a material called a *photoresist*, place a *mask* on top, and illuminate it by ultraviolet light [Fig. 9.53(b)]. The mask has opaque and transparent areas, so it can define the region of the photoresist upon which the ultraviolet will fall. It is usually a photographic mask drawn originally at a size that may be 1000 times larger than the one required finally, and reduced using a rather sophisticated 'enlarger' backwards.

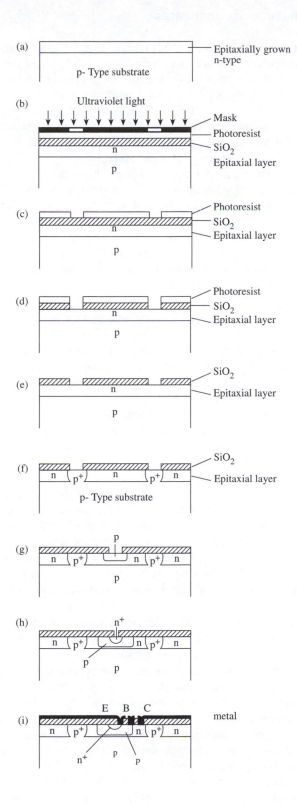

Fig. 9.53
Stages in the production of microelectronic circuits.

Photoresists are organic compounds whose solubility is affected by exposure to ultraviolet light. We are using here a *positive* photoresist in which the exposed areas can be washed away by a suitable developer.* After this operation, we are left with two windows in the photoresist, as shown in Fig. 9.53(c). The one chemical that readily attacks SiO_2 is hydrofluoric acid (HF), but it does not dissolve the photoresist. Hence, the windows in the resist can be turned into windows in the SiO_2 by etching with HF [Fig. 9.53(d)] and the remaining photoresist may then be removed [Fig. 9.53(e)].

The purpose of all these operations, starting with oxidation, was to get access to the epitaxial layer at selected places. The next operation that makes use of these windows is *diffusion*. The silicon is sealed into a clean furnace tube, containing a volatile form of the required doping material. It is then heated for a prescribed time, and the impurity diffuses into the surface. This is a solid-state diffusion process, and the important thing is that it is completely inhibited in the oxide covered regions. For the p-type doping we want here we could use boron bromide (BBr_3) heated to about $1100°C$. The emerging p^+ region is shown in Fig. 9.53(f).

Note that there is lateral diffusion as well, so the p^+ region extends somewhat under the oxide layer. The aim of this diffusion is to *isolate* the present transistor from others made on the same chip.

The oxide layer has now done its duty, so it may be removed, and we are ready to perform the next operation, which is to provide a window for base diffusion. The steps are again oxidation, photoresist coating, masking, illumination by ultraviolet light, removal of the exposed photoresist, and removal of the oxide layer underneath. Then comes the p-type diffusion with which the stage shown in Fig. 9.53(g) is reached.

A further repetition of the technique leads to a window for a diffusion of phosphorus, which forms the n^+ emitter region [Fig. 9.53(h)]. The n^+–p–n transistor is now ready, though it still needs to be connected to other elements on the same chip; so we need some electrodes. This may be done by forming three more windows and evaporating a metal, usually aluminium, for the emitter, base, and collector contacts. The finished transistor is shown in Fig. 9.53(i).

In practice, the above structure is rarely used because of two major disadvantages, first the parasitic p–n–p transistor (formed by the base, collector, and substrate regions) may draw away current to the substrate, and second there is a long path of high resistance from the emitter to the collector. The remedy is to diffuse an n^+ *buried layer* into the p-type substrate prior to the epitaxial growth of the n-layer. Thus, the starting point is as shown in Fig. 9.54(a) instead of that in Fig. 9.53(a).

There is also an additional n^+ diffusion, following the emitter diffusion, leading to the final product shown in Fig. 9.54(b). Note that there is a more modern technique of doping called *ion implantation*. As the name implies, this involves the implantation (in fact, shooting them in with high energy) of ions to wherever the impurities are needed.

Now we know how to make one transistor. The beauty of the technique is that it can make simultaneously millions or billions of transistors. The information where the circuits reside is contained in the corresponding photographic mask. So how many transistors of the type shown in Fig. 9.54(b) can be produced on a chip that is, say, of the size of $1\,cm^2$? Let us do a very, very simple calculation which will give us a very rough answer. To make the calculation even simpler let us consider the less elaborate structure of an inversion type MOSFET shown in Fig. 9.55. The crucial quantity that will determine the

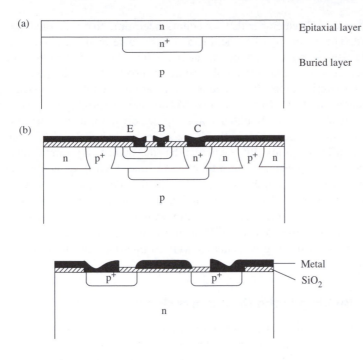

Fig. 9.54

(a) Buried layer diffusion prior to epitaxial growth. (b) The completed transistor differing from that of Fig. 9.53 by having an additional n^+ region.

Fig. 9.55

A simple MOST with contacts.

density of the components is the so called *minimum feature size* which depends on the wavelength of the light used to produce the mask. With an excimer laser of 156 nm and using a technique called 'phase shifting masks'* the minimum feature size, comparable to the wavelength, might be $a = 120$ nm. This would correspond to the minimum distance in Fig. 9.55 which is about half the length of the p^+ region or the distance between the metal electrodes. The length of the MOSFET is then about $9a$. Taking the width of the device as $4a$ and the distance between two devices as a the area required for one MOSFET is $50a^2$ whence the number of components on a chip of 1 cm^2 comes to 140 million—a figure which is in the right ballpark.

We have to note here that not only is the science difficult, but increasing resolution by a factor n, and consequently the component density by n^2, involves formidable problems in costs and man-years, which also go up by n^2. The costs begin to look like the national debt, and the personnel involve management problems of large teams, to reduce (say) 500 man-years to a development period that keeps you ahead of the opposition. This has led to international collaboration involving all the major players. One of the results coming out is a massive document called the International Technology Roadmap for Semiconductors. It sets industry standards and predicts the main trends in the semiconductor industry up to the year 2016. It is quite optimistic. The minimum feature size is supposed to decline by a factor of about 5 during the period ahead. Clearly, the experts believe that saturation is still far away. Can that be true? Saturation is bound to come eventually. The same technique cannot go on for ever. So what are the limitations? First, the number of electrons: as it declines a stage is reached when the fluctuations will cause unacceptable degradation in performance. Another obvious factor as size decreases is the increase in

* The detailed technology of phase shifting masks is formidable, but we can appreciate the principle by considering a masking pattern consisting of an opaque array of bars interspersed by transparent regions. The shadow of one bar (whose width is about the same as the wavelength) will be made fuzzy by diffraction. Now if alternate 'windows' are 'glazed' with a half wavelength phase shifter, the light through adjacent windows is out of phase, so the light diffracting around the bar will interfere destructively, reducing the fuzziness and making the bar's shadow sharp, as if it were a larger object, away from the diffraction limit.

capacitance, and remember that capacitors need time to be charged and discharged. The oncoming of tunnelling across the gate oxide or from source to drain is also a serious limitation. What else? Heat. As the density of components increases heat dissipation will become an insolvable problem. Finally, we should not overlook some less obvious factors, as for example the constant bombardment of our devices by cosmic rays. As device size shrinks a single incident cosmic ray might cause enough damage appreciably to influence performance.

Let us finish this section with Moore's famous law that has been quoted in many different forms. Its essence is that the number of components on a chip increases exponentially as a function of time. Figures quoted have been doubling every year, doubling every 18 months, doubling every two years depending on the mood at the time (optimistic or pessimistic). The Roadmap, mentioned above, predicts a commercially available minimum feature size of about 20 nm by 2016. What will happen afterwards? Will new technologies come? We shall try to give some answers to this question in the next three sections.

9.23   Building in the third dimension

We have talked about microelectronic circuits in the previous section. It was conceived as a two-dimensional technique (for a while it was called the planar technique) and it is still a two-dimensional technique. It is invariably a plane surface that is manipulated for producing a host of devices. It is like Flatland,* a world of two dimensions. Can we get out of Flatland and start building circuits in the third dimension? There are some people who claim they can do it, and an increasing number of people who believe that three-dimensional structures have become a practical possibility.

How would we attempt to build them? Having completed our circuits in two dimensions we would carefully put an insulating oxide layer on the top and start afresh. Alas, we no longer have our nice, epitaxial layer of silicon: the crystalline regularity has been lost. There is no problem depositing silicon on the top of the insulator but it will be an amorphous layer and everyone knows that amorphous materials are not good for building high quality transistors. This has certainly been the state of the art until recently. What has changed is the ability to produce a 'good' amorphous layer by depositing the silicon at the right temperature to be followed by the right heat treatment. Good in this context means that the single crystal grains, of which all amorphous materials are made, can now be quite large, large enough to accommodate a fair number of transistors. One more problem that had to be solved was the presence of irregularities, hills and valleys, after each deposition process. A technique to eliminate them, called chemical-mechanical polishing, has also been perfected. So the road to three-dimension-land is open.

What are the advantages? The main advantage, clearly, is higher packing density: to gain a factor of 10 is not to be sniffed at. The devices being closer to each other also means that the signals have shorter paths to travel, and that increases speed. Unfortunately, there are still a number of disadvantages which will exclude them for the moment from flooding the market. The greatest disadvantage is of course a straight consequence of the polycrystalline nature of the layers. Devices that lie on the grain boundaries will not work. Therefore

Flatland: a romance of many dimensions is a short novel by Edwin Abbot published in London in 1884. It is about life how it is lived in two dimensions and how the inhabitants can deduce evidence about the existence of a third dimension.

error detection and correction techniques must be an integral part of the system. Speed may also suffer. The advantage of shorter paths is offset by the slower switching speed of amorphous devices. And then comes the problem of heat. It is difficult enough to avoid overheating in a two-dimensional structure. It is much more difficult to do so in three-dimension. The answer is to reduce voltages or simply cool the system. Will it be economic to do so? For some applications, for example, for using them as simple memory cells, the answer is already yes. They will very likely replace photographic film at the lower end of the digital recording market.

9.24   Microelectro-Mechanical Systems (MEMS)

Up to now everything has been immobile. Well, nearly. Electrons had a licence to roam about and the lattice was allowed to vibrate. The difference is that from now on part of a structure can mechanically move to perform some useful function. This is a big subject to which I am unable to do justice in the few lines I can devote to it but we shall try to convey the essence of the idea by going in some detail through one example, that of a movable mirror. The optical aspects will be discussed in Chapter 13. In the present section I shall talk about the construction of the mirror. The various steps in the construction process bear strong similarity to those I have discussed in relation to microelectronic circuits.

I shall start with a silicon wafer with a SiO_2 insulator on the top. We could deposit polysilicon on the insulator, as outlined in the previous section, but if we need a thicker layer and higher quality then another technique, called Bonded Silicon-on-Insulator, is used. It involves the bonding of another silicon wafer to the oxidised silicon substrate. The initial bonding is carried out under ultra-clean conditions, and the assembly is then heated in a furnace to strengthen the bond by inter-diffusion. The bonded layer may then be ground and polished, to leave a high-quality single crystal Si layer which can be of virtually any desired thickness.

To fabricate the mirror, the bonded layer is first metal-coated, typically with Cr to improve adhesion and then Au to improve reflectivity. This is shown in Fig. 9.56(a). We have five layers on top of each other: silicon, silicon oxide, silicon, chromium and gold. The next problem is to shape both the mirror and its elastic torsion suspension in the bonded layer. As you may guess, the bonded layer is coated with photoresist which is then patterned with the mechanical shape of the mirror, elastic suspension and surround. After that come two different kinds of etching, the first one to transfer the pattern to the metal and the second one to transfer it to the silicon layer. We arrive then to the situation shown in Fig. 9.56(b) (cross section) and 9.56(c) (top view). The next step is to remove the photoresist after which the whole thing is turned upside down, the substrate side is coated with photoresist and patterned to define a clearance cavity (we need the cavity for the mirror to be able to move). Two further etchings are needed now, one to remove the silicon and the next one to remove the silicon oxide. At this point the mirror is free to rotate on its suspension [Fig. 9.56(d) and (e)].

The released structure is then turned upside down once more and attached to a second wafer, which carries a pair of patterned metal electrodes on an insulating oxide layer [Fig. 9.56(f)]. The mirror is now complete [Fig. 9.56(g) and (h)],

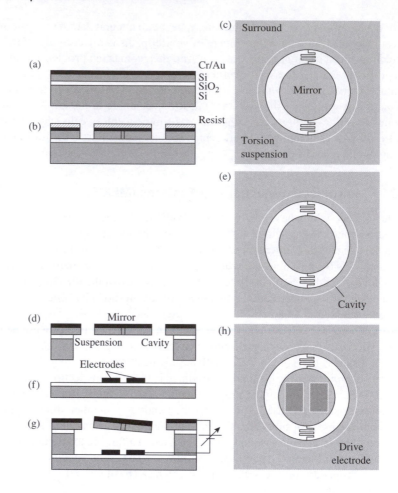

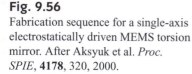

Fig. 9.56

Fabrication sequence for a single-axis electrostatically driven MEMS torsion mirror. After Aksyuk et al. *Proc. SPIE*, **4178**, 320, 2000.

and may be rotated by applying a voltage between the upper electrode and one lower electrode. The mirror will rotate until the attractive electrostatic force between the electrodes is balanced by the restoring force provided by the elastic suspension. The elastic qualities of silicon are surprisingly good, and there are few problems with fatigue and brittle fracture if the assembly is packaged and carefully handled.

You will realize that the aim was to show the basic principles by giving an example of practical significance. There are better solutions but the present one will also work well for a limited angular range. Turn angles of a few degrees may be achieved with drive voltages of 100–200 V. Two-axis mirrors may be constructed using similar principles, by mounting the mirror in a gymbal with two orthogonal elastic suspensions and two orthogonal sets of drive electrodes.

9.25 Nanoelectronics

This is a highly controversial subject and is likely to remain so for a couple of decades. The question is what happens when one or more of the dimensions of a device fall between 1 and 1000 nm. A fair amount is known about the upper end of the range from 100 nm upwards. Mass produced devices in the next few years might even breach the 100 nm limit. A lot is known about the other end as well. Current technology is capable of depositing one atomic layer at a time. So why is the subject controversial? Partly due to some exaggerated claims made by science fiction writers aided and abetted by a few scientists. It is the dream of nanobots, little robots that will do everything conceivable producing food from basic elements and when needed scrambling up in your vein in order to repair a clot. However, the main reason for the controversy is that nobody knows in what direction to push ahead or rather every participant in the game, which is amply financed, is pushing ahead in a different direction.

By the nature of the problem there are two basic approaches: top-down and bottom-up. In the former case one proceeds like a sculptor chiselling away unwanted material and adding bits here and there. This is the approach of microelectronics, the familiar approach. The bottom-up approach is the new one. It can be done. It is not impossible, just pretty difficult. Let me give you some indication of how it has been done. The technique is that of Micro-Electro-Mechanical Systems, discussed earlier, but on a much smaller scale so that we can replace the 'Micro' by 'Nano' arriving at the field of NEMS. The bottom-up approach is based on a small cantilever* that can actually capture an atom off the surface of some material and deposit it at another place. The principles followed are that of the atomic force microscope.

There is no doubt that successful experiments have been done and a lot has been learned but that is still a far cry from buidling useful devices in a reliable manner. Do we know the laws governing the nanometre scale? Yes, of course, they are the laws of quantum mechanics. We know the basic equations, but computers are just not powerful enough to get even near to solving them for practical situations. So far in this course we have been able to manage by injecting no more than a small amount of quantum mechanics. We needed some basic tenets in order to explain the mechanism of conduction, the role of the periodic structure of atoms, the concept of tunnelling, etc. But having accepted the notion of conduction and valence bands, the presence of two kinds of carriers, energy gaps, impurity levels and so on, we could really use the familiar classical picture. It did not really stretch our magination to the limits to 'see' holes diffusing across the base region. We could legitimately boast to have tamed quantum mechanics when dimensions are above about 100 nm. For structures smaller than that the taming has just began. One hopes that it will successfully continue.

I shall now discuss in a little more detail one of the devices that need some structure on the nanoscale. It bears some resemblance to the High-Electron-Mobility-Transistor discussed in Section 9.15. The only essential difference, as shown in Fig. 9.57(a), is that the gate electrode is now split. There are now three finger-gates placed close to each other, where each finger-gate can be biased independently. Let us have reverse bias on the outer gates, so there is a depletion layer below them. This means that the charge sheet sticking to the AlGaAs/GaAs boundary will have discontinuities. If there are no charges, there is no current. So how will this device work? Let me quickly add that the inner gate is forward-biased, so that the potential distribution between source and drain will have the approximate shape shown in Fig. 9.57(b). Well, we have a lower potential in the middle, but will that help? It will if the dimensions are sufficiently small—then electrons may tunnel through the barriers. Does this

* The small cantilever may also serve as the basis of a new type of memory promising gigabytes of information on a few square centimetres. Several companies, including IBM, made progress in that direction. The information is written by a sharp tip perched at the end of the cantilever dipping into a polymer and creating a pit. The presence of a pit may then be regarded as a one and the absence of a pit as a zero. Reading is also done by a tip relying on a change of electrical resistance when it enters the pit.

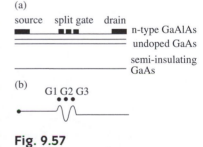

Fig. 9.57
(a) Schematic representation of a Lateral Resonant Tunnelling Transistor; (b) Energy diagram influenced by the voltages on the split electrodes.

mean that the current will flow as in a tunnel diode? In one sense yes, because tunnelling is necessary for the existence of electron flow. It is different, though, in another sense. In the tunnel diode the energy levels of the electrons were nigh infinitely close to each other. The current depended on the density of states. With gate fingers around 20–50 nm the electrons are confined to such a small range that the individual energy levels can be distinguished.

One mode of operation is where the potentials at the outer gates and between the source and the drain are fixed and the inner gate potential is varied, that is the depth of the potential well is controlled. The energy levels are determined by the confinement, so their positions are fixed relative to the bottom of the well. Hence, when the depth of the well is changed, the energy levels move up and down. There will be current flowing whenever a given energy level inside the well matches that of the electrons outside the barrier. The name given to this phenomenon is resonant tunnelling, and the device shown schematically in Fig. 9.57(a) is known as a Lateral Resonant Tunnelling Transistor. Lateral, because the electrons move in the lateral (horizontal) direction, and resonant because current flows only at certain resonant values of the inner gate voltage. One can see that here is a device which can be switched by very small changes (of the order of millivolts) in voltage. In practice, at least for the present, the energy levels are not so well resolved, mainly due to electron scattering in the well which will spread the electron energies. Cooling the device will help, but we are still very far away from measuring delta functions in current. What we do see are little current maxima at the right voltages. But these are early days.

There are a number of other devices too which have been shown to work in the laboratory. Let me mention three of them.

Nanotube transistors. In order to make nanotube transistors we first need nanotubes. What are they? They are thin-walled cylinders of about 1.5 nm diameter and 25 nm long bearing strong resemblance to the fullerenes discussed in Section 5.3.6. The wall is usually made up of one single atomic layer of carbon atoms. The process of preparing them is relatively easy. An arc needs to be struck between graphite electrodes at the right pressure. They are made up of graphite sheets and have their desirable properties (strong covalent bonds) without the disadvantages (weak bonds between the layers). They are very strong mechanically, they have both metallic and semiconductor varieties. At the moment (writing in the spring of 2003) there is no way to decide in advance that this batch will be semiconducting and the next one metallic. Both types appear in the same batch with about equal chance. The only way to have only the semiconducting variety is to destroy the metallic ones, and there are ways of doing so by sending for example an electric shockwave.

What do they look like? They may be tubes hollow all the way or they might be nicely, hemispherically terminated. Their shape also depends on the way the graphite sheet is rolled up. If one of the axes of the hexagonal set coincides with the axis of the nanotube, they take the form shown in Fig. 9.58(a), if it is at an angle we obtain Fig. 9.58(b). They certainly give pretty patterns.

A property that is important for applications is that due to van der Waals forces they stick to the surfaces they are deposited on but they do not stick too firmly, so they can be shifted about, turned around, can be bent and can be cut. When we want to make FETs out of them the substrate can serve as the gate electrode, the tube provides the semiconducting path that is affected by the gate

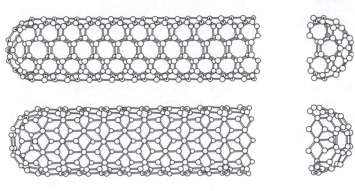

Fig. 9.58
Graphite sheets rolled up into
nanotubes at different angles.

voltage, and the ends of the tube serve as the source and the drain with metal
electrodes deposited upon them.

The second device is the *Single Electron Transistor* which, strictly speaking, does not belong to this chapter since the materials involved are metals and
insulators not semiconductors. On the other hand they can only work when the
dimensions are in the nanometre region so it is not unreasonable to discuss them
here. The effect upon which these devices are built comes from a combination
of electrostatics and tunnelling. The basic configuration is a Metal–Insulator–
Metal–Insulator–Metal (MIMIM) junction. The metal in the middle is called a
Coulomb island. The aim is to show that a single electron can make a difference. This may occur when the electrostatic energy due to a single electron,
$\left(\frac{1}{2}\right)e^2/C$, exceeds the thermal energy $\left(\frac{1}{2}\right)kT$ where C is the capacitance. When
the dimensions are sufficiently small* this capacitance is also small allowing
a high enough electrostatic energy. When a voltage is applied and an electron
tunnels across to the Coulomb island, the resulting change in energy is sufficiently large to forbid any further flow until the voltage is raised to such a value
as to overcome this barrier. The argument can be made a little more precise
by using an energy diagram. At thermal equilibrium [Fig. 9.59(a)], due to the
presence of a significant electrostatic energy (denoted here by E_s), there are no
states available to tunnel into in the vicinity of the Fermi level. The potential
barrier E_s is partly below and partly above the Fermi level. Clearly, no current
can flow in response to a small voltage. This is called a Coulomb blockade.
However when the applied voltage is sufficiently large to overcome the barrier
[Fig. 9.59(b)] there is an opportunity for a single electron, to tunnel across.
But only for a single electron because as soon as it tunnels across from right to
left a new barrier is erected which can only be overcome by increasing again
the voltage above the next threshold. Hence, the current voltage characteristics
consists of a series of steps known as a Coulomb staircase.

Having discussed the basic phenomena it is now easy to imagine how they
can be utilized in a three-terminal device. We need an additional gate electrode
to control the flow of electrons as shown schematically in Fig. 9.60(a), or two
gates and two islands if we want more sophisticated control [Fig. 9.60(b)]. The
latter arrangment, a little similar to that used for CCDs (Section 9.16), permits
the transfer of a single electron from source to drain by choosing a suitable
sequence of gate voltages. In a practical case one should of course choose
a planar configuration. The Coulomb island(s) and the two metal electrodes

* One must be a little careful here. One
cannot just say that small dimensions lead
to small capacitance. In fact, a small
intermetallic distance, needed for tunnelling to take place, leads to a high
capacitance. It needs to be emphasised
then that the metallic areas facing each
other must be very small. Taking the insulator as air, the inter-metallic distance as
1.5 nm and the cross-sections facing each
other as circles of 10 nm radius we end up
with a capacitance of 1.8×10^{-18} F and an
electrostatic energy of 7×10^{-21} J. Note
that this is about 3.5 times higher than the
thermal energy at room temperature.

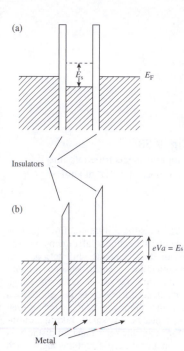

Fig. 9.59

Energy diagram for a Single Electron Transistor (a) in thermal equilibrium, (b) when a voltage Va is applied.

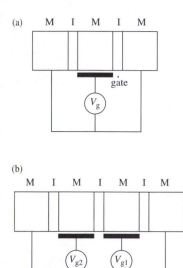

Fig. 9.60

Single electron transfer controlled by (a) a gate electrode in the MIMIM, and (b) by two gate electrodes in the MIMIMIM configuration.

would be evaporated upon one side of an insulator and the gate electrode(s) upon the other side.

If one thinks about it one must acknowledge that this is an amazing feat of science: the control of current down to a single electron. Will these devices ever reach the market place or will they remain a scientfic curiosity? I think they will—in the fullness of time. The idea is so revolutionary, so challenging that sooner or later the necessary effort will be invested into it. What can one hope for? The advent of an entirely new family of logic circuits.

The third new type of device I wish to mention here but only very briefly is the *Molecular Transistor*. It is made of Rotoxane—a molecule that can switch from a conducting to a not-so-well-conducting state by the application of a small voltage.

Most of the ideas behind devices on the nanometric scale have been around for quite some time and experimental results showing the feasibility of the ideas have also been available. A more detailed investigation of the relevant phenomena is however quite recent for the simple reason that it took time to develop the technology. The main motivation has been to put more devices on a mm^2 but many of the experiments conducted have also led to some new physics, as for example to the discovery that in a sufficiently narrow bridge both electrical and thermal conductivity are quantized.

Nanostructures have been made by a variety of methods. It is obviously beyond the scope of the present course to enumerate them. The one that is worth mentioning is the obvious one, electron beam machining, that can produce the required accuracy due to the very short wavelength of accelerated electrons (see examples in Chapter 2). It can write features on an atomic scale, although that method is not free of some technical difficulties either, for example spurious effects due to the electrons bouncing about in the photoresist. The biggest problem however is cost. In microelectronics one can simultaneously produce the pattern for a million elements. If we use electron beams, the pattern must be written serially, and that takes time and effort.

9.26 Social implications

Do great men change the world? They surely do. History is full of them. But, to use an engineering term, they are randomly distributed in space and time so their effect on the whole cancels out. They can, admittedly, cause significant local perturbations, but the associated time constants are invariably small.

Technology is in a different class. Whatever is learnt is rarely forgotten. The interactions are cumulative. So it is not unreasonable to assume that when they exceed a critical value society is no longer able to escape their effect. We may roughly say (only a first order approximation, mind you) that present-day society is determined by the invention and by subsequent improvements in the performance of the steam engine. With the same degree of approximation, we may predict that our future society will be determined by the invention and by subsequent improvements in the performance of semiconductor technology.

The consequence of the steam engine was the First Industrial Revolution, which came to assist our muscles. The consequence of semiconductor techno-logy is the Second Industrial Revolution, which assists our brains. The time is not very far when a man in an armchair will be able to look after a factory. Will it lead to unemployment? There are no signs of it happening. Will it lead to shorter working hours? There are no signs of that happening either. Why not? The explanation is to be thought in one of Parkinson's laws*. People love to create work for each other. Administrative apparatuses grow and grow.

*C. Northcote Parkinson, *Parkinson's Law*. London, John Murray, 1958.

Exercises

9.1. Show that the 'built-in' voltage in a p–n junction is given by

$$U_0 = \frac{kT}{e} \log_e \frac{N_{en}}{N_{ep}} = \frac{kT}{e} \log_e \frac{N_{hp}}{N_{hn}}$$

where N_{en}, N_{hn} and N_{ep}, and N_{hp} are the carrier densities beyond the transition region in the n and p-type materials respectively.
[Hint: Use eqns (8.17) and (8.20) and the condition that the Fermi levels must agree.]

9.2. If both an electric field and concentration gradients are present the resulting current is the sum of the conduction and diffusion currents. In the one-dimensional case it is given in the form

$$J_e = e\mu_e N_e \mathcal{E} + eD_e \frac{dN_e}{dx},$$

$$J_h = e\mu_h N_h \mathcal{E} + eD_h \frac{dN_h}{dx},$$

When the p–n junction is in thermal equilibrium there is no current flowing, i.e. $J_e = J_h = 0$. From this condition show that the 'built-in' voltage is

$$U_0 = \frac{D_e}{\mu_e} \log_e \frac{N_{en}}{N_{ep}}.$$

Compare this with the result obtained in example 9.1, and prove the 'Einstein relationship'

$$\frac{D_e}{\mu_e} = \frac{D_h}{\mu_h} = \frac{kT}{e}.$$

Prove that $N_e N_h = N_i^2$ everywhere in the two semiconductors, including the junction region.

9.3. Owing to a density gradient of the donor impurities there is a built-in voltage of 0.125 V at room temperature between the ends of a bar of germanium. The local resistivity at the high impurity density end is 10 Ωm. Find the local resistivity at the other end of the bar. Assume that all donor atoms are ionized. [Hint: Use the same argument for the conduction and diffusion currents cancelling each other as in the previous example.]

9.4. In a metal–insulator–n-type semiconductor junction the dielectric constants are ϵ_i and ϵ_s for the insulator and the semiconductor respectively. Taking the width of the insulator (sufficient to prevent tunnelling) to be equal to d_i, determine the width of the depletion region as a function of reverse voltage.

9.5. The doping density across a p–n junction is of the form shown in Fig. 9.61. Both the donor and acceptor densities increase linearly in the range $|x| < d_0/2$ and are constant for $|x| > d_0/2$. Determine d, the width of the depletion region when the 'built-in' voltage is such that (i) $d < d_0$, (ii) $d > d_0$.

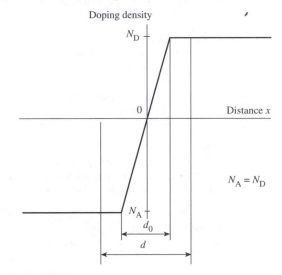

Fig. 9.61
The variation of doping density in a p–n junction.

9.6. Use the expression given in Exercise 9.1 to evaluate the 'built-in' voltage for junctions in germanium and silicon. At room temperature the following data may be assumed

	μ_h (m² V⁻¹ s⁻¹)	μ_e (m² V⁻¹ s⁻¹)	σ_p (Ω⁻¹ m⁻¹)	σ_n (Ω⁻¹ m⁻¹)	N_i (m⁻³)
Ge	0.17	0.36	10^4	100	2.4×10^{19}
Si	0.04	0.18	10^4	100	1.5×10^{16}

where σ_p and σ_n are the conductivities of the p- and n-type materials respectively.

9.7. If a forward bias of 0.1 V is applied to the germanium p–n junction given in example 9.6, what will be the density of holes injected into the n-side and the density of electrons injected into the p-side?

9.8. (i) In a certain n-type semiconductor a fraction α of the donor atoms is ionized. Derive an expression for $E_F - E_D$, where E_F and E_D are the Fermi level and donor level respectively.

(ii) In a certain p-type semiconductor a fraction β of the acceptor atoms is ionized. Derive an expression for $E_F - E_A$ where E_A is the acceptor level.

(iii) Assume that both of the above materials were prepared by doping the same semiconductor, and that $E_D = 1.1$ eV and $E_A = 0.1$ eV, where energies are measured from the top of the valence band. By various measurements at a certain temperature T we find that $\alpha = 0.5$ and $\beta = 0.05$. When a p–n junction is made, the built-in voltage measured at the same temperature is found to be 1.05 eV. Determine T.

9.9. Calculate the reverse breakdown voltage in an abrupt Ge p–n junction for $N_A = 10^{23}\,\mathrm{m}^{-3}$, $N_D = 10^{22}\,\mathrm{m}^{-3}$, $\epsilon_r = 16$, and breakdown field $\mathcal{E}_{br} = 2 \times 10^7\,\mathrm{V\,m}^{-1}$.

9.10. Determine the density distribution of holes injected into an n-type material. Assume that $\partial/\partial t = 0$ (d.c. solution), and neglect the conduction current in comparison with the diffusion current.
[Hint: solve the continuity equation subject to the boundary conditions, $N_h(x = 0) =$ injected hole density, $N_h(x \to \infty) =$ equilibrium hole density in the n-type material.

9.11. Determine from the solution of example 9.10 the distance at which the injected hole density is reduced by a factor e. Calculate this distance numerically for germanium where $D_h = 0.0044\,\mathrm{m}^2\,\mathrm{s}^{-1}$ and the lifetime of holes is 200 μ.

9.12. Determine the spatial variation of the hole current injected into the n-type material.
[Hint: Neglect again the conduction current in comparison with the diffusion current.]

9.13. Express the constant I_0 in the rectifier equation

$$I = I_0[\exp(eU_1/kT) - 1]$$

in terms of the parameters of the p- and n-type materials constituting the junction.

9.14. Owing to the dependence of atomic spacing on pressure the bandgap E_g of silicon decreases under pressure at the rate of 2×10^{-3} eV per atmosphere from its atmospheric value of 1.1 eV.

(i) Show that in an intrinsic semiconductor the conductivity may be expressed with good approximation as

$$\sigma = \sigma_0 \exp(-E_g/2kT)$$

and find an expression for σ_0.

(ii) Calculate the percentage change in conductivity of intrinsic silicon at room temperature for a pressure change of 10 atmospheres.

(iii) Show that I_0, the saturation current of a diode (found in example 9.13), is to a good approximation proportional to $\exp(-E_g/kT)$.

(iv) Compare the pressure sensitivity of σ with that of I_0.

9.15. Take two identical samples of a semiconductor which are oppositely doped so as the number of electrons in the n-type material (N) is equal to the number of holes in the p-type material. Denoting the lengths of the samples by L and the cross-sections by A, the number of electrons and holes are:

p-type	number of holes	LAN
	number of electrons	LAN_i^2/N
n-type	number of holes	LAN_i^2/N
	number of electrons	LAN.

Thus the total number of carriers in the two samples is

$$2LAN\left(1 + \frac{N_i^2}{N^2}\right).$$

Assume now that we join together (disregard the practical difficulties of doing so) the two samples. Some holes will cross into the n-type material and some electrons into the p-type material until finally an equilibrium is established.

Show that the total number of mobile carriers is reduced when the two samples are joined together (that is some electrons and holes must have been lost by recombination).

9.16. A pn junction LED made of GaAsP emits red light (approximate wavelength 670 nm). When forward biased it takes 0.015 mA at 1.3 V rising to 17 mA at 1.6 V. When reverse biased its capacity is 83.1 pF at 1 V and 41.7 pF at 10 V. Avalanche breakdown occurs at 20 V.

(i) Deduce I_0 and the effective device temperature from the rectifier equation.

(ii) Assuming that the junction is very heavily doped on the n side, find the carrier density on the p side.

(iii) Deduce the junction area and the built in voltage.

Data for GaAsP: Permittivity $= 12\epsilon_0$; breakdown field $= 8 \times 10^7\,\mathrm{V\,m}^{-1}$.

Dielectric materials

Le flux les apporta, le reflux les emporte.
Corneille *Le Cid*

10.1 Introduction

In discussing properties of metals and semiconductors we have seen that, with
a little quantum mechanics and a modicum of common sense, a reasonable
account of experiments involving the transport (the word meaning motion in
the official jargon) of electrons emerges. As a dielectric is an insulator, by
definition, no transport occurs. We shall see that we can discuss the effects
of dielectric polarization adequately in terms of electromagnetic theory. Thus,
all we need from band theory is an idea of what sort of energy gap defines an
insulator.

Suppose we consider a material for which the energy gap is 100 times the
thermal energy at 300 K, that is 2.5 eV. Remembering that the Fermi level is
about halfway across the gap in an intrinsic material, it is easily calculated that
the Fermi function is about 10^{-22} at the band edges. With reasonable density
of states, this leads to less than 10^6 mobile electrons per cubic metre, which is
usually regarded as a value for a good insulator. Thus, because we happen to
live at room temperature, we can draw the boundary between semiconductors
and insulators at an energy gap of about 2.5 eV.

Another possible way of distinguishing between semiconductors and insulators is on the basis of optical properties. Since our eyes can detect electromagnetic radiation between the wavelengths of 400 nm and 700 nm, we attribute
some special significance to this band, so we may define an insulator as a
material in which electron–hole pairs are *not* created by visible light. Since a
photon of 400 nm wavelength has an energy of about 3 eV, we may say that an
insulator has an energy gap in excess of that value.

10.2 Macroscopic approach

This is really the subject of electromagnetic theory, which most of you already
know, so I shall briefly summarize the results.

A dielectric is characterized by its dielectric constant ϵ, which relates the
electric flux density to the electric field by the relationship,

$$D = \epsilon \mathcal{E}. \qquad (10.1)$$

In the SI system ϵ is the product of
ϵ_0 (permittivity of free space) and
ϵ_r (relative dielectric constant).

The basic experimental evidence (as discovered by Faraday some time ago)
comes from the condenser experiment in which the capacitance increases by
a factor, ϵ_r, when a dielectric is inserted between the condenser plates. The

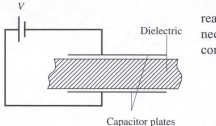

Fig. 10.1
Inserting a dielectric between the
plates of a capacitor increases the
surface charge.

reason is the appearance of charges on the surface of the dielectric (Fig. 10.1) necessitating the arrival of fresh charges from the battery to keep the voltage constant.

In vacuum the surface charge density on the condenser plates is

$$Q = \epsilon_0 \frac{V}{d}, \tag{10.2}$$

where d is the distance between the plates. In the presence of the dielectric the surface charge density increases to

$$Q' = \epsilon_0 \epsilon_r \frac{V}{d}. \tag{10.3}$$

Remember now from electromagnetic theory that the dielectric displacement, D, is equal to the surface charge on a metal plate. Denoting the increase in surface charge density by P, and defining the 'dielectric susceptibility' by

$$\chi = \epsilon_r - 1, \tag{10.4}$$

we may get from eqns (10.2) and (10.3) the relationships

$$P = D - \epsilon_0 \mathcal{E} \quad \text{and} \quad P = \epsilon_0 \chi \mathcal{E}. \tag{10.5}$$

10.3   Microscopic approach

We shall now try to explain the effect in terms of atomic behaviour, seeing how individual atoms react to an electric field, or even before that recalling what an atom looks like. It has a positively charged nucleus surrounded by an electron cloud. In the absence of an electric field the statistical centres of positive and negative charges coincide. (This is actually true for a class of molecules as well.) When an electric field is applied, there is a shift in the charge centres, particularly of the electrons. If this separation is δ, and the total charge is q, the molecule has an *induced dipole* moment,

$$\mu = q\delta. \tag{10.6}$$

Let us now switch back to the macroscopic description and calculate the amount of charge appearing on the surface of the dielectric. If the centre of electron charge moves by an amount δ, then the total volume occupied by these electrons is $A\delta$, where A is the area. Denoting the number of molecules per unit volume by N_m and taking account of the fact that each molecule has a charge q, the total charge appearing in the volume $A\delta$ is $A\delta N_m q$, or simply $N_m q\delta$ per unit area—this is what we mean by surface charge density.

It is interesting to notice that this polarized surface charge density (denoted previously by P, known also as induced polarization or simply polarization) is exactly equal to the amount of dipole moment per unit volume, which from eqn (10.6) is also $N_m q\delta$, so we have obtained our first relationship between the microscopic and macroscopic quantities

$$P = N_m \mu. \tag{10.7}$$

For low electric fields, we may assume that the dipole moment is proportional to the local electric field, $\mathcal{E}'$:

$$\mu = \alpha\mathcal{E}'. \tag{10.8}$$

α is a constant called the *polarizability*.

Notice that the presence of dipoles increases the local field (Fig. 10.2), which will thus always be larger than the applied electric field.

10.4   Types of polarization

Electronic   All materials consist of ions surrounded by electron clouds. As electrons are very light, they have a rapid response to field changes; they may even follow the field at optical frequencies.

Molecular   Bonds between atoms are stretched by applied electric fields when the lattice ions are charged. This is easily visualized with an alkali halide crystal (Fig. 10.3), where small deformations of the ionic bond will occur when a field is applied, increasing the dipole moment of the lattice.

Orientational   This occurs in liquids or gases when whole molecules, having a permanent or induced dipole moment, move into line with the applied field. You might wonder why in a weak static field all the molecules do not eventually align just as a weather vane languidly follows the direction of a gentle breeze. If they did, that would be the lowest energy state for the system, but we know from Boltzmann statistics that in thermal equilibrium the number of molecules with an energy E is proportional to $\exp(-E/kT)$; so at any finite temperature other orientations will also be present.

Physically, we may consider the dipole moments as trying to line up but, jostled by their thermal motion, not all of them succeed. Since the energy of a dipole in an electric field, $\mathcal{E}$ is (Fig. 10.4)

$$E = -\mu\mathcal{E}\cos\theta, \tag{10.9}$$

the number of dipoles in a solid angle, $d\Omega$, is

$$A\exp\left(\frac{\mu\mathcal{E}\cos\theta}{kT}\right)2\pi\sin\theta\,d\theta. \tag{10.10}$$

A is a constant.

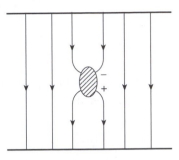

Fig. 10.2
Presence of an electric dipole increases the local electric field.

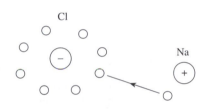

Fig. 10.3
The inter-atomic bond in NaCl is caused by Coulomb attraction. An external electric field will change the separation, thus changing the dipole moment.

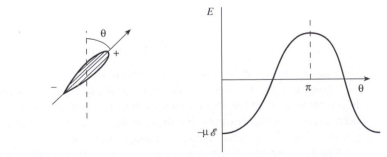

Fig. 10.4
Energy of a dipole in an electric field.

Hence, the average dipole moment is given as

$$\langle \mu \rangle = \frac{\text{net moment of the assembly}}{\text{total number of dipoles}}$$

$$= \frac{\int_0^\pi A \exp\left(\frac{\mu \mathcal{E} \cos\theta}{kT}\right)(\mu \cos\theta) 2\pi \sin\theta \, d\theta}{\int_0^\pi A \exp\left(\frac{\mu \mathcal{E} \cos\theta}{kT}\right) 2\pi \sin\theta \, d\theta}. \tag{10.11}$$

Equation (10.11) turns out to be integrable, yielding

$a = (\mu \mathcal{E}/kT)$, and $L(a)$ is called the Langevin function.

$$\frac{\langle \mu \rangle}{\mu} = L(a) = \coth a - \frac{1}{a}. \tag{10.12}$$

If a is small, which is true under quite wide conditions, eqn (10.12) may be approximated by

$$\langle \mu \rangle = \frac{\mu^2 \mathcal{E}}{3kT}. \tag{10.13}$$

That is, the polarizability is inversely proportional to the absolute temperature.

10.5 The complex dielectric constant and the refractive index

In engineering practice the dielectric constant is often divided up into real and imaginary parts. This can be derived from Maxwell's equations by rewriting the current term in the following manner:

* The complex dielectric constant used by electrical engineers is invariably in the form $\epsilon = \epsilon_0(\epsilon' - j\epsilon'')$. We found a different sign because we had adopted the physicists' time variation, $\exp(-i\omega t)$.

$$J - i\omega\epsilon\mathcal{E} = \sigma\mathcal{E} - i\omega\epsilon\mathcal{E}$$

$$= -i\omega\left(\epsilon + i\frac{\sigma}{\omega}\right)\mathcal{E}, \tag{10.14}$$

where the term in the bracket is usually referred to as the complex dielectric constant. The usual notation is*

The *loss tangent* is defined as $\tan\delta \equiv \epsilon''/\epsilon'$.

$$\epsilon \equiv \epsilon'\epsilon_0 \quad \text{and} \quad \frac{\sigma}{\omega} = \epsilon''\epsilon_0. \tag{10.15}$$

The refractive index is defined as the ratio of the velocity of light in a vacuum to that in the material,

$$n = \frac{c}{v}$$

$$= \sqrt{\epsilon_r \mu_r} = \sqrt{\epsilon'} \tag{10.16}$$

since $\mu_r = 1$ for most materials that transmit light.

Conventionally, we talk of 'dielectric constant' (or permittivity) for the lower frequencies in the electromagnetic spectrum and of refractive index for light. Equation (10.16) shows that they are the same thing—a measure of the polarizability of a material in an alternating electric field.

A fairly recent and important application of dielectrics to optics has been that of multiply-reflecting thin films. Consider the layered structure represented in Fig. 10.5 with alternate layers of transparent material having refractive indices

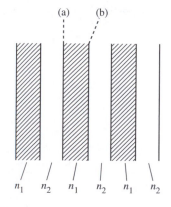

(a)   (b)

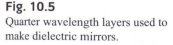

$n_1 \quad n_2 \quad n_1 \quad n_2 \quad n_1 \quad n_2$

Fig. 10.5

Quarter wavelength layers used to make dielectric mirrors.

n_1 and n_2 respectively. At each interface there will be some light reflected and some transmitted. The *reflection coefficient*, from electromagnetic theory, at an interface like (a) in Fig. 10.5 is

$$r_a = \frac{n_2 - n_1}{n_2 + n_1}. \tag{10.17}$$

By symmetry, the reflection coefficient at (b) will be the reverse of this

$$r_b = \frac{n_1 - n_2}{n_1 + n_2} = -r_a. \tag{10.18}$$

Now suppose that all the layers are a quarter wavelength thick—their actual thickness will be $n_1(\lambda/4)$ and $n_2(\lambda/4)$ respectively. Then the wave reflected back from (b) will be π radians out of phase with the wave reflected back from (a) because of its extra path length, and another π radians because of the phase difference in eqn (10.18). So the two reflected waves are 2π radians different; that is, they add up in phase. A large number of these layers, often as many as 17, makes an excellent mirror. In fact, provided good dielectrics (ones with low losses, that is), are used, an overall reflection coefficient of 99.5% is possible, whereas the best metallic mirror is about 97–98% reflecting. This great reduction in losses with dielectric mirrors has made their use with low-gain gas lasers almost universal. I shall return to this topic when discussing lasers.

Another application of this principle occurs when the layer thickness is one *half* wavelength. Successive reflections then cancel, and we have a reflectionless or 'bloomed' coating, much used for the lenses of microscopes and binoculars. A simpler form of 'blooming' uses only one intermediate layer on the glass surface (Fig. 10.6) chosen so that

$$n_1 = \sqrt{n_2}. \tag{10.19}$$

The layer of the material of refractive index n_1 is this time one quarter wavelength, as can be seen by applying eqn (10.17).

10.6 Frequency response

Most materials are polarizable in several different ways. As each type has a different frequency of response, the dielectric constant will vary with frequency in a complicated manner; for example at the highest frequencies (light waves) only the electronic polarization will 'keep up' with the applied field. Thus, we may measure the electronic contribution to the dielectric constant by measuring the refractive index at optical frequencies. An important dielectric, water has a dielectric constant of about 80 at radio frequencies, but its refractive index is 1.3, not $(80)^{1/2}$. Hence we may conclude that the electronic contribution is about 1.7, and the rest is probably due to the orientational polarizability of the H_2O molecule.

The general behaviour is shown in Fig. 10.7. At every frequency where ϵ' varies rapidly, there tends to be a peak of the ϵ'' curve. In some cases this is analogous to the maximum losses that occur at resonance in a tuned circuit: the molecules have a natural resonant frequency because of their binding in the crystal, and they will transfer maximum energy from an electromagnetic wave

The two reflections have a phase difference of π radians.

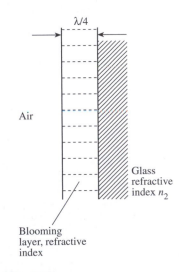

Fig. 10.6
Simple coating for a 'bloomed' lens.

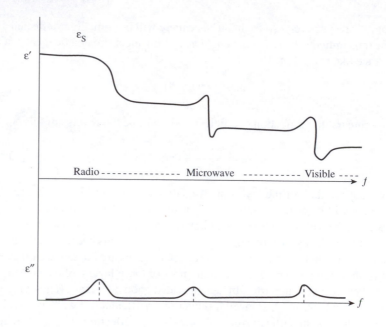

Fig. 10.7
Typical variation of ϵ' and ϵ'' with frequency.

at this frequency. Another case is the 'viscous lag' occurring between the field and the polarized charge which is described by the Debye equations, which we shall presently consider. A consequence of all this is that materials that transmit light often absorb strongly in the ultraviolet and infrared regions, for example most forms of glass. Radio reception indoors is comparatively easy because (dry) bricks transmit wireless waves but absorb light; we can listen in privacy. The Earth's atmosphere is a most interesting dielectric. Of the fairly complete spectrum radiated by the Sun, not many spectral bands reach the Earth. Below 10^8 Hz the ionosphere absorbs or reflects; between 10^{10} and 10^{14} Hz there is molecular resonance absorption in H_2O, CO_2, O_2, and N_2; above 10^{15} Hz there is a very high scattering rate by molecules and dust particles. The visible light region (about 10^{14}–10^{15} Hz) has, of course, been of greatest importance to the evolution of life on Earth. One wonders what we would all be like if there had been just a little more dust around, and we had had to rely on the 10^8–10^{10} Hz atmosphere window for our vision.

10.7   Polar and non-polar materials

This is a distinction that is often made for semiconductors as well as dielectrics. A non-polar material is one with no permanent dipoles. For example, Si, Ge, and C (diamond) are non-polar. The somewhat analogous III–V compounds, such as GaAs, InSb, and GaP, share their valency electrons, so that the ions forming the lattice tend to be positive (group V) or negative (group III). Hence, the lattice is a mass of permanent dipoles, whose moment changes when a field is applied. As well as these ionic bonded materials, there are two other broad classes of polar materials. There are compounds, such as the hydrocarbons (C_6H_6 and paraffins) that have permanent dipole arrangements but still have a net dipole moment of zero (one can see this very easily for the benzene ring).

Table 10.1 *Dielectric constant and refractive index of polar and non-polar materials*

Material	Refractive index	(Refractive index)^2	Dielectric constant measured at 10^3 Hz
Non-polar			
C (diamond)	2.38	5.66	5.68
H$_2$ (liquid)	1.11	1.232	1.23
Weakly polar			
polythene	1.51	2.28	2.30
ptfe (poly-tetra fluoro-ethylene)	1.37	1.89	2.10
CCl$_4$ (carbon-tetrachloride)	1.46	2.13	2.24
Paraffin	1.48	2.19	2.20
Polar			
NaCl (rocksalt)	1.52	2.25	5.90
TiO$_2$ (rutile)	2.61	6.8	94.0
SiO$_2$ (quartz)	1.46	2.13	3.80
Al$_2$O$_3$ (ceramic)	1.66	2.77	6.5
Al$_2$O$_3$ (ruby)	1.77	3.13	4.31
Sodium carbonate	1.53	2.36	8.4
Ethanol	1.36	1.85	24.30
Methanol	1.33	1.76	32.63
Acetone	1.357	1.84	20.7
Soda glass	1.52	2.30	7.60
Water	1.33	1.77	80.4

Then there are molecules such as water and many transformer oils that have permanent dipole moments, and the total dipole moment is determined by their orientational polarizability.

A characteristic of non-polar materials is that, as all the polarization is electronic, the refractive index at optical wavelengths is approximately equal to the square root of the relative dielectric constant at low frequencies. This behaviour is illustrated in Table 10.1.

From Table 10.1 (more comprehensive optics data would show the same trend) you can see that most transparent dielectrics, polar or not, have a refractive index of around 1.4–1.6; only extreme materials like liquid hydrogen, diamond, and rutile (in our list) show appreciable deviation. Let us look for an explanation of this remarkable coalescence of a physical property; starting with our favourite (simplest) model of a solid, the cubic lattice of Fig. 1.1, with a lattice spacing a. Suppose the atoms are closely packed, each having a radius r, so that

$$a = 2r. \qquad (10.20)$$

For an optical property we need to consider only electronic polarizability as ionic, and molecular responses are too slow. Let us suppose that each atomic volume, $(4/3)\pi r^3$ is uniformly occupied by the total electronic charge, Ze. When an electric field, $\mathcal{E}$, is applied, the centre of charge of the electronic cloud shifts a distance d to C^- from the nucleus at C (Fig. 10.8). To find

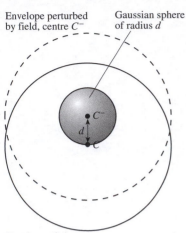

Envelope perturbed by field, centre C^-

Gaussian sphere of radius d

Envelope of electron cloud about centre C

Fig. 10.8

Displacement of electron cloud about an atom centred on C by a distance d, by an applied electric field.

The LHS is the quantity defined as the dipole moment, μ, in eqn (10.6).

the restoring force attracting the electrons back towards the nucleus, we can construct a Gaussian surface of radius d about C^-, so that C is just excluded. The negative charge inside the Gaussian sphere is, according to the uniform charge approximation, equal to $Ze(d/r)^3$. So the attractive force, F, towards the nucleus is

$$F = \frac{(Ze)^2 d}{4\pi\epsilon_0 r^3}. \tag{10.21}$$

This must be balanced by the field force causing the charge displacement,

$$F = Ze\mathcal{E}. \tag{10.22}$$

So by eqns (10.21) and (10.22)

$$Zed = 4\pi r^3 \epsilon_0 \mathcal{E}. \tag{10.23}$$

By eqn (10.8) we see that the polarizability is

$$\alpha = 4\pi r^3 \epsilon_0. \tag{10.24}$$

We can now return to eqn (10.7), $P = N_m \mu$, to find the induced polarization. The density of atoms per unit volume is

$$N_m = \frac{1}{a^3} = \frac{1}{(2r)^3}, \tag{10.25}$$

leading to

$$P = \frac{1}{(2r)^3} 4\pi r^3 \epsilon_0 \mathcal{E} = \frac{\pi}{2}\epsilon_0 \mathcal{E}. \tag{10.26}$$

From eqns (10.4) and (10.5) it follows then that

$$\chi = \frac{P}{\epsilon_0 \mathcal{E}} = \frac{\pi}{2}, \tag{10.27}$$

whence

$$\epsilon_r = 1 + \chi \cong 2.57, \tag{10.28}$$

and

$$n = \sqrt{\epsilon_r} \cong 1.6. \tag{10.29}$$

Thus, our very approximate estimate is at the high end of our small sample in Table 10.1. We could refine this model to give a different fit by remarking that less close packing would give $N_m < 1/(2r)^3$, and we could also take into account quantum orbits. This would change things slightly at the expense of considerable calculations. But the main point of this aside is that a simple approach can sometimes give a reasonable answer and at the same time enhance insight into the phenomenon.

10.8 The Debye equation

We have seen that frequency variation of relative permittivity is a complicated affair. There is one powerful generalization due to Debye of how materials with orientational polarizability behave in the region where the dielectric polarization is 'relaxing', that is the period of the a.c. wave is comparable to the alignment time of the molecule. When the applied frequency is much greater than the reciprocal of the alignment time, we shall call the relative dielectric constant ϵ_∞ (representing atomic and electronic polarization). For much lower frequencies it becomes ϵ_s, the static relative dielectric constant. We need to find an expression of the form

$$\epsilon(\omega) = \epsilon_\infty + f(\omega), \tag{10.30}$$

for which $\omega \to 0$ reduces to

$$f(0) = \epsilon_s - \epsilon_\infty. \tag{10.31}$$

Now suppose that a steady field is applied to align the molecules and then switched off. The polarization and hence the internal field will diminish. Following Debye, we shall assume that the field decays exponentially with a time constant, τ, the characteristic relaxation time of the dipole moment of the molecule,

$$P(t) = P_0 \exp(-t/\tau). \tag{10.32}$$

You know that time variation and frequency spectrum are related by the Fourier transform. In this particular case it happens to be true that the relationship is

$$\begin{aligned} f(\omega) &= K \int_0^\infty P(t) e^{i\omega t} \, dt \\ &= \frac{K P_0}{-i\omega \times 1/\tau}, \end{aligned} \tag{10.33}$$

K is a constant ensuring that $f(\omega)$ has the right dimension.

using the condition (10.31) for the limit when $\omega = 0$, we obtain

$$K P_0 \tau = \epsilon_s - \epsilon_\infty. \tag{10.34}$$

Hence, eqn (10.30) becomes

$$\epsilon(\omega) = \epsilon_\infty + \frac{\epsilon_s - \epsilon_\infty}{-i\omega\tau + 1}, \tag{10.35}$$

which, after the separation of the real and imaginary parts, reduces to

$$\epsilon' = \epsilon_\infty + \frac{\epsilon_s - \epsilon_\infty}{1 + \omega^2 \tau^2} \tag{10.36}$$

$$\epsilon'' = \frac{\omega\tau}{1 + \omega^2\tau^2}(\epsilon_s - \epsilon_\infty). \tag{10.37}$$

These equations agree well with experimental results. Their general shape is shown in Fig. 10.9. Notice particularly that ϵ'' has a peak at $\omega\tau = 1$, where the slope of the ϵ' curve is a maximum.

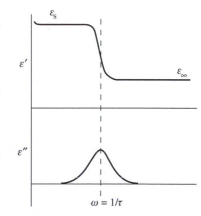

Fig. 10.9
Frequency variation predicted by the Debye equations.

10.9   The effective field

We have remarked that the effective or local field inside a material is increased above its value in free space by the presence of dipoles. Generally, it is difficult to calculate this increase, but for a non-polar solid, assumptions can be made that give reasonable agreement with experiment and give some indication of how the problem could be tackled for more complicated materials. Consider the material to which an external field is applied. We claim now that the local electric field at a certain point is the same as that inside a spherical hole. In this approximate picture the effect of all the 'other' dipoles is represented by the charges on the surface of the sphere. Since in this case the surface is not perpendicular to the direction of the polarization vector, the surface charge is given by the scalar product (Fig. 10.10)

$$\mathbf{P} \cdot d\mathbf{A} = P\, dA\, \cos\theta, \tag{10.38}$$

giving a radial electric field,

$$d\mathcal{E}_r = \frac{P\, dA\, \cos\theta}{4\pi\epsilon_0 r^2}, \tag{10.39}$$

in the middle of the sphere. Clearly, when we sum these components the net horizontal field in our drawing will be zero, and we have to consider only the vertical field. We get this field, previously called $\mathcal{E}'$, by integrating the vertical component and adding to it the applied field, that is

$$\mathcal{E}' = \iint\limits_{\text{surface}} d\mathcal{E}_r \cos\theta + \mathcal{E}, \tag{10.40}$$

whence

$$\begin{aligned}
\mathcal{E}' - \mathcal{E} &= \iint \frac{P\cos^2\theta\, dA}{4\pi\epsilon_0 r^2} \\
&= \int_0^\pi \frac{P\cos^2\theta}{4\pi\epsilon_0 r^2} r^2 2\pi \sin\theta\, d\theta \\
&= \frac{P}{3\epsilon_0}.
\end{aligned} \tag{10.41}$$

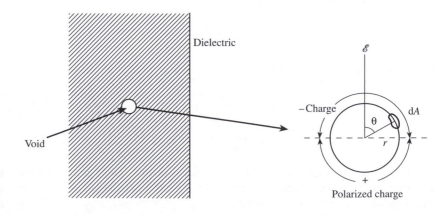

Fig. 10.10
Calculation of effective internal field.

Substituting for P from eqn (10.5) and solving for $\mathcal{E}'$ we get

$$\mathcal{E}' = \tfrac{1}{3}(\epsilon' + 2)\mathcal{E}. \tag{10.42}$$

This result is clearly acceptable for $\epsilon' = 1$; and it is also consistent with our assumption that $\mathcal{E}'$ is proportional to $\mathcal{E}$.

We can now derive an expression for the polarizability α as well, by combining our expression for the local field with eqns (10.7) and (10.8), yielding

$$\alpha = \frac{\epsilon' - 1}{\epsilon' + 2} \cdot \frac{3\epsilon_0}{N_{\mathrm{m}}}. \tag{10.43}$$

This is known as the Clausius–Mosotti equation. It expresses the microscopically defined quantity α in terms of measurable macroscopic quantities.

10.10 Dielectric breakdown

Electric breakdown is a subject to which it is difficult to apply our usual scientific rigour. A well-designed insulator (in the laboratory) breaks down in service if the wind changes direction or if a fog descends. An oil-filled high-voltage condenser will have bad performance, irrespective of the oil used, if there is 0.01% of water present. The presence of grease, dirt, and moisture is the dominant factor in most insulator design. The flowing shapes of high-voltage transmission line insulators are not entirely due to the fact that the ceramic insulator firms previously made chamberpots; the shapes also reduce the probability of surface tracking. The onset of dielectric breakdown is an important economic as well as technical limit in capacitor design. Generally, one wishes to make capacitors with a maximum amount of stored energy. Since the energy stored per unit volume is $\tfrac{1}{2}\epsilon\mathcal{E}^2$, the capacitor designers value high breakdown strength even more highly than high dielectric constant.

In general, breakdown is manifested by a sudden increase in current when the voltage exceeds a critical value U_{b}, as shown in Fig. 10.11. Below U_{b} there is a small current due to the few free electrons that must be in the conduction band at finite temperature. When breakdown occurs it does so very quickly, typically in 10^{-8} s in a solid.

There are three main mechanisms that are usually blamed for dielectric breakdown: (1) intrinsic, (2) thermal, and (3) discharge breakdown.

10.10.1 Intrinsic breakdown

When the few electrons present are sufficiently accelerated (and lattice collisions are unable to absorb the energy) by the electric field, they can ionize lattice atoms. The minimum requirement for this is that they give to the bound (valence) electron enough energy to excite it across the energy gap of the material. This is, in fact, the same effect that we mentioned before in connection with avalanche diodes.

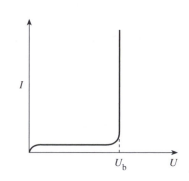

Fig. 10.11
Current voltage characteristics for an insulator. The current increases very rapidly at the breakdown voltage, U_{b}.

10.10.2 Thermal breakdown

This occurs when the operating or test conditions heat the lattice. For example, an a.c. test on a material in the region of its relaxation frequency, where ϵ'' is large would cause heating by the lossy dipole interaction rather than by accelerating free electrons. The heated lattice ions could then be more easily ionized by free electrons, and hence the breakdown field could be less than the intrinsic breakdown field measured with d.c. voltages. The typical polymer, polyethylene, has a breakdown field of 3 to 5×10^8 V m^{-1} for very low frequencies, but this falls to about 5×10^6 V m^{-1} around 10^6 Hz, where a molecular relaxation frequency occurs. Ceramics such as steatite and alumina exhibit similar effects.

If it were not for dielectric heating effects, breakdown fields would be lower at high frequencies simply because the free electrons have only half a period to be accelerated in one direction. I mentioned that a typical breakdown time was 10^{-8} s; so we might suppose that at frequencies above 10^8 Hz breakdown would be somewhat inhibited. This is true, but a fast electron striking a lattice ion still has a greater speed after collision than a slow one, and some of these fast electrons will be further accelerated by the field. Thus, quite spectacular breakdown may sometimes occur at microwave frequencies (10^{10} Hz) when high power densities are passed through the ceramic windows of klystrons or magnetrons. Recent work with high-power lasers has shown that dielectric breakdown still occurs at optical frequencies. In fact the maximum power available from a solid-state laser is about 10^{12} W from a series of cascaded neodymium glass amplifier lasers. The reason why further amplification is not possible is that the optical field strength disrupts the glass laser material.

10.10.3 Discharge breakdown

In materials such as mica or porous ceramics, where there is occluded gas, the gas often ionizes before the solid breaks down. The gas ions can cause surface damage, which accelerates breakdown. This shows up as intermittent sparking and then breakdown as the test field is increased.

10.11 Piezoelectricity

* Derived from the Greek word meaning 'to press'.

It is easy to describe the piezoelectric* effect in a few words: a mechanical strain will produce dielectric polarization and, conversely, an applied electric field will cause mechanical strain. Which crystals will exhibit this effect? Experts say that out of the 21 classes of crystals that lack a centre of symmetry, 20 are piezoelectric. Obviously, we cannot go into the details of all these crystals structures here, but one can produce a simple argument showing that lack of a centre of symmetry is a necessary condition. Take for example the symmetric cubic structures shown in Fig. 10.12(a). If mechanical force is applied [Fig. 10.12(b)] then the dimensions change, but no net electric dipole moment is created. If we take, however, a crystal which clearly lacks a centre of symmetry (Fig. 10.13) and apply a mechanical force, then the centres of positive and negative charge no longer coincide, and a dipole moment is produced. For small deformations and small electric fields, the relationships are linear and may be described by the two equations,

T=stress, S=strain, c=elastic constant, e=piezoelectric constant.

$$T = cS - e\mathcal{E} \tag{10.44}$$

and

$$D = \epsilon\mathcal{E} + eS. \tag{10.45}$$

With $e = 0$, the above equations reduce to Hooke's law and to the $D = \epsilon\mathcal{E}$ relationship, respectively. It may be seen that when $\mathcal{E} = 0$, the electric flux

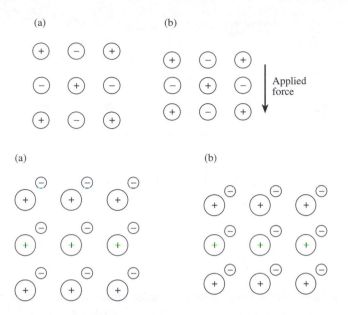

Fig. 10.12
Schematic representation of a symmetric crystal; (a) in the absence of applied force, (b) in the presence of applied force.

Fig. 10.13
Schematic representation of a crystal that lacks a centre of symmetry; (a) in the absence of applied force, (b) in the presence of applied force.

density, D, may be finite in spite of zero electric field. Similarly, an electric field sets up a strain without a mechanical stress being applied.

In general, all the constants in eqns (10.44) and (10.45) are tensors. In the worst case there are 45 independent coefficients comprising 21 elastic constants, 6 dielectric constants, and 18 piezoelectric constants. In practical cases the situation is not so complicated because we would apply an electric field in some particular direction and would make use of the mechanical displacement in some other specified direction, so that you can safely think in terms of those three scalar constants in the above equations.

It follows from the properties of piezoelectrics that they are ideally suited to play the role of electromechanical transducers. Common examples are the microphone, where longitudinal sound vibrations in the air are the mechanical driving force, and the gramophone pick-up, which converts into electrical signals the mechanical wobbles in the groove of a record—for these applications Rochelle salt has been used. More recently, ceramics of the barium titanate type, particularly lead titanate, are finding application. They have greater chemical stability than Rochelle salt but suffer from larger temperature variations.

A very important application of piezoelectricity is the quartz (SiO_2) stabilized oscillator, used to keep radio stations on the right wavelength, with an accuracy of about one part in 10^8 or even 10^9. The principle of operation is very simple. A cuboid of quartz (or any other material for that matter) will have a series of mechanical resonant frequencies of vibration whenever its mechanical length, L, is an odd number of half wavelengths. Thus, the lowest mode will be when $L = \lambda/2$. The mechanical disturbance will travel in the crystal with the velocity of sound, which we shall call v_s. Hence, the frequency of mechanical oscillation will be $f = v_s/\lambda = v_s/2L$. If the ends of the crystal are metallized, it forms a capacitor that can be put in a resonant electrical circuit, having the same resonant frequency, f. The resonant frequency of a transistor oscillator circuit depends a little on things outside the inductor and capacitor

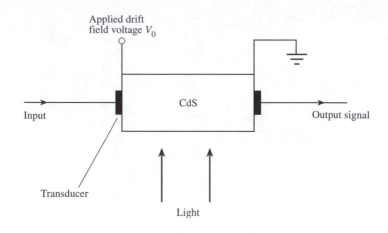

Fig. 10.14

General arrangement of an acoustic amplifier. The applied electric field causes the electrons (photoelectrically generated by the light) to interact with the acoustic wave in the crystal.

Quartz-controlled master oscillators, followed by stages of power amplification, are used in all radio and television transmitters, from the most sophisticated, down to the humblest 'ham'.

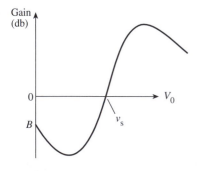

Fig. 10.15

The gain of an acoustic amplifier as a function of the applied voltage. At the voltage where the electron drift velocity is equal to the sound velocity (v_s) the gain changes from negative to positive.

of the resonant circuit. Usually these are small effects that can be ignored, but if you want an oscillator that is stable in frequency to one part in 10^8, things like gain variation in the amplifier caused by supply voltage changes or ageing of components become important; on this scale they are virtually uncontrollable. This is where the mechanical oscillation comes in. We have seen it is a function only of the crystal dimension. Provided the electrical frequency is nearly the same, the electrical circuit will set up mechanical as well as electrical oscillations, linked by the piezoelectric behaviour of quartz. The mechanical oscillations will dominate the frequency that the whole system takes up, simply because the amplifier part of the oscillator circuit works over a finite band of frequencies that is greater than the frequency band over which the mechanical oscillations can be driven. A circuit engineer would say that the $'Q'$ of the mechanical circuit is greater than that of the electrical circuit. The only problem in stabilizing the frequency of the quartz crystal-controlled oscillator is to keep its mechanical dimension, L, constant. This of course, changes with temperature, so we just have to put the crystal in a thermostatically controlled box. This also allows for slight adjustment to the controlled frequency by changing the thermostat setting by a few degrees.

The reverse effect is used in earphones and in a variety of transducers used to launch vibrations in liquids. These include the 'echo-sounder' used in underwater detection and ultrasonic washing and cleaning plants.

I should like to discuss in a little more detail another application in which a piezoelectric material, cadmium sulphide (CdS) is used. The basic set-up is shown in Fig. 10.14. An input electric signal is transformed by an electromechanical transducer into acoustic vibrations that are propagated through the crystal and are converted back into electric signals by the second transducer. Assuming for simplicity that the transducers are perfect (convert all the electric energy into acoustic energy) and the acoustic wave suffers no losses, the gain of the device is unity (0 db if measured in decibels). If the crystal is illuminated, that is mobile charge carriers are created, the measured gain is found to decrease to B (Fig. 10.15). If further a variable d.c. voltage is applied across the CdS crystal, the gain varies as shown.

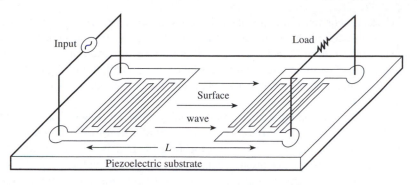

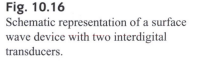

Fig. 10.16
Schematic representation of a surface wave device with two interdigital transducers.

These experimental results may be explained in the following way. The input light creates charge carriers that interact with the acoustic waves via the piezoelectric effect. If the carriers move slower than the acoustic wave, there is a transfer of power from the acoustic wave to the charge carriers. If, on the other hand, the charge carriers move faster than the acoustic wave, the power transfer takes place from the carriers to the acoustic wave, or in other words the acoustic wave is amplified.

Hence, there is the possibility of building an electric amplifier relying on the good services of the acoustic waves. Since we can make electric amplifiers without using acoustic waves, there is not much point using this acoustic amplifier unless it has some other advantages. The main advantage is compactness. The wavelength of acoustic waves is smaller by five orders of magnitude than that of electromagnetic waves, and this makes the acoustic amplifier much shorter than the equivalent electromagnetic travelling wave amplifier (the travelling wave tube). It is unlikely that this advantage will prove very important in practical applications, but one can certainly regard the invention of the acoustic amplifier as a significant step for the following two reasons: (i) it is useful when acoustic waves need to be amplified, and (ii) it has created a feeling (or rather expectation) that whatever electromagnetic waves can do, acoustic waves can do too. So it stimulates the engineer's brains in search of new devices, and a host of new devices may one day appear.

Since the first edition of this book was published, a new device of considerable importance has, in fact, evolved from these ideas. It is the Surface Acoustic Wave device (acronym SAW). We are all familiar with the ripples and surface waves that form on the surfaces of disturbed liquids. It was shown long ago by Lord Rayleigh that analogous waves rippled across the surface of solids, travelling with a velocity near to the velocity of sound in the bulk material.

Why do we want to use surface waves? There are two reasons: (1) they are there on the surface so we can easily interfere with them, and (2) the interfering structures may be produced by the same techniques as those used for integrated circuits. Why do we want to interfere with the waves? Because these devices are used for signal processing which, broadly speaking, requires that the signal which arrives during a certain time interval at the input should be available some time later in some other form at the output.

Let us first discuss the simplest of these devices which will produce the same waveform at the output with a certain delay. The need for such a device is obvious. We want to compare a signal with another one, which is going to arrive

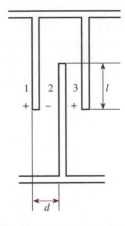

Fig. 10.17

A section of the interdigital transducer.

a bit later. A schematic drawing of the device is shown in Fig. 10.16. The electric signal is transformed by the input transducers into an acoustic one, which travels slowly to the output transducer, where it is duly reconverted into an electric signal again. The input and output transducers are so called interdigital lines. The principles of this transducer's operations may be explained with the aid of Fig. 10.17.

The electric signal appearing simultaneously between fingers 1 and 2, and 2 and 3 excites acoustic waves. These waves will add up in a certain phase depending on the distance d between the fingers. If d is half a wavelength (wavelength and frequency are now related by the acoustic velocity!) and considering that the electric signals are in opposite phase (one is plus–minus, the other one is minus–plus) then the effect from each finger pair adds up, and we have maximum transmission. At twice the frequency, where d is a whole wavelength, the contribution from neighbouring finger pairs will be opposite, and the transducer will produce no net acoustic wave. This shows that the delay line cannot be a very broad-band device, but it shows in addition that by having a frequency-dependent output, we might be able to build a filter. The parameters at our disposal are the length of overlap between fingers (l on Fig. 10.17) and the relative position of the fingers, the former controlling the strength of coupling and the latter the relative phase. It turns out that excellent filters can be produced which are sturdy, cheap, and better than anything else available in the MHz region. In the stone-age days of radio, when we were younger, such a band-pass filter was accomplished by a series of transformers tuned by capacitors. These would appear ridiculously bulky compared to today's microelectronic amplifiers, so a SAW band-pass filter on a small chip of piezoelectric ceramic with photoengraved transducers is compatible in bulk and in technology with integrated circuits, and much cheaper than the old hardware. If you happen to own a colour television set it will most likely have such a filter in the IF amplifier.

What else can one do with SAW devices? One of the applications is quite a fundamental one related to signal processing. The problem to be solved is posed as follows. If a signal with a known waveform and a lot of noise arrives at the input of a receiver, how can one improve the chances of detecting the signal? The answer is that a device exists which will so transform the signal as to make it *optimally* distinguishable from noise. The device is called, rather inappropriately, a matched filter. It turns out that SAW devices are particularly suitable for their realization. They are vital parts of certain radar systems.

We cannot resist (this is now 2003) to mention one more application that will soon appear on the market: a device that monitors the pressure of car tyres while on the road. The heart of the device is a SAW resonator inserted into the tyre. The useful information is contained in its resonant frequency which is pressure dependent. The resonator is interrogated by a pulse from the transmitter attached to the wheel arch. It re-radiates at its resonant frequency providing the pressure information that is then displayed on the instrument panel.

I shall finish this section by mentioning *electrostriction*, a close relative of piezoelectricity. It is also concerned with mechanical deformation caused by an applied electric field, but it is not a linear phenomenon. The mechanical deformation is proportional to the square of the electric field, and the relationship applies to all crystals whether symmetric or not. It has no inverse effect.

The mechanical strain does not produce an electric field via electrostriction. *Biased* electrostriction is, however, very similar to piezoelectricity. If we apply a large d.c. electric field $\mathcal{E}_0$, and a small a.c. electric field, $\mathcal{E}_1$, then the relationship is

$$S = \gamma(\mathcal{E}_0 + \mathcal{E}_1)^2 \cong \gamma\mathcal{E}_0^2 + 2\gamma\mathcal{E}_0\mathcal{E}_1, \qquad (10.46)$$

γ is a proportionality factor.

and we find that the a.c. strain is linearly proportional to the amplitude of the a.c. field.

10.12 Ferroelectrics

There is one more class of dielectrics I should like to mention, which, as well as being piezoelectric, have permanent dipole moments and a polarization that is not necessarily zero when there is no electric field. In fact, they get their name by analogy with ferromagnetics, which have a $B-H$ loop, hysteresis, and remanent magnetism. Ferroelectrics have a $P-E$ loop, hysteresis, and remanent polarized charge as shown in Fig. 10.18. They are very interesting scientifically but so far have not found much application. The high relative dielectric constants of the titanates ($BaTiO_3$ is one of them; it is the example usually given of a ferroelectric material) are used in capacitor-making, where the essential ferroelectric effects of voltage and temperature changes of capacity are usually an embarrassment. At one time it seemed as if the voltage change of dielectric constant would find application in voltage-tunable capacitors, but varactor diodes won the race. Another potential application (as an externally tunable phase-shifter) has not materialized either because of the high hysteresis losses at high frequencies. Both these situations could be changed by improved materials.

There is one ferroelectric device (well, the essential element in it is ferroelectric), which may hit the big time selling in billions and billions in the future. It is the Ferroelectric Random Access Memory known as FRAM. The operational principles are quite simple. There is a FET with source, drain and gate electrodes and a capacitor in the drive line containing a lead-zirconium titanate (PZT) ferroelectric (Fig. 10.19). With the aid of the drive line the capacitor

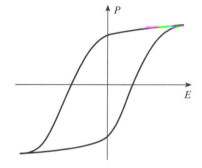

Fig. 10.18
Ferroelectric hysteresis loop.

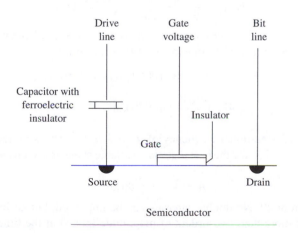

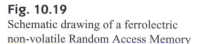

Fig. 10.19
Schematic drawing of a ferroelectric non-volatile Random Access Memory

is biased in one or the other direction causing the domains point up or down storing thereby a logical 0 or 1. Reading is done by applying a short voltage pulse to the capacitor. If the applied field is in the same direction as the domain orientation, then a small current pulse will appear in the bit line. If the applied field is in the opposite direction, the current pulse is much larger, hence the state of the memory can be ascertained. The main merit of this memory cell is that when the pulse is removed, the domains retain their orientation. This is a non-volatile memory greatly in demand. Another contender for the market's attention, providing also non-volatile memory is the Magnetic Random Access Memory that will be briefly discussed in Section 11.9.2.

10.13 Optical fibres

I have tried to show that dielectric properties have importance in optics as well as at the more conventional electrical engineering frequencies. That there are no sacred boundaries in the electromagnetic spectrum is shown very clearly by a fairly recent development in communications engineering. This involves the transmission (guiding) of electromagnetic waves. The principle of operation is very simple. The optical power remains inside the fibre because the rays suffer total internal reflection at the boundaries. This could be done at any frequency, but dielectric waveguides have distinct advantages only in the region around μm wavelengths. The particular configuration used is a fibre of rather small diameter (say $5-50\,\mu$m) made of glass or silica. Whether this transmission line is practical or not will clearly depend on the attenuation. Have we got the formula for the attenuation of a dielectric waveguide? No, we have not performed that specific calculation, but we do have a formula for the propagation coefficient of a plane wave in a lossy medium, and that gives a sufficiently good approximation.

Recall eqn (1.38)

$$k = (\omega^2\mu\epsilon + i\omega\mu\sigma)^{1/2},$$ (10.47)

and assuming this time that

$$\omega\epsilon \gg \sigma,$$ (10.48)

we get the attenuation coefficient

$$k_{\text{imag}} = \frac{1}{2}\frac{\omega\sqrt{\epsilon'}}{c}\frac{\sigma}{\omega\epsilon} = \frac{1}{2}\frac{\omega\sqrt{\epsilon'}}{c}\tan\delta.$$ (10.49)

The usual measure is the attenuation in decibels for a length of one kilometre, which may be expressed from eqn. (10.49) as follows:

$$A = 20\log_{10}\exp(1000\,k_{\text{imag}}) = 8680\,k_{\text{imag}}$$

$$= 4340\frac{\omega\sqrt{\epsilon'}}{c}\tan\delta\,\text{db km}^{-1}.$$ (10.50)

For optical communications to become feasible, A should not exceed $20\,\text{db km}^{-1}$ first pointed out by Kao and Hockham in 1966.

Taking an operational frequency of $f = 3 \times 10^{14}\,$Hz, a typical dielectric constant $\epsilon' = 2.25$, and the best material available at the time with $\tan\delta \approx 10^{-7}$, we get

$$A \approx 4 \times 10^3\,\text{db km}^{-1},$$ (10.51)

a far cry from 20. No doubt materials can be improved, but an improvement in $\tan\delta$ of more than two orders of magnitude looked at the time somewhat

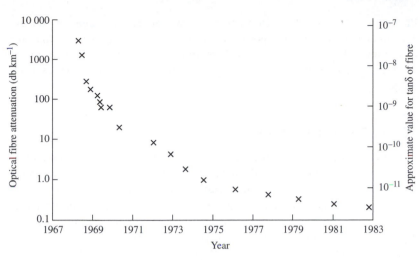

Fig. 10.20
Improvement in optical fibre
attenuation.

beyond the realm of practical possibilities. Nevertheless, the work began, and
Fig. 10.20 shows the improvement achieved. The critical 20 db was reached at
the end of 1969, and by 1983 the figure was down to $0.2\,\text{db}\,\text{km}^{-1}$, an amazing
improvement on a difficult enough initial target.

The most usual material used for these successful fibres has been purified silica (SiO_2) with various dopants to produce the refractive index profile
across the diameter to contain the light ray in a small tube along the axis,
with total internal reflection from the lower refractive index cladding. To
get very low attenuations, the wavelength of the light has to be carefully
controlled. Even with highly purified silica, there are some hydroxyl (OH)
impurity radicals, which are stimulated into vibrations and hence absorb bands
of frequencies. There are 'windows' in this absorption spectrum, one of which
between 1.5 and $1.7\,\mu\text{m}$ wavelength was used to obtain the $0.2\,\text{db}\,\text{km}^{-1}$ result.
A new impetus to the in any case fast-growing optical fibre communications
has been given by the invention of a fibre amplifier that makes orthodox repeaters superfluous. The principles of operation will be explained in Section 12.10
after some acquaintance with lasers.

The lowest attenuation available at the time of writing (spring 2003) is
$0.15\,\text{db/km}^{-1}$ at a wavelength of 1568 nm, not much less than that achieved in
1983. The latest fibres, free of OH absorption, can cover the wavelength range
from about 1275–1625 nm. The amount of information one can get through
these fibres, using Wavelength Division Multiplex, is enormous, much above
present demand. Demultiplexing is usually done by Bragg reflection filters (An
example of such a filter has already been given in Section 10.5. Two further
realizations in integrated optics form will be discussed in Section 13.7).

Up to now about 6×10^8 km of fibre has been laid. The rate of increase has
halved in 2002 but that is believed to be due to the recession rather than to the
coming saturation of the market.

In conclusion, it is worth mentioning that the reduction of attenuation
by four orders of magnitude in less than two decades is about the same
feat as was achieved by strenuous efforts between the first attempts of the

Phoenicians (around 2000 BC) and the dawn of the fibre age. If you ever encounter a problem that appears to be too daunting, remember the story of the optical fibres. It is an excellent illustration of the American maxim (born in the optimism of the post-war years) that the impossible takes a little longer.

10.14   The Xerox process

This great development of the past three decades enables the production of high quality reproductions of documents quickly and easily. This has in turn made decision making more democratic, bureaucrats more powerful, and caused vast forests of trees to disappear to provide the extra paper consumed. Scientifically, the principles are simple. The heart of the Xerox machine is a plate made of a thin layer of amorphous semiconductor on a metal plate. The semiconductor is a compound of As, Se, and Te. It is almost an insulator, so that it behaves like a dielectric, but it is also photo-conductive; that is, it becomes more conducting in the light (remember Section 8.6). The dielectric plate is highly charged electrostatically by brushing it with wire electrodes charged to about 30 kV. The document to be copied is imaged onto the plate. The regions that are white cause the semiconductor to become conducting, and the surface charge leaks away to the earthed metal backing plate. However, where the dark print is imaged, charge persists. The whole plate is dusted with a fine powder consisting of grains of carbon, silica, and a thermosetting polymer. Surplus powder is shaken off, and it adheres only to the highly charged dark regions. A sheet of paper is then pressed onto the plate by rollers. It picks up the dust particles and is then treated by passing under an infra-red lamp. This fuses polymeric particles, which subsequently set, encasing the black C and SiO_2 dust to form a permanent image of the printed document. To clear the plate, it is illuminated all over so that it all discharges, the ink is shaken off, and it is ready to copy something else.

A very simple process scientifically, it works so well because of the careful and very clever technological design of the machine.

10.15   Liquid crystals

I suppose we have heard so much about liquid crystal displays (LCD) in the last couple of decades that we tend to ignore the implied contradiction. Is it a liquid, or is it a crystal? Well, it can be both, and the fact has been known for nearly a hundred years. This particular set of viscous liquids happens to have anisotropic properties due to ordering of long rod-like molecules. If you would like to visualize the short range order of long rods have a look at the photograph shown in Fig. 10.21 taken in Canada by Dr Raynes, an expert on liquid crystals. Mind you, the analogy is not exact. In contrast to logs liquid crystals have in addition strong electric dipole moments and easily polarizable chemical groups, so there is a voltage dependent anisotropy.

There are three types of liquid crystal structures, nematic, cholesteric, and smectic. The one most often used for display purposes belongs to the nematic

Fig. 10.21

type and is known as the twisted nematic display for reasons which will become obvious in a moment.

The liquid crystal is held between two glass plates. If the glass plate is suitably treated, then the molecules next to the surface will align in any desired direction. We can thus achieve that the molecules on the two opposite surfaces will lie at right angles to each other, and those in between change gradually from one orientation to the other [Fig. 10.22(a)]. It is not too difficult to imagine now that light incident with a polarization perpendicular to the direction of the molecules will be able to follow the twist and will emerge with a polarization twisted by 90°. Thus, in the configuration of Fig. 10.22(a), with the two polarizers in place, light will be easily transmitted. When a voltage is applied, the molecules line up in parallel with the electric field [Fig. 10.22(b)], the incident light no longer twists its polarization, and consequently no light transmission takes place.

You may legitimately ask at this stage, why is this a display device? The answer is that this is only a valve, but it can be turned into a display device by placing a mirror behind it. With the voltage off the display is bright because it reflects ambient light. With the voltage on there is no light reflection.

Other types of liquid crystal displays also exist. A colour response can be obtained by the so called guest–host effect, which relies on an anisotropic dye aligning with the liquid crystal molecules.

The biggest advantage of liquid crystal displays is that the voltages they need are compatible with those used for semiconductor devices. At the low character density end, they are gradually moving into the market held by cathode ray tube video display units. Will they ever replace our bulky cathode ray tubes in television receivers? They might. The race for a commercially available flat screen television is on but liquid crystals are not very highly placed at the moment.

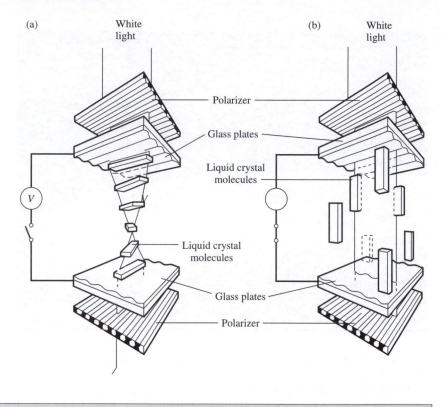

Fig. 10.22

A schematic representation of the operational principles of a twisted nematic display device.

Exercises

10.1. Sketch qualitatively how you would expect the permittivity and loss tangent to vary with frequency in those parts of the spectrum that illustrate the essential properties, limitations, and applications of the following materials; window glass, water, transformer oil, polythene, and alumina.

10.2. What is the atomic polarizability of argon? Its susceptibility at 273 K and 1 atm is 4.35×10^{-4}.

10.3. A long narrow rod has an atomic density 5×10^{28} m^{-3}. Each atom has a polarizability of 10^{-40} farad m^2. Find the internal electric field when an axial field of 1 V/m is applied.

10.4. The energy of an electric dipole in an electric field is given by eqn (10.9). Derive this expression by finding the work done by the electric field when lining up the dipole.

10.5. The tables[†] show measured values of dielectric loss for thoria (ThO$_2$) containing a small quantity of calcium. For this material the static and high frequency permittivities have been found from other measurements to be

$$\epsilon_s = 19.2\epsilon_0, \qquad \epsilon_\infty = 16.2\epsilon_0.$$

$f = 695$ Hz		$f = 6950$ Hz	
T (K)	$\tan \delta$	T (K)	$\tan \delta$
555	0.023	631	0.026
543	0.042	621	0.036
532	0.070	612	0.043
524	0.086	604	0.055
516	0.092	590	0.073
509	0.086	581	0.086
503	0.081	568	0.086
494	0.063	543	0.055
485	0.042	518	0.025
475	0.029	498	0.010

[†] Data taken from PhD thesis of J. Wachtman, University of Maryland, 1962, quoted in *Physics of solids*, Wert and Thomson (McGraw–Hill), 1964.

Assume that orientational polarization is responsible for the variation of tan δ. Use the Debye equations to show that by expressing the characteristic relaxation time as

$$\tau = \tau_0 \exp(H/kT)$$

(where τ_0 and H are constants) both of the experimental curves can be approximated. Find τ_0 and H.

If a steady electric field is applied to thoria at 500 K, then suddenly removed, indicate how the electric flux density will change with time.

10.6. A more general time-varying relationship between the electric displacement D and the electric field $\mathcal{E}$ may be assumed to have the form

$$D + a\frac{dD}{dt} = b\mathcal{E} + c\frac{d\mathcal{E}}{dt}$$

where a, b, c are constants. Determine the values of these constants in terms of ϵ_s, the static dielectric constant, ϵ_∞, the high frequency dielectric constant and τ, the relaxation time for dipoles under constant electric field conditions.

10.7. Figure 10.23 shows two types of breakdown that can occur in the reverse characteristic of a p–n junction diode. The 'hard' characteristic is the desired avalanche breakdown discussed in Chapter 9. The 'soft' characteristic is a fault that sometimes develops with disastrous effect on the rectification efficiency. It has been suggested that this is due to precipitates of metals such as copper or iron in the silicon, leading to local breakdown in high field regions. [Goetzberger and Shockley (1960). 'Metal precipitates in p–n junctions', *J. Appl. Phys.*, **31**, 1821.] Discuss briefly and qualitatively this phenomenon in terms of the simple theories of breakdown given in this chapter.

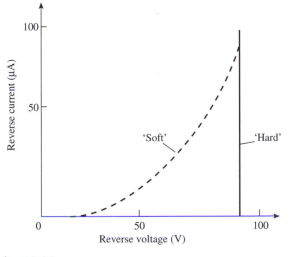

Fig. 10.23

10.8. A capacitor is to be made from a dielectric having a breakdown field strength $\mathcal{E}_b$ and a relative permittivity ϵ_r. The electrodes are metal plates fixed to the sides of a slab (thickness 0.5 mm) of the dielectric. Owing to a slight distortion of one of the plates, one-third of its area is separated from the dielectric

by an air-filled gap of thickness $1\,\mu\text{m}$. The remaining two-thirds of the plate and the whole of the second plate are in intimate contact with the dielectric. The breakdown field $\mathcal{E}_b$ is $2.0\,\text{MV m}^{-1}$ for the dielectric and $3.0\,\text{MV m}^{-1}$ for air, ϵ_r is 1000. Discuss the effect of the gap (compared with a gap-free capacitor), on (a) the capacitance and (b) the breakdown voltage.

10.9. Derive an expression for the gain of the piezoelectric ultrasonic amplifier (Hutson, McFee, and White (1961), *Phys. Rev. Lett.*, 7, 237).
[Hint: In the one-dimensional case we can work in terms of scalar quantities. Our variables are: ϵ, D, T, S, J, N_e. The equations available are: eqns (10.34) and (10.35) for the relationship between the mechanical and electrical quantities, the equation for the electron current including both a conduction and a diffusion term (given in exercise 9.2), the continuity equation for electrons, and one of Maxwell's equations relating D to N_e. Altogether there are five equations and six variables. The missing equation is the one relating strain to stress for an acoustic wave.

It is of the form

$$\frac{\partial^2 T}{\partial z^2} = \rho_m \frac{\partial^2 S}{\partial t^2},$$

where ρ_m is the density of the piezoelectric material.

The gain may be derived in a manner analogous to that adopted in Chapter 1 for the derivation of the dispersion relations for electromagnetic and plasma waves. The steps are as follows:

(i) Assume that the a.c. quantities are small in comparison with the d.c. quantities (e.g. the a.c. electric field is much smaller than the applied d.c. electric field) and neglect the products of a.c. quantities.

(ii) Assume that the a.c. quantities vary as $\exp[-i(\omega t - kz)]$ and reduce the linear differential equation system to a set of linear equations.

(iii) Derive the dispersion equation from the condition that the linear equation system must have a non-trivial solution.

(iv) Substitute $k = \omega/v_s + \delta$ [where $v_s = (c/\rho_m)^{1/2}$ is the velocity of sound in the medium] into the dispersion equation and neglect the higher powers of δ.

(v) Calculate the imaginary part of δ which will determine (by its sign) the growing or attenuating character of the wave. Show that there is gain for $v_0 > v_s$, where v_0 is the average velocity of the electrons.]

10.10. Find the frequency dependence of the complex permittivity due to electronic polarizability only.
[Hint: Write down the equation of motion for an electron, taking into account viscous friction (proportional to

velocity) and restoring force (proportional to displacement). Solve the equations for a driving force of $e\mathcal{E}\exp(-i\omega t)$. Remember that polarization is proportional to displacement and finally find the real and imaginary parts of the dielectric constant as a function of frequency.]

10.11. Using the data for KCl given in Exercise 5.4 estimate a frequency at which the dielectric constant of KCl will relax to a lower value.

[Hint: Around the equilibrium value the force is proportional to displacement (to a first approximation), whence a characteristic frequency can be derived.]

Magnetic materials

Quod superest, agere incipiam quo foedere fiat
naturae, lapis hic ut ferrum ducere possit,
quem Magneta vocant partio de nomine Grai,
Magnetum quia fit patriis in finibus ortus
hunc homines lapidem mirantur; . . .
. . . .
Hoc genus in rebus firmandumst multa prius quam
ipsius rei rationem reddere possis,
et nimium longis ambagibus est adeundum;

Lucretius *De Rerum Natura*

To pass on, I will begin to discuss by what law of nature it comes about that
iron can be attracted by that stone which the Greeks call magnet from the name
of its home, because it is found within the national boundaries of the Magnetes.
This stone astonishes men . . .

In matters of this sort many principles have to be established before you can
give a reason for the thing itself, and you must approach by exceedingly long
and roundabout ways;

11.1   Introduction

There are some curious paradoxes in the story of magnetism that make the
topic of considerable interest. On the one hand, the lodestone was one of the
earliest known applications of science to industry—the compass for shipping;
and ferromagnetism is of even more crucial importance to industrial society
today than it was to early navigators. On the other hand, the origin of magnetism
eluded explanation for a long time, and the theory is still not able to account
for all the experimental observations.

It is supposed that the Chinese used the compass around 2500 BC. This may
not be true, but it is quite certain that the power of lodestone to attract iron was
known to Thales of Miletos in the sixth century BC. The date is put back another
two hundred years by William Gilbert (the man of science in the court of Queen
Elizabeth the First), who wrote in 1600 that 'by good luck the smelters of iron or
diggers of metal discovered magnetite as early as 800 BC.' There is little doubt
about the technological importance of ferromagnetism today. In the United
Kingdom as much as 7×10^{11} W of electricity are generated at times; electrical
power in this quantity would be hopelessly impractical without large quantities
of expertly controlled ferromagnetic materials. Evidence for the statement that
the theory is not fully understood may be obtained from any honest man who
has done some work on the theory of magnetism.

11.2 Macroscopic approach

By analogy with our treatment of dielectrics, I shall summarize here briefly the main concepts of magnetism used in electromagnetic theory. As you know, the presence of a magnetic material will enhance the magnetic flux density. Thus the relationship

$$\mathbf{B} = \mu_0 \mathbf{H}, \tag{11.1}$$

M is the magnetic dipole moment per unit volume, or shortly, magnetization.

valid in a vacuum, is modified to

$$\mathbf{B} = \mu_0 (\mathbf{H} + \mathbf{M}) \tag{11.2}$$

in a magnetic material. The magnetization is related to the magnetic field by the relationship

χ_m is the magnetic susceptibility.

$$\mathbf{M} = \chi_m \mathbf{H}. \tag{11.3}$$

Substituting eqn (11.3) into eqn (11.2) we get

μ_r is called the *relative permeability*.

$$\mathbf{B} = \mu_0 (1 + \chi_m) \, \mathbf{H} = \mu_0 \mu_r \mathbf{H}. \tag{11.4}$$

11.3 Microscopic theory (phenomenological)

Our aim here is to express the macroscopic quantity, M, in terms of the properties of the material at atomic level. Is there any mechanism at atomic level that could cause magnetism? Reverting for the moment to the classical picture, we can say yes. If we imagine the atoms as systems of electrons orbiting round protons, they can certainly give rise to magnetism. We know this from electromagnetic theory, which maintains that an electric current, I, going round in a plane will produce a magnetic moment*

* It is an unfortunate fact that the usual notation is μ both for the permeability and for the magnetic moment. I hope that, by using the subscripts 0 and r for permeability and m for magnetic moment, the two things will not be confused.

$$\mu_m = IS, \tag{11.5}$$

where S is the area of the current loop. If the current is caused by a single electron rotating with an angular frequency ω_0, then the current is $e\omega_0/2\pi$, and the magnetic moment becomes

$$\mu_m = \frac{e\omega_0}{2} r^2, \tag{11.6}$$

where r is the radius of the circle. Introducing now the angular momentum

$$\Pi = mr^2 \omega_0 \tag{11.7}$$

Remember that the charge of the electron is negative; the magnetic moment is thus in a direction opposite to the angular momentum.

we may rewrite eqn (11.6) in the form

$$\mu_m = \frac{e}{2m} \Pi. \tag{11.8}$$

We now ask what happens when an applied magnetic field is present. Consider a magnetic dipole that happens to be at an angle θ to the direction of the magnetic field (Fig. 11.1). The magnetic field produces a torque $\boldsymbol{\mu}_m \times \mathbf{B}$ on the magnetic dipole. Since the torque is perpendicular both to $\boldsymbol{\mu}_m$ and $\mathbf{B}$, the change in the angular momentum will also be perpendicular, causing the magnetic dipole to precess around the magnetic field. From kinetics we can easily show that the frequency of precession is

$$\omega_L = \frac{eB}{2m}, \tag{11.9}$$

which is usually called the *Larmor frequency*. If the magnetic dipole precesses, some electric charge must go round. So we could use eqn (11.6) to calculate the magnetic moment due to the precessing charge. Replace ω_0 by ω_L; we get

$$(\mu_m)_{ind} = \frac{Br^2 e^2}{4m}, \tag{11.10}$$

where r is now the radius of the precessing orbit. The sign of this induced magnetic moment can be deduced by remembering Lenz's law. It must oppose the magnetic field responsible for its existence.

We are now in a position to obtain M from the preceding microscopic considerations. If there are N_a atoms per unit volume and each atom contains Z electrons, the total induced magnetic dipole moment is

$$M = N_a Z (\mu_m)_{ind}. \tag{11.11}$$

Hence, the magnetic susceptibility is

$$\chi_m = \frac{M}{H} = -\frac{N_a Z e^2 r^2 \mu_0}{4m}. \tag{11.12}$$

Rarely exceeding 10^{-5}, χ_m given by the above equation is a small number, but the remarkable thing is that it is negative. This is in marked contrast with the analogous case of electric dipoles, which invariably give a positive contribution.* The reason for this is that the electric dipoles line up, whereas the magnetic dipoles precess in a field. Magnetic dipoles can line up as well. The angle of precession will stay constant in the absence of losses but not otherwise. In the presence of some loss mechanism the angle of precession gradually becomes smaller and smaller as the magnetic dipoles lose energy; in other words, the magnetic dipoles do line up. They will not align completely because they occasionally receive some energy from thermal vibrations that frustrates their attempt to line up. This is exactly the same argument we used for dielectrics, and we can therefore use the same mathematical solution. Replacing the electric energy in eqn (10.13) by the magnetic energy, we get the average magnetic moment in the form

$$\langle \mu_m \rangle = \mu_m L(a), \qquad a = \frac{\mu_m \mu_0}{kT} H. \tag{11.13}$$

Denoting by N_m the number of magnetic dipoles per unit volume, we get for the total magnetic moment

$$M = N_m \langle \mu_m \rangle. \tag{11.14}$$

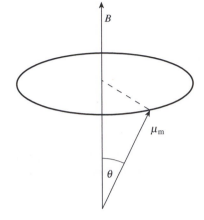

Fig. 11.1
A magnetic dipole precessing around a static magnetic field.

* The electric susceptibility can also be negative, but that is caused by free charges and not by electric dipoles.

At normal temperatures and reasonable magnetic fields, $a \ll 1$ and eqn (11.14) may be expanded to give

$$M = \frac{N_m \mu_m^2 \mu_0 H}{3kT}, \qquad (11.15)$$

leading to

$$\chi_m = \frac{N_m \mu_m^2 \mu_0}{3kT}. \qquad (11.16)$$

Here χ_m is definitely positive and varies inversely with temperature.

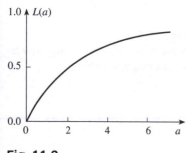

Fig. 11.2
The Langevin function, $L(a)$.

At the other extreme of very low temperatures all the magnetic dipoles line up; this can be seen mathematically from the fact that the function $L(a)$ (plotted in Fig. 11.2) tends to unity for large values of a. The total magnetic moment is then

$$M_s = N_m \mu_m, \qquad (11.17)$$

which is called the *saturation magnetization* because this is the maximum contribution the material can provide.

We have now briefly discussed two distinct cases: (i) when the induced magnetic moment opposes the magnetic field, called *diamagnetism*; and (ii) when the aligned magnetic moments strengthen the magnetic field, called *paramagnetism*. Both phenomena give rise to small magnetic effects that are of little use when the aim is the production of high magnetic fluxes. What about our most important magnetic material, iron? Can we explain its properties with the aid of our model? Not in its present state. We have to modify our model by introducing the concept of the internal field. This is really the same sort of thing that we did with dielectrics. We said then that the local electric field differs from the applied electric field because of the presence of the electric dipoles in the material. We may argue now that in a magnetic material the local magnetic field is the sum of the applied magnetic field and the internal magnetic field, and we may assume (as Pierre Weiss did in 1907) that this internal field is proportional to the magnetization, that is

$$H_{int} = \lambda M. \qquad (11.18)$$

λ is called the *Weiss constant*.

Using this newly introduced concept of the internal field, we may replace H in eqn (11.13) by $H + \lambda M$ to obtain for the magnetization:

$$\frac{M}{N_m \mu_m} = L \left\{ \frac{\mu_m \mu_0}{kT} (H + \lambda M) \right\}. \qquad (11.19)$$

Thus for any given value of H we need to solve eqn (11.19) to get the corresponding magnetization. It is interesting to note that eqn (11.19) still has solutions when $H = 0$. To prove this, let us introduce the notations,

$$b = \mu_m \mu_0 \lambda M / kT, \qquad \theta = N_m \mu_m^2 \mu_0 \lambda / 3k, \qquad (11.20)$$

and

$$\frac{M}{N_m \mu_m} = \frac{T \mu_m \mu_0 \lambda M / kT}{3 N_m \mu_m^2 \mu_0 \lambda / 3 k} = \frac{Tb}{3\theta}. \qquad (11.21)$$

We can now rewrite eqn (11.19), for the case $H = 0$, in the form

$$\frac{T}{3\theta}b = L(b). \qquad (11.22)$$

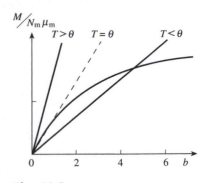

Plotting both sides of the above equation (Fig. 11.3) it becomes apparent that there is a solution when $T < \theta$ but no solution when $T > \theta$.

What does it mean if there is a solution? It means that M may be finite for $H = 0$; that is, the material can be magnetized in the absence of an external magnetic field. This is a remarkable conclusion. We have managed to explain, with a relatively simple model, the properties of permanent magnets.

Note that there is a solution only below a certain temperature. Thus,if our theory is correct, permanent magnets should lose their magnetism above this temperature. Is this borne out by experiment? Yes, it is a well-known experimental fact (discovered by Gilbert) that permanent magnets cease to function above a certain temperature. What happens when $T > \theta$? There is no magnetization for $H = 0$, but we do get some magnetization for finite H.

The mathematical solution may be obtained from eqn (11.19), noting that the argument of the Langevin function is small, and we may use again the approximation $L(a) \cong a/3$, leading to

Fig. 11.3
Graphical solution of eqn (11.22). There is no solution, that is the curves do not intersect each other, for $T > \theta$.

$$\frac{M}{N_m\mu_m} = \frac{\mu_m\mu_0}{3kT}(H + \lambda M), \qquad (11.23)$$

which may be solved for M to give

$$M = \frac{N_m\mu_m^2\mu_0/3k}{T - N_m\mu_m^2\mu_0\lambda/3k}H = \frac{C}{T - \theta}H, \qquad (11.24)$$

where

$$C = \frac{N_m\mu_m^2\mu_0}{3k} \qquad (11.25)$$

is called the *Curie constant* and θ is the *Curie temperature*. The $M-H$ relationship is linear, and the susceptibility is given by

$$\chi_m = \frac{C}{T - \theta}. \qquad (11.26)$$

Thus, we may conclude that our ferromagnetic material (the name given to materials like iron that exhibit magnetization in the absence of applied magnetic fields) becomes paramagnetic above the Curie temperature.

We have now explained all the major experimental results on magnetic materials. We can get numerical values if we want to. By measuring the temperature where the ferromagnetic properties disappear, we get θ (it is 1043 K for iron) and by plotting χ_m above the Curie temperature as a function of $1/(T - \theta)$ we get C (it is about unity for iron). With the aid of eqns (11.20) and (11.25) we can now express the unknowns μ_m and λ as follows:

$$\lambda = \frac{\theta}{C} \quad \text{and} \quad \mu_m = \left(\frac{3\,kC}{N_m\mu_0}\right)^{1/2}. \qquad (11.27)$$

Taking $N_m = 8 \times 10^{28} m^{-3}$ and the above-mentioned experimental results, we get for iron:

$$\lambda \cong 1000 \quad \text{and} \quad \mu_m \cong 2 \times 10^{-23} \, Am^2. \tag{11.28}$$

The value for the magnetic dipole moment of an atom seems reasonable. It would be produced by an electron going round a circle of 0.1 nm radius about 10^{15} times per second. One can imagine that, but it is much harder to swallow a numerical value of 1000 for λ. It means that the internal field is 1000 times as large as the magnetization. When all the magnetic dipoles line up, M comes to about $10^6 A \, m^{-1}$, leading to a value for the internal flux density $B_{int} = \mu_0 \lambda M = 10^3$ T, which is about an order of magnitude higher than the highest flux density ever produced. Where does such an enormous field come from? It is a mysterious problem, and we shall leave it at that for the time being.

11.4   Domains and the hysteresis curve

We have managed to explain the spontaneous magnetization of iron, but as a matter of fact, freshly smelted iron does not act as a magnet. How is this possible? If, below the Curie temperature, all the magnetic moments line up spontaneously, how can the outcome be a material exhibiting no external magnetic field? Weiss, with remarkable foresight, postulated the existence of a domain structure. The magnetic moments do line up within a domain, but the magnetizations of the various domains are randomly oriented relative to each other, leading to zero net magnetism.

The three most important questions we need to answer are as follows:

1. Why does a domain structure exist at all?
2. How thick are the domain walls?
3. How will the domain structure disappear as the magnetic field increases?

It is relatively easy to answer the first question. The domain structure comes about because it is energetically unfavourable for all the magnetic moments to line up in one direction. If it were so then, as shown in Fig. 11.4(a), there would be large magnetic fields and, consequently, a large amount of energy outside the material. This magnetic energy would be reduced if the material would break up into domains as shown in Fig. 11.4(b)–(e). But why would this process ever stop? Should not the material break up into as many domains as it possibly could, down to a single atom? The reason why this would not happen is because domains must have boundaries and, as everyone knows, it is an expensive business to maintain borders of any kind. Customs officials must be paid, not mentioning the cost of guard towers and barbed wires, with which some borders are amply decorated. Thus some compromise is necessary. The more domains there are, the smaller will be the magnetic energy outside, but the more energy will be needed to maintain the boundary walls. When putting up one more wall needs as much energy as the achieved reduction of energy outside, an equilibrium is reached, and the energy of the system is minimized.

We have now managed to provide a reasonable answer to question (1). It is much more difficult to describe the detailed properties of domains, and their

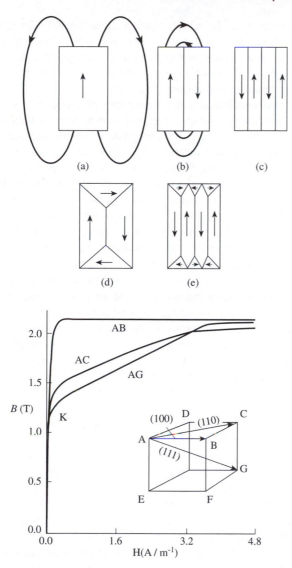

Fig. 11.4
The formation of domains (from
C. Kittel, *Introduction to solid state
physics*, John Wiley, New York).

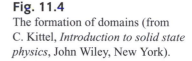

Fig. 11.5
Magnetization curve of single
crystal iron in three different
crystallographical directions.

dependence on applied magnetic fields. We must approach the problem, as
Lucretius said, 'by exceedingly long and roundabout ways'.

However much I dislike talking about crystal structure, there is no escape
now because magnetic properties do depend on crystallographic directions. I
am not suggesting that magnets are ever made of single-crystal materials, but
in order to interpret some of the properties of ordinary polycrystalline magnets,
we have to know the magnetic properties of the single crystals.

In Fig. 11.5 the magnetization curve (B against H) of iron is plotted for
three different directions in the crystal. It may be seen that magnetization is
relatively easy in the AB direction and harder in the AC and AG directions, or
in other words, it is easier to magnetize iron along a cube edge than along a face
or a body diagonal. This does not mean, of course, that all magnetic materials

follow the same pattern. In nickel, another cubic crystal, the directions of
easy and difficult magnetization are the other way round. What matters is that
in most materials magnetization depends on crystallographic directions. The
phenomenon is referred to as anisotropy, and the internal forces which bring
about this property are called anisotropy forces.

Let us now see what happens at the boundary of two domains, and choose
for simplicity two adjacent domains with opposite magnetizations as shown in
Fig. 11.6. Note that the magnitude of the magnetic moments is unchanged during
the transition, but they rotate from an 'up' position into a 'down' position. Why
is the transition gradual? The forces responsible for lining up the magnetic
moments (let us call them for the time being 'lining up' forces) try to keep
them parallel. If we wanted a sudden change in the direction of the magnetic
moments, we should have to do a lot of work against the 'lining up' forces, and
consequently there would be a lot of 'lining up' energy present. The anisotropy
forces would act the opposite way. If 'up' is an easy direction (the large majority
of domains may be expected to line up in an easy direction), then 'down' must
also be an easy direction. Thus, most of the directions in between must be looked
upon unfavourably by the anisotropy forces. If we want to rotate the magnetic
moments, a lot of work needs to be done against the anisotropy forces, resulting
in large anisotropy energy. According to the foregoing argument, the thickness
of the boundary walls will be determined by the relative magnitudes of the
'lining up' and the anisotropy forces in the particular ferromagnetic material.
If anisotropy is small, the transition will be slow and the boundary wall thick,
say $10\,\mu\text{m}$. Conversely, large anisotropy forces lead to boundary walls which
may be as thin as $0.3\,\mu\text{m}$.

We are now ready to explain the magnetization curves of Fig. 11.5. When
the piece of single-crystal iron is unmagnetized, we may assume that there are
lots of domains, and the magnetization in each domain is in one of the six easy
directions. Applying now a magnetic field in the AB direction, we find that,
as the magnetic field is increased, the domains which lay originally in the AB
direction will increase at the expense of the other domains, until the whole
material contains only one single domain. Since the domain walls move easily,
the magnetic field required to reach saturation is small.

What happens if we apply to the same single-crystal iron a magnetic field in
the AG direction? First, the domain walls will move just as in the previous case

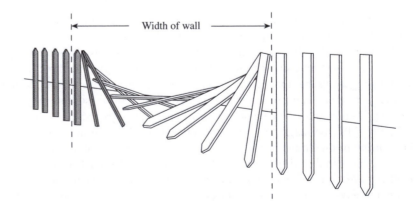

Fig. 11.6
Rotation of magnetic moments at a
domain wall.

until only three easy directions are left, namely AB, AD, and AE, that is those with components in the AG direction. This may be achieved with very little magnetic field, but from then on (K in Fig. 11.5) the going gets hard. In order to increase the magnetization further, the magnetic moments need to change direction, which can only happen if the internal anisotropy forces are successfully overcome. This requires more effort, hence the slope of the magnetization curve changes, and saturation will only be achieved at greater magnetic fields.

Is this explanation still correct for polycrystalline materials? Well, a polycrystalline material contains lots of single crystal grains, and the above argument applies to each of the single crystals; thus the magnetization curve of a polycrystalline material should look quite similar to that of a single-crystal material in a difficult direction. As you know from secondary school, this is not the case. Figure 11.5 does not tell the whole story. The magnetization curve of a typical ferromagnetic material exhibits *hysteresis*, as shown in Fig. 11.7. Starting with a completely demagnetized material, we move up the curve along 2, 3, 4, 5 as the magnetic field is increased. Reducing then the magnetic field, we get back to point 6, which is identical with point 4, but further decrease takes place along a different curve. At 7 there is no applied magnetic field, but B is finite. Its value, $B = B_r$, is the so called *remanent* flux density. Reducing further, the magnetic field B takes the values along 8, 9, 10. Returning from 10, we find that 11 is identical with 9 and then proceed further along 12 and 13 to reach finally 4.

The loop 4, 7, 8, 9, 12, 13, 4 is referred to as the hysteresis loop. It clearly indicates that the magnetization of iron is an irreversible phenomenon.

The paths 4, 5 and 9, 10 suggest that rotation from easy into difficult directions is reversible, thus the causes of irreversibility should be sought in domain movement. Because of the presence of all sorts of defects in a real material, the domain walls move in little jerks, causing the magnetization to increase in a discontinuous manner (region 2, 3 magnified in Fig. 11.7). The walls get stuck once in a while and then suddenly surge forward, setting up in the process some eddy currents and sound waves, which consume energy. If energy is consumed, the process cannot be reversible, and that is the reason for the existence of the hysteresis loop.

Note that the value of H at 13 is called the *coercivity*, denoted by H_c. It represents the magnetic field needed for the flux density to vanish.

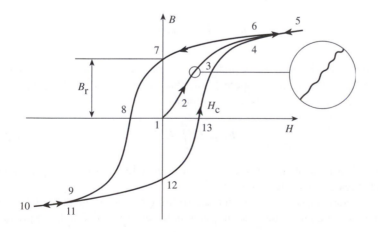

Fig. 11.7
The magnetization curve of a typical ferromagnetic material.

Is it possible to describe more accurately the movement of domains? One can go indeed a little further by taking into account the effect of *magnetostriction*, which, as you may guess, is the magnetic counterpart of electrostriction. Strictly speaking, one should distinguish between magnetostriction and piezomagnetism, the magnetic counterpart of piezoelectricity. But biased magnetostriction (see discussion on biased electrostriction in Section 10.11) is phenomenologically equivalent to piezomagnetism, and piezomagnetism has not been much investigated anyway; thus most authors just talk about magnetostriction. Disregarding the problem of nomenclature, the relevant fact is that when a magnetic field is applied, the dimensions of the material change,* and conversely, strain in the material leads to changes in magnetization and may also affect the directions of easy magnetization. Now if the material exhibits a large anisotropy and stresses are present as well, then there will be *local* easy directions resisting the movement of domain walls everywhere. The stresses may be caused by the usual defects in crystals and particularly by impurities. In addition, a cluster of non-magnetic impurity atoms might be surrounded by domains (see Fig. 11.8). This is a stable configuration which cannot be easily changed.

How can we classify magnetic materials? There is a simple division into soft and hard magnetic materials. Why soft and hard? Well, the hard materials are those which are hard to magnetize and demagnetize. So materials which are easy to magnetize and demagnetize should be called *easy* materials. In fact, they are called soft materials, and there is nothing we can do about that. We have to remember, though, that these are only very tenuously related to mechanical properties, which may also be hard and soft.

* It is, incidentally the cause of the humming noise of transformers.

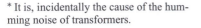

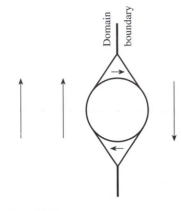

Fig. 11.8

Non-magnetic impurity surrounded by a domain.

11.5　Soft magnetic materials

Their main role is to enhance the magnetic effect produced by a current carrying coil. So, obviously, they should have large saturation magnetization and large permeability. If the material is subjected to alternating voltages, then an important consideration is to reduce losses caused by the induced eddy currents, which can be done by increasing resistivity. What else is needed in order to reduce losses? A narrow hysteresis loop is needed as shown below.

The energy dissipated in a coil for a period T may be expressed with the aid of the current and voltage as

$$E_{\mathrm{d}} = \int_0^T V(t)i(t)\,\mathrm{d}t. \tag{11.29}$$

Now, using Faraday's law (that the voltage is proportional to the derivative of the flux density) and Ampère's law (that the magnetic field is proportional to current) eqn (11.29) may be rewritten as

$$E_{\mathrm{d}} = C\int H\,\mathrm{d}B, \tag{11.30}$$

where C is a constant. Thus, clearly, the energy loss per cycle is proportional to the area of the hysteresis loop.

The most important parameter determining the desirable properties of soft magnetic materials is the frequency at which they are used. For d.c. applications the best material is the one with the largest saturation magnetization. As the frequency increases, it is still important to have large saturation magnetization, but low coercitivity is also a requirement. At high frequencies, considering that eddy current losses are proportional to the square of the frequency, the most important property is high resistivity.

Do losses matter? In practical terms this is probably the most important materials science problem that we have touched upon. Something like many millions of megawatts of electricity is being generated around the world, all by generators with hysteresis losses of order 0.5–1.0%. Then a large fraction of this electricity goes into motors and transformers with more iron losses. If all inventors were paid a 1% royalty on what they saved the community, then a good way to become rich would be to make a minute improvement to magnetic materials. Is there any good scientific way to set about this? Not really. We know that anisotropy, magnetostriction, and local stresses are bad, but we cannot start from first principles and suggest alloys which will have the required properties. The considerable advances that have been made in magnetic materials have largely been achieved by extensive and expensive trial and error. To Gilbert's seventeenth-century crack about 'good luck' we must add *diligence* for the modern smelters of iron. The currently used phrase is actually 'enlightened empiricism'.

Iron containing silicon is used in most electrical machinery. An alloy with about 2% silicon, a pinch of sulfur, and critical cold rolling and annealing processes is used for much rotating machinery. Silicon increases the resistivity, which is a good thing because it reduces eddy-current losses. Iron with a higher silicon content is even better and can be used in transformer laminations, but it is mechanically brittle and therefore no good for rotating machinery. Where small quantities of very low-loss material are required and expense is not important, as for radio-frequency transformers, Permalloy [78.5% Ni, 21.5% Fe] is often used. A further improvement is achieved in the material called 'Supermalloy' which contains a little molybdenum and manganese as well. It is very easily magnetizable in small fields [Fig. 11.9(a)] and has no magnetostriction.

We may now mention a fairly new and rather obvious trick. If anisotropy is bad, and anisotropy is due to crystal structure, then we should get rid of the crystal structure. What we obviously need is an amorphous material. How can we produce an amorphous material? We can produce it by cooling the melt rapidly, so that the liquid state disorder is frozen in. The key word is 'rapidly'. In fact, the whole process is called Rapid Solidification Technology, abbreviated as RST. The cooling should proceed at a speed of about a million degrees per second, so the technological problems have not been trivial. In the first successful commercial solution a stream of molten metal is squirted on a cooled rotating drum, followed usually by a stress relief anneal at about 300°C. The resulting magnetic material has the form of long thin ribbons typically about 50 μm thick and a few millimetres wide. New production methods, for example the planar flow casting method, in which a stable rectangular melt 'puddle' feeds the material into the drum, have led to further improvements. It is now possible to obtain a uniform ribbon with a thickness of 20–30 μm and up to 20 cm wide. The main advantage of amorphous materials is that they can be produced easily and relatively cheaply, with magnetic properties

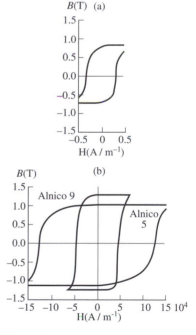

Fig. 11.9

Hysteresis loops of (a) Supermalloy and (b) Alnico 5 and 9. Note the factor 10^5 between the horizontal scales of (a) and (b).

nearly as good as those of commercial alloys, which require careful melting and elaborate sequences of rolling and annealing. The presently available amorphous materials have not quite reached the quality of supermalloy, but they are quite close. The cobalt-based commercially available 2714 A material has a saturation magnetization of 0.5 T with a maximum permeability of a million. Another one, known as 2605S-3 A made of iron and chromium has a saturation magnetization of 1.4 T and a maximum permeability over a quarter of a million.

The latest line of soft magnetic materials are the nanocrystalline alloys with grain sizes of the order of 10 nm. They have been around for about 10 years. Typical representatives are Fe–B–Si–Cu–Nb alloys, which may reach relative permeabilities over 100 000. The excellent soft magnetic properties may be explained by the reduction in effective crystal anisotropy expected when grain sizes are reduced below the bulk-domain wall thickness.

The situation is somewhat different in power applications, such as transformers. There the traditional materials are cheaper, but amorphous materials may still represent the better choice on account of lower losses; their higher cost may be offset in the long term by lower power consumption (or even possible future legislation in some countries requiring higher efficiency in electrical equipment).

At higher frequencies, as mentioned before, high resistivities are required for which a family of ferrites with chemical formula $MO \cdot Fe_2O_3$ (where M is a metal, typically Ni, Al, Zn, or Mg) is used. If the metal M is iron, the material is iron ferrite, Fe_3O_4, the earliest-known magnetic material.

Ferrites are usually manufactured in four stages. In the first stage the material is produced in the form of a powder with the required chemical composition. In the second stage the powder is compressed, and the third stage is sintering to bind the particles together. The fourth stage is machining (grinding, since the material is brittle) to bring the material to its final shape.

For the properties of a number of soft magnetic materials see Table 11.1.

11.6   Hard magnetic materials (permanent magnets)

What kind of materials are good for permanent magnets? Well, if we want large flux density produced, we need a large value of B_r. What else? We need a large H_c. Why? A rough answer is that the high value of B_r needs to be protected. If for some reason we are not at the $H = 0$ point, we do not want to lose much flux, therefore the $B - H$ curve should be as wide as possible.

A more rigorous argument in favour of large H_c can be produced by taking account of the so-called demagnetization effect, but in order to explain that, I shall have to make a little digression and go back to electromagnetic theory. First of all, note that in a ring magnet [Fig. 11.10(a)] $B = B_r = $ constant everywhere in the material to a very good approximation. Of course, such a permanent magnet is of not much interest because we cannot make any use of the magnetic flux. It may be made available, though, by cutting a narrow gap in the ring, as shown in Fig. 11.10(b). What will be the values of B and H in the gap? One may argue from geometry that the magnetic lines will not spread out (this is why we chose a narrow gap, so as to make the calculations simpler) and the flux density in the gap will be the same as in the magnetic material.

Table 11.1 *Major families of soft magnetic materials with typical properties*

Category	B_s (T)	ρ ($\mu\Omega - m$)	μ_{max}	Typical core loss, W kg^{-1} measured at f (Hz)	Applications, notes
A. Steels					
lamination (low C)	2.1–2.2	0.4		2.0 (60)	Inexpensive fractional hp motors
non-oriented (2% Si)	2.0–2.1	0.35		2.7 (60)	High efficiency motors
convent. grain oriented (CGO M-4)	2.0	0.48	5 000	0.9 (60)	50/60 Hz distribution transformers
high grain oriented (HGO)	2.0	0.45		1.2 (60)	50/60 Hz DTs: high design B_{max}
B. Fe–(Ni, Co) alloys					
40–50 Ni	1.6	0.48	150 000	110 (50 k)	
77–80 Ni (square permalloy)	1.1	0.55	150 000	40 (50 k)	High μ, used as thin ribbon
79 Ni–4 Mo (4–79 Mo permalloy, supermalloy)	0.8	0.58	10^6	33 (50 k)	Highest μ/lowest core loss of any metallic material
49 Co–2 V (permendur, supermendur)	2.3	0.35	50 000	2.2 (60)	Highest B_s of commercial soft magnetic material
C. Ferrites					
MnZn	0.5	2×10^6	6 000	35 (50 k)	Power supply inductors, transformers
NiZn	0.35	10^{10}	4 000		MHz applications

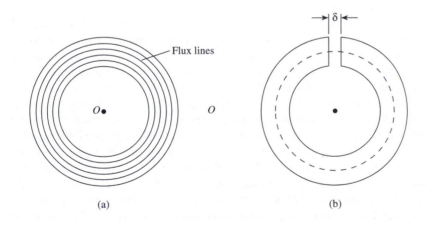

(a)   (b)

Fig. 11.10
(a) Magnetic field lines inside a permanent magnet. (b) The same magnet with a narrow gap.

But, and this is the question of interest, will the flux density be the same in the presence of the gap as in its absence? Without the gap, $B = B_r$ (Fig. 11.11). If the value of flux density is denoted by $B = B_{r1}$ in the presence of the gap, the magnetic field in the gap will be $H_g = B_{r1}/\mu_0$.

If you can remember Ampère's law, which states that the line integral of the magnetic field in absence of a current must vanish for a closed path, it follows that

$$H_g \delta + H_m l = 0 \qquad (11.31)$$

H_m is the magnetic field in the material, and δ and l are the lengths of the paths in the gap and in the material, respectively.

From the above equations we get

$$B_{r1} = -\frac{\mu_0 l}{\delta} H_m. \qquad (11.32)$$

But remember that the relationship between B_{r1} and H_m is given also by the hysteresis curve. Hence, the value of B_{r1} may be obtained by intersecting the hysteresis curve by the straight line of eqn (11.32) as shown in Fig. 11.11. The aforegoing construction depends on the particular geometry of the permanent magnet we assume, but similar 'demagnetization' will occur for other geometries as well. Hence, we may conclude in general that in order to have a large, useful flux density, the $B - H$ curve must be wide. We may therefore adopt, as a figure of merit, the product $B_r H_c$ or, as it is more usual, the product $(BH)_{max}$ in the second quadrant.

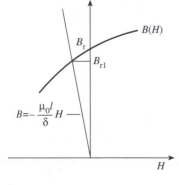

Fig. 11.11
Construction for finding B_{r1}.

How can one achieve a large value of H_c? It is relatively easy to give an answer in principle. All the things which caused the quality of soft materials to deteriorate are good for permanent magnets. In particular, when a domain gets stuck on an impurity, that is bad for a soft magnetic material but good for the hard variety. An obvious way to include impurities is to add some carbon. High-carbon steels were indeed *the* permanent magnet materials in the nineteenth century until displaced by tungsten steels towards the end of the century.

The simplest permanent magnet one could conceive in principle would be a single crystal of a material that has a large anisotropy and has only one axis of easy magnetization. The anisotropy may be characterized by an effective field H_a, which attempts to keep the magnetization along the axis. If a single crystal material is magnetized along this axis, and a magnetic field is applied in the opposite direction, nothing should happen in principle until the field H_a is reached, and then, suddenly, the magnetization of the whole crystal should reverse. Going one step further in this direction, one could claim that any collection of anisotropic particles that are too small to contain a domain wall (having a diameter of the order of 20 nm) will have large coercivity. This idea, due to Stoner and Wohlfarth, was the inspiration behind many attempts to make better permanent magnets. In particular, the so-called Elongated Single Domain (ESD) magnets owe their existence to the above concept. It is also likely that elongated particles play a significant role in the properties of the Alnico series of alloys, which contain aluminium, nickel, and cobalt besides iron. They first appeared in the early 1930s but have been steadily improving ever since. A major early advance was the discovery that cooling in a magnetic field produced anisotropic magnets with improved properties in the field-annealed direction. The hysteresis curves of their best-known representatives (Alnico 5 and 9) are shown in Fig. 11.9(b).

Ferrites are also used for hard magnetic materials in the form $MO \cdot (Fe_2O_3)_6$ (M = Ba, Sr, or Pb). They were introduced in the 1950s. They have been steadily growing in tonnage ever since, overtaking the Alnico alloys in the late 1960s and rising in the late 1980s to 97.4% of world production (note that in value they represent only about 60%). Their high coercivity derives from the high anisotropy of the hexagonal phase of the materials. They have many advantages: they are cheap, easily manufactured, chemically stable, and have low densities. Their disadvantages are the relatively low remanence and declining performance for even moderate rises in temperature.

One might be forgiven for believing that the late entry of rare-earth magnets into the market place was due to their rarity. In fact, rare-earth elements are not particularly rare, but they occur in mixtures with each other which cannot easily be separated owing to their similar chemical properties. However, once the problem of separation was satisfactorily solved (early 1970s) these magnets could be produced at an economic price. Their first champion was the samarium–cobalt alloy $SmCo_5$, produced by powdering and sintering. The next major advance owed its existence to political upheavals in Africa. Uncertainties in the supply of cobalt, not to mention a five-fold price increase, lent some urgency to the development of a cobalt-free permanent magnet. Experiments involving boron led to new (occasionally serendipitous) discoveries, culminating in the development of the $Nd_2Fe_{14}B$, which became known as 'neo' magnets, referring not so much to their novelty (although new they were) but to their neodymium content. They hold the current record of $(BH)_{max} = 400 \, kJ \, m^{-3}$ obtained under laboratory conditions. The commercially available value is about $300 \, kJ \, m^{-3}$, as may be seen in Table 11.2. They have, though, the major disadvantage of a fairly low Curie temperature. Note that these new materials have radically different looking hysteresis curves as shown in Fig. 11.12 for the second quadrant only.

Let us see now two rather revealing indications of progress. As shown in Fig. 11.13, the introduction of new magnetic materials led to quite significant shrinking of the magnetic circuit of a moving-coil meter. Our second example is the historical development of the maximum value of $(BH)_{max}$ shown in Fig. 11.14. The points labelled 1–3 are steels, 4–8 are alnicos, and 9–12 are rare-earth magnets. The increase may be seen to be roughly exponential, a factor of 200 in a century—not as spectacular as the improvement in the attenuation of optical fibres, but one certainly gains the impression of steady advance.

For the properties of a number of hard magnetic materials, see Table 11.2.

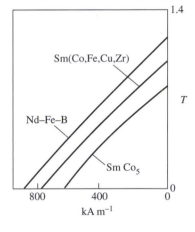
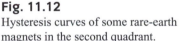

Fig. 11.12
Hysteresis curves of some rare-earth magnets in the second quadrant.

Table 11.2 *Hard magnetic materials*

Material	H_c (A m^{-1})	B_r (T)	$(BH)_{max}$ J m^{-3}
Carbon steel 0.9%C, 1% Mn	4.0×10^3	0.9	8×10^2
Alnico 5			
8% Al, 24% Co, 3% Cu, 14% Ni	4.6×10^4	1.25	2×10^4
'Ferroxdur' $(BaO)(Fe_2O_3)_6$	1.6×10^5	0.35	1.2×10^4
ESD Fe–Co	8.2×10^4	0.9	4×10^4
Alnico 9	1.3×10^5	1.05	10^5
$SmCo_5$	7×10^5	0.8	2×10^5
$Nd_2Fe_{14}B$	8.8×10^5	1.2	3×10^5

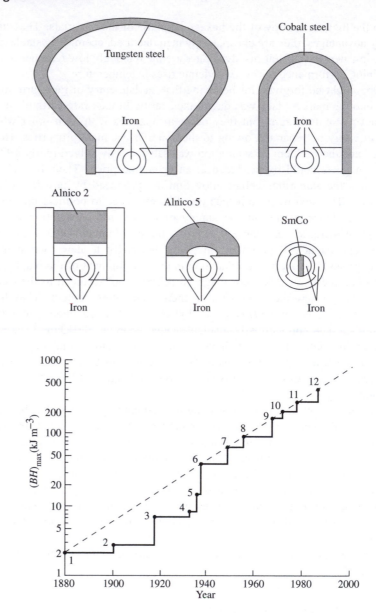

Fig. 11.13
The evolution of the magnetic circuit of moving-coil meters reflects the progress in magnet materials development. Coil size and magnetic field are equal in all five sketches.

Fig. 11.14
The achieved optimal value of $(BH)_{max}$ against time.

* The theory we have discussed so far is not really consistent because classical theory cannot even justify the existence of atoms and so cannot provide any good reasons for the presence of circulating electronic currents in a material.

11.7   Microscopic theory (quantum-mechanical)

Classical theory gives a reasonable physical picture of what is happening in a magnetic material and does give some guidance to people searching for new materials.* The question arises whether we should discuss quantum theory as well. I would like to advise against excessive optimism. Do not expect too much; the situation is not as cheerful as for semiconductors, where the injection of a tiny dose of quantum theory sufficed to explain all the major phenomena. The same is not true for magnetism. The quantum theory of magnetism is much more complicated and much less useful to an engineer. The most important activity, the search for better magnetic materials, is empirical anyway, and there are not many magnetic devices clamouring for quantum theory to solve the riddle of their operation.

I do think, however, that a brief look into the quantum theory of magnetism will yield some dividends. It is worth learning, for example, how quantum numbers come into the picture. We have, after all, come across them when studying the hydrogen atom, so it is not unreasonable to expect them to be able to say something about magnetic properties. It is also worth knowing that there is a very simple experiment showing the quantized nature of magnetic moments, and there are a few devices which need quantum theory for their description. So let me describe the basic concepts.

First, we should ask how much of the previously outlined theory remains valid in the quantum-mechanical formulation. Not a word of it! There is no reason whatsoever why a classical argument (as, for example, the precession of magnetic dipoles around the magnetic field) should hold water. When the resulting formulae turn out to be identical (as, for example, for the paramagnetic susceptibility at normal temperatures), it is just a lucky coincidence.

So we have to start from scratch.

Let us first talk about the single electron of the hydrogen atom. As we mentioned before, the electron's properties are determined by the four quantum numbers n, l, m_l, and s, which have to obey certain relationships between themselves; as for example, that l must be an integer and may take values between 0 and $n - 1$. Any set of these four quantum numbers will uniquely determine the properties of the electron. As far as the specific magnetic properties of the electron are concerned, the following rules are relevant:

1. The total angular momentum is given by

$$\Pi = \hbar \{ j (j + 1) \}^{1/2}, \tag{11.33}$$

   where $j = l + \frac{1}{2}$, that is a combination of the quantum numbers l and s.
2. The possible components of the angular momentum along *any* specified direction* are determined by the combination of m_l (which may take on any integral value between $-l$ and $+l$) and s, yielding

$$j, j - 1, \ldots, -j + 1, -j.$$

   Taking as an example a d-electron, for which $l = 2$, the total angular momentum is

$$\Pi = \hbar \left(\frac{5}{2} \cdot \frac{7}{2} \right)^{1/2} = \frac{\hbar}{2} \sqrt{35}, \tag{11.34}$$

   and its possible components along (say) the z-axis are

$$\frac{5}{2}\hbar, \frac{3}{2}\hbar, \frac{\hbar}{2}, -\frac{\hbar}{2}, -\frac{3}{2}\hbar, -\frac{5}{2}\hbar$$

   as shown in Fig. 11.15.
3. The quantum-mechanical relationship between magnetic moment and angular momentum is nearly the same as the classical one, represented by eqn (11.8)

$$\mu_m = g \frac{e}{2m} \Pi. \tag{11.35}$$

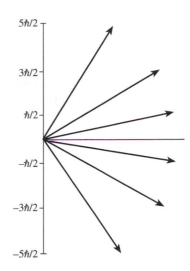

Fig. 11.15
The possible directions of the angular momentum vector for a d-electron.

* This is sheer nonsense classically because, according to classical mechanics, once the angular momentum is known about three axes perpendicular to each other, it is known about any other axes (and it will not therefore take necessarily integral multiples of a certain unit). In quantum mechanics we may know the angular momentum about several axes but *not simultaneously*. Once the angular momentum is measured about one axis, the measurement will alter the angular momentum about some other axis in an unpredictable way. If it were otherwise, we would get into trouble with the uncertainty relationship. Were we to know the angular momentum in all directions, it would give us the plane of the electron's orbit. Hence, we would know the electron's velocity in the direction of the angular momentum vector (it would be zero), and also the position (it would be in the plane perpendicular to the angular momentum in line with the proton). But this is forbidden by the uncertainty principle, which says that it is impossible to know both the velocity and the position coordinate in the same direction as the velocity.

The only difference is the factor g (admirably called the g-factor). For pure orbital motion its value is 1; for pure spin motion its value is 2; otherwise it is between 1 and 2.

4. The energy of a magnetic dipole in a magnetic field H (taken in the z-direction) is

The term $-e\hbar/2m$ is called a *Bohr magneton* and denoted by μ_{mB}.

$$E_{mag} = -(\mu_m)_z \mu_0 \, H = -ge\Pi_z \mu_0 \frac{H}{2m}. \tag{11.36}$$

We may rewrite eqn (11.36) in the form

$$E_{mag} = g\mu_{mB} \Pi_z \mu_0 \frac{H}{\hbar}, \tag{11.37}$$

where $\Pi_z/\hbar$, as we have seen before, may take the values $j, j - 1$, etc. down to $-j$.

We know now everything about the magnetic properties of an electron in the various states of the hydrogen atom. In general, of course, the hydrogen atom is in its ground state, for which $l = 0$ and $m_l = 0$, so that only the spin of the electron counts. The new quantum number j comes to $\frac{1}{2}$, and the possible values of the angular momentum in any given direction are $\hbar/2$ and $-\hbar/2$. Furthermore, $g = 2$, and the magnetic moment is

The magnetic moment of hydrogen happens to be one Bohr magneton.

$$\mu_m = \mu_{mB}. \tag{11.38}$$

We can get the magnetic properties of more complicated atoms by combining the quantum numbers of the individual electrons. There exist a set of rules (known as Hund's rules) that tell us how to combine the spin and orbital quantum numbers in order to get the resultant quantum number J. The role of J for an atom is exactly the same as that of j for an electron. Thus, for example, the total angular momentum is given by

$$\Pi = \hbar \left\{ J(J + 1) \right\}^{1/2}, \tag{11.39}$$

and the possible components of the angular momentum vector along any axis by

$$\hbar J, \hbar (J - 1), \ldots, \hbar (-J + 1), -\hbar J.$$

The general rules are fairly complicated and can be found in text books on magnetism. I should just like to note two specific features of the magnetic properties of atoms:

1. Atoms with filled shells have no magnetic moments (this is because the various electronic contributions cancel each other);
2. The spins arrange themselves so as to give the maximum possible value consistent with the Pauli principle.

It follows from (1) that helium and neon have no magnetic moments; and stretching the imagination a little one may also conclude that hydrogen, lithium, and silver, for example, possess identical magnetic properties (because all of them have one outer electron).

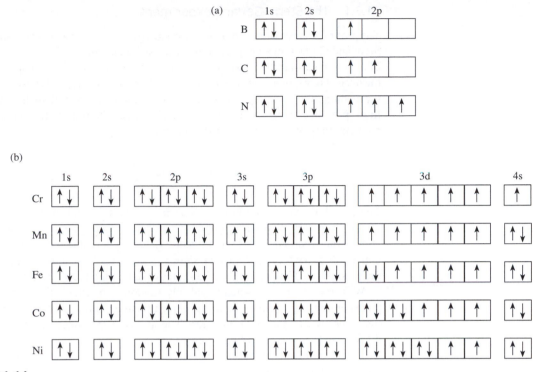

Fig. 11.16

The electron configurations of (a) boron, carbon, and nitrogen and (b) chromium, manganese, iron, cobalt, and nickel.

The consequences of (2) are even more important. It follows from there that states with identical spins are occupied first. Thus, boron with a configuration $1s^2 2s^2 2p^1$ has one electron with spin 'up' in the outer shell [see Fig. 11.16(a)]; carbon has two electrons with spin up, and nitrogen has three. Similarly all five electrons of chromium and manganese in the 3d shell have spins up, and the states with opposite spins start to fill up only later, when there is no alternative. This is shown in Fig. 11.16(b), where the electronic configurations are given for chromium, manganese, iron, cobalt, and nickel.

We shall return to the spins of the 3d electrons a little later; first let me summarize the main points of the argument. The most important thing to realize is that electrons in an atom do not act individually. We have no right to assume (as we did in the classical treatment) that all the tiny electronic currents are randomly oriented. They are not. They must obey Pauli's principle, and so within an atom they all occupy different states that do bear some strict relationships to each other. The resultant angular momentum of the atom may be obtained by combining the properties of the individual electrons, leading to the quantum number J, which may also be zero. Thus an atom that contains many 'magnetic' electrons may end up without any magnetic moment at all.

You may ask at this stage what is the evidence for these rather strange tenets of quantum theory? Are the magnetic moments of the atoms really quantized? Yes, they are. The experimental proof actually existed well before the theory was properly formulated.

11.7.1 The Stern–Gerlach experiment

The proof for the existence of discrete magnetic moments was first obtained by Stern and Gerlach in an experiment shown schematically in Fig. 11.17. Atoms of a chosen substance (it was silver in the first experiment) are evaporated in the oven. They move then with the average thermal velocity, and those crossing the diaphragms S_1 and S_2 may be expected to reach the target plane in a straight line—provided they are non-magnetic. If, however, they do possess a magnetic moment, they will experience a force expressed by

$$F = (\mu_{\mathrm{m}})_z \mu_0 \frac{\partial H}{\partial z}. \tag{11.40}$$

Thus, the deflection of the atoms in the vertical plane depends on the magnitude of this force. $\partial H / \partial z$ is determined by the design of the magnet [a strong variation in the z-component of the magnetic field may be achieved by making the upper pole piece wedge-shaped as shown in Fig. 11.17(b)] and is a constant in the experiment. Hence, the actual amount of deflection is a measure of $(\mu_{\mathrm{m}})_z$.

Were the magnetic moments entirely randomly oriented, the trace of the atoms on the target plane would be a uniform smear along a vertical line. But that is not what happens in practice. The atoms in the target plane appear in distinct spots as shown in Fig. 11.17(c).

For silver $J = \frac{1}{2}$, and the beam is duly split into two, corresponding to the angular momenta $\Pi_z = \hbar/2$ and $-\hbar/2$. If the experiment is repeated with other substances, the result is always the same. One gets a discrete number of beams, corresponding to the discrete number of angular momenta the atom may have.

11.7.2 Paramagnetism

We are now in a position to work out, with the aid of quantum theory, the paramagnetic susceptibility of a substance containing atoms with quantum numbers, $J \neq 0$. When we apply a magnetic field, all the atoms will have some magnetic moments in the direction of the magnetic field. The relative number of atoms, possessing the same angular momentum, is determined again by Boltzmann statistics. The mathematical procedure for obtaining the average magnetic moment is analogous to the one we used for electric dipoles but must now be applied to a discrete distribution.

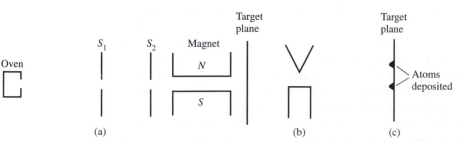

Fig. 11.17
Schematic representation of the Stern–Gerlach experiment.

The possible magnetic moments are

$$M_J g \mu_{mB}, \quad \text{where } M_J = J, J-1, \ldots, -J+1, -J.$$

Hence, their energies are

$$E_{mag} = -M_J g \mu_{mB} \mu_0 H, \tag{11.41}$$

and the average magnetic moment may be obtained in the form

$$\langle \mu_m \rangle = \frac{\sum_{-J}^{J} M_J g \mu_{mB} \exp(M_J g \mu_{mB} \mu_0 H/kT)}{\sum_{-J}^{J} \exp(M_J g \mu_{mB} \mu_0 H/kT)}. \tag{11.42}$$

The macroscopic magnetic moment may now be calculated by multiplying $\langle \mu_m \rangle$ by the number of atoms per unit volume.

Equation (11.42) turns out to be a very accurate formula* for describing the average magnetic moment as shown in Fig. 11.18, where it is compared with the experimental results of Henry on potassium chromium alum. The vertical scale is in Bohr magnetons per *ion*. Note that experimental results for paramagnetic properties are often given for ions embedded in some salt. The reason is that in these compounds the ions responsible for magnetism (Cr^{3+} in the case of potassium chromium alum) are sufficiently far from each other for their interaction to be disregarded.

If the exponent is small enough, the exponential function may be expanded to give

$$\langle \mu_m \rangle = -g \mu_{mB} \frac{\sum_{-J}^{J} M_J (1 - M_J g \mu_{mB} \mu_0 H/kT)}{\sum_{-J}^{J} (1 - M_J g \mu_{mB} \mu_0 H/kT)}$$

$$= \frac{g^2 \mu_{mB}^2 \mu_0 H}{(2J+1)kT} \sum_{-J}^{J} M_J^2, \tag{11.43}$$

because

$$\sum_{-J}^{J} M_J = 0. \tag{11.44}$$

* We need not be too much impressed by these close agreements between theory and experiments. The theoretical curve was *not* calculated from first principles, in the sense that the value of J was arrived at by semi-empirical considerations. The problem is far too difficult to solve exactly. The usual approach is to set up a simple model and modify it (e.g. by taking account of the effect of neighbouring atoms) until theory and experiment agree. It is advisable to stop rather abruptly at that point because further refinement of the model might increase the discrepancy.

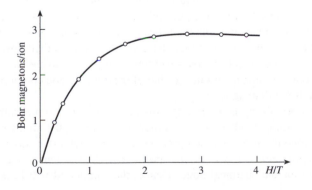

Fig. 11.18
The magnetic moment as a function of H/T for potassium chromium alum (after Henry).

The summation in eqn (11.43) is one of the simpler ones to perform, yielding

$$\tfrac{1}{3}J(J+1)(2J+1),$$

which gives finally

$$\langle \mu_m \rangle = g^2 \mu_{mB}^2 J(J+1)\mu_0 H/3kT. \tag{11.45}$$

We may now express the above equation in terms of the *total* angular momentum

$$\Pi = \hbar\{J(J+1)\}^{1/2} \tag{11.46}$$

and total magnetic momentum

$$\mu_m = ge\Pi/2m \tag{11.47}$$

to get

$$\langle \mu_m \rangle = \mu_m^2 \mu_0 H/3\,kT, \tag{11.48}$$

in agreement with the classical result.*

* This perhaps shows the power of human imagination. If one has a fair idea how the final conclusion should look, one can get a reasonable answer in spite of following a false track.

11.7.3   Paramagnetic solids

As we have seen, the magnetic properties of electrons combine to produce the magnetic properties of atoms. These properties can be measured in a Stern–Gerlach apparatus, where each atom may be regarded as a separate entity. This is because the atoms in the vapour are far enough from each other not to interact. However, when the atoms aggregate in a solid, the individual magnetic properties of atoms combine to produce a resultant magnetic moment. The electrons that are responsible for chemical bonding are usually responsible for the magnetic properties as well. When, for example, sodium atoms and chlorine atoms combine to make up the ionic solid, NaCl, then the valence electron of the sodium atom moves over to the chlorine atom and fills up the shell. Hence, both the sodium and the chlorine ions have filled shells, and consequently, solid NaCl is non-magnetic. A similar phenomenon occurs in the covalent bond, where electrons of opposite spin strike up a durable companionship, and as a result, the magnetic moments cancel again.

How then can solids have magnetic properties at all? Well, there is first the metallic bond, which does not destroy the magnetic properties of its con-stituents. It is true that the immobile lattice ions have closed shells and hence no magnetic properties, but the pool of electrons do contribute to magnetism, owing to their spin. Some spins will be 'up' (in the direction of the magnetic field); others will be 'down'. Since there will be more up than down, the sus-ceptibility of all metals has a paramagnetic component, of the order of 10^{-5}. This is about the same magnitude as that of the diamagnetic component; hence some metals are diamagnetic.

Another possibility is offered by salts of which potassium chromium alum is a typical example. There again, as mentioned above, the atoms responsible for the magnetic properties, being far away from each other, do not interact. In these compounds, however, the atoms lose their valence electrons; they are needed for the chemical bond. Hence, the compound will have magnetic properties only if some of the ions remain magnetic. This may happen in the so called 'transition elements', which have unfilled inner shells. The most notable of them is the 3d shell, but Table 4.1 shows that the 4d, 4f, 5d, and 5f shells have similar properties.

Taking chromium again as an example, it has a valency of two or three; hence, in a chemical bond it must lose its 4s electron [see Fig. 11.16(b)] and one or two of its 3d electrons. The important thing is that there are a number of 3d electrons left that have identical spins, being thus responsible for the paramagnetic properties of the salt.

11.7.4 Antiferromagnetism

Let us now study the magnetic properties of solid chromium. From what we have said so far it would follow that chromium is a paramagnetic solid with a susceptibility somewhat larger than that of other metals because free electrons contribute to it, and the lattice ions are magnetic as well. These expectations are not entirely false, and this is what happens above a certain temperature, the *Néel temperature* (475 K for chromium). Below this temperature, however, a rather odd phenomenon occurs. The spins of the neighbouring atoms suddenly acquire an ordered structure; they become antiparallel as shown in Fig. 11.19. This is an effect of the 'exchange interaction', which is essentially just another name for Pauli's principle. According to Pauli's principle, two electrons cannot be in the same state unless their spins are opposite. Hence, two electrons close to each other have a tendency to acquire opposite spins. Thus, the electron-pairs participating in covalent bonds have opposite spins, and so have the electrons in neighbouring chromium atoms. Besides chromium, there are a number of compounds like MnO, MnS, FeO, etc. and another element, manganese (Néel temperature 100 K) that have the same antiferromagnetic properties.

Antiferromagnetics display an ordered structure of spins; so in a sense, they are highly magnetic. Alas, all the magnetic moments cancel each other (in practice *nearly* cancel each other) and there are therefore no external magnetic effects.

11.7.5 Ferromagnetism

Leaving chromium and manganese, we come to iron, cobalt, and nickel, which are ferromagnetic. In a ferromagnetic material the spins of neighbouring atoms are parallel to each other [Fig. 11.19(b)]. Nobody quite knows why. There seems to be general agreement that the exchange interaction is responsible for the lining-up of the spins (as suggested first by Heisenberg in 1928) but there is no convincing solution yet. The simplest explanation (probably as good as any other) is as follows.

Electrons tend to line up with their spins antiparallel. Hence, a conduction electron passing near a 3d electron of a certain iron atom will acquire a tendency to line up antiparallel. When this conduction electron arrives at the next iron ion, it will try to make the 3d electron of that atom antiparallel to itself; that is, parallel to the 3d electron of the previous iron atom. Hence, all the spins tend to line up.

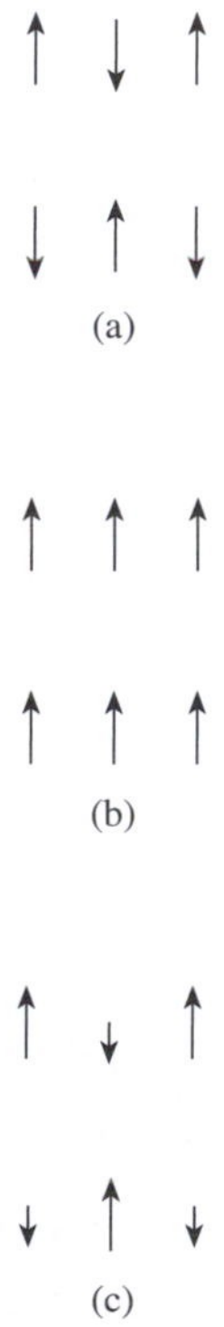

(a)

(b)

(c)

Fig. 11.19
The angular momentum vector for (a) antiferromagnetic, (b) ferromagnetic and (c) ferrimagnetic materials.

* F. Keffer, Magnetic properties of materials, *Scientific American*, September 1967.

In Weiss's classical picture the magnetic moments are lined up by a long-range internal field. In the quantum picture they are lined up owing to nearest-neighbour interaction. 'One is reminded,' writes Keffer* 'of the situation when, as the quiet of evening descends, suddenly all the dogs in a town get to barking together, although each dog responds only to the neighbouring dogs.'

11.7.6 Ferrimagnetism

This type of magnetism occurs in compounds only, where the exchange interaction causes the electrons of each set of atoms to line up parallel, but the two sets are antiparallel to each other. If the magnetic moments are unequal, then we get the situation shown in Fig. 11.17(c), where the resultant magnetic moment may be quite large. For most practical purposes ferrimagnetic materials behave like ferromagnetics but have a somewhat lower saturation magnetization.

11.7.7 Garnets

This is the name for a class of compounds crystallizing in a certain crystal structure. As far as magnetic properties are concerned, their most interesting representative is yttrium–iron garnet ($Y_3Fe_5O_{12}$), which happens to be ferromagnetic for a rather curious reason. The spin of the yttrium atoms is opposite to the spin of the iron atoms, so the magnetic moments would line up alternately—if the orbital magnetic moments were small. But for yttrium the orbital magnetic moment is large, larger actually than the spin, and is in the opposite direction. Hence, the total magnetic moment of the yttrium atom is in the same direction as that of iron, making the compound ferromagnetic.

11.7.8 Helimagnetism

You may wonder why the magnetic moments of neighbouring atoms in an ordered structure are either parallel or antiparallel. One would expect quantum mechanics to produce a larger variety. In actual fact, there *are* some materials in which the spins in a given atomic layer are all in the same direction, but the spins of adjacent layers lie at an angle (e.g. 129° in MnO_2 below a certain temperature), producing a kind of helix. For the moment this is a scientific curiosity with no practical application.

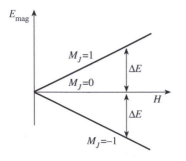

Fig. 11.20

The energy of an atom as a function of magnetic field for $J = 1$.

11.8 Magnetic resonance

11.8.1 Paramagnetic resonance

The possible energies of an atom in a magnetic field are given by eqn (11.41). There are $2J + 1$ energy levels, with separations of $\Delta E = g\mu_{mB}\mu_0 H$, as shown in Fig. 11.20 for $J = 1$.

We now put a sample containing magnetic atoms (e.g. a paramagnetic salt) into a waveguide and measure the transmission of the electromagnetic waves as a function of frequency. When $f = \Delta E/h$, the incident photon has just the right energy to excite the atom from a lower energy level into a higher energy level. Thus, some of the photons transfer their energies to the atomic system; this means loss of photons, or in other words, absorption of electromagnetic energy.

Hence, there is a dip in the transmission spectrum as shown in Fig. 11.21. Since the absorption occurs rather sharply in the vicinity of the frequency $\Delta E/h$, it is referred to as *resonant absorption*, and the whole phenomenon is known as *paramagnetic resonance*.

In practice the energy diagram is not quite like the one shown in Fig. 11.20 (because of the presence of local electric fields) and a practical measuring apparatus is much more complicated than our simple waveguide (in which the absorption would hardly be noticeable) but the principle is the same.

11.8.2   Electron spin resonance

This is really a special case of paramagnetic resonance, when only the spin of the electron matters. It is mainly used by organic chemists as a tool to analyse chemical reactions. When chemical bonds break up, electrons may be left unpaired, that is the 'fragments' may possess a net spin (in which case they are called free radicals). The resonant absorption of electromagnetic waves indicates the presence of free radicals, and the magnitude of the response can serve as a measure of their concentration.

11.8.3   Ferromagnetic, antiferromagnetic, and ferrimagnetic resonance

When a crystal with ordered magnetic moments is illuminated by an electro-magnetic wave, the mechanism of resonant absorption is quite complicated, owing to the interaction of the magnetic moments. The resonant frequencies cannot be predicted from first principles (though semiclassical theories exist) but they have been measured under various conditions for all three types of materials.

11.8.4   Nuclear magnetic resonance

If electrons, by virtue of their spins, can cause resonant absorption of electro-magnetic waves, one would expect protons to behave in a similar manner. The main difference between the two particles is in mass and in the sign of the electric charge; so the analogous formula,

$$f = \frac{1}{2\pi} g \frac{e}{2m_\mathrm{p}} \mu_0 H \tag{11.49}$$

should apply.

The linear dependence on magnetic field is indeed found experimentally, but the value of g is not 2 but 5.58, indicating that the proton is a more complex particle than the electron.

Neutrons also possess a spin, so they can also be excited from spin 'down' into spin 'up' states. Although they are electrically neutral, the resonant frequency can be expressed in the same way, and the measured g-factor is 3.86.

The resonance is sharp in liquids but broader, by a few orders of magnitude, in solids. The reason for this is that the nuclear moments are affected by the local fields, which may vary in a solid from place to place but average to zero in a liquid.

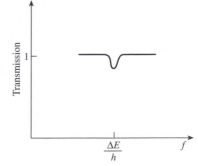

Fig. 11.21
Transmission of electromagnetic waves as a function of frequency through a paramagnetic material. There is resonant absorption where $hf = \Delta E$.

m_p is the mass of the proton.

Since both the shape of the resonant curve and the exact value of the resonant frequency depend on the environment in which a nucleus finds itself, nuclear magnetic resonance can be used as a tool to investigate the properties of crystals*. An important application in a different direction is the precision measurement of magnetic fields. The proton resonance of water is generally used for this purpose. The accuracy that can be achieved is about 1 part in 10^6.

* See also Section 11.9.5

11.8.5   Cyclotron resonance

We have already discussed the phenomenon of cyclotron resonance from a classical point of view, and we shall now consider it quantum mechanically. For resonant absorption one needs at least two energy levels or, even better, many energy levels equally spaced from each other. What are the energy levels of an electron in a solid? Remember that in our earlier model we neglected the interaction between electrons and simply assumed that the solid may be regarded as an infinte potential well. The possible energy levels were then given by eqn (6.2),

$$E = \frac{\hbar^2}{2m}(k_x^2 + k_y^2 + k_z^2) = \frac{\hbar^2}{8\,m\,(2a)^2}(n_x^2 + n_y^2 + n_z^2),$$

where n_x, n_y, n_z are integers.

When a magnetic field is applied in the z-direction, then the above equation modifies to*

* The effect of the magnetic field may be taken into account by replacing p^2 by $(p - eA)^2$ in the Hamiltonian of Schrödinger's equation (where A is the vector potential).

$$E = \left(\lambda + \frac{1}{2}\right)\hbar\omega_c + \frac{\hbar^2}{2m}k_z^2, \tag{11.50}$$

where λ is an integer and ω_c is the cyclotron frequency. For constant k_z, the difference between the energy levels (called *Landau levels*) is $\hbar\omega_c$. Hence, we may look upon cyclotron resonance as a process in which electrons are excited by the incident electromagnetic wave from one energy level to the next.

11.8.6   The quantum Hall effect

Strictly speaking this does not belong to magnetic resonance (although Landau levels are involved) and may be a little out of place in an engineering textbook. The argument for including it is that there might be some relationship to high temperature superconductivity (see Section 14.9) which is of great practical significance, and it is also true that the effect would have never been discovered had not engineers invented field effect transistors, whose operation depended on a two-dimensional electron gas (see Section 9.15).

You may remember the discussion of the ordinary Hall effect in Chapter 1. The experimental set-up for the quantum Hall effect is exactly the same. The only difference is that the dimension of the current channel perpendicular to the applied magnetic field is now comparable with the electron wavelength. The requirements for observing the effect are high magnetic fields ($B \cong 10$T) and low temperatures, say a few K. The measured value is the so-called Hall resistance, which relates the measured transverse voltage (Hall voltage) to the longitudinal current. Since the Hall voltage is known to be proportional to the applied magnetic field [eqn (1.20)] we would expect the Hall resistance versus longitudinal current curve to vary linearly with B. The striking result is that the Hall resistance turns out to be independent of the magnetic flux density within certain intervals as shown in Fig. 11.22. It looks as if the Hall resistance was quantized.

How can we explain these results? Surely, if something is quantized, we need quantum theory to explain it. Unfortunately, quantum theories are complicated, so one tries to avoid them. That is what we did in Section 8.4, where relationships for the mobilities of semiconductors were derived. In order to explain the present results there is, however, no reprieve. We have to approach the concept of resistance from an entirely different viewpoint, from that of quantum mechanics.

Classically, a piece of resistive material always leads to power absorption. In quantum mechanics we have to ask the question whether an electron is capable of absorbing the energy available. It can only do so if there are empty states at a higher energy into which the electrons can scatter, so we need to find out whether there are any empty states available.

Let us assume that the temperature is low enough and the magnetic field is high enough, so that only the two lowest Landau levels are occupied. The lowest energy level is completely filled, the second energy level is partially filled, and the third level is empty. If the magnetic field is reduced, then the energy difference between the second and first Landau levels is reduced, consequently some electrons must move up from the first level to the second level. That means that there are now fewer states at the second level, which an electron can scatter into, hence the probability of transition is smaller, and the resistance (we are talking about *longitudinal* resistance not Hall resistance) decreases. If the magnetic field is further reduced, then at a certain stage the second Landau level will be completely filled. The only way an electron in the ground state can now absorb energy is by scattering into the third Landau level, but that is too far away. Hence, the probability of scattering into that level is extremely low, which means that the resistance is extremely low. In practice, this resistance would be low indeed, lower than that of copper.

Let us now slightly complicate our model and assume that there are some impurity states just below the third Landau level, as shown in Fig. 11.23. The argument for the longitudinal resistance is unaffected: the impurity states are still very far away from the second Landau level. But let us return to the Hall resistance. How will it vary as the magnetic field is reduced? The electrons moving into the impurity levels will no longer be available for deflection by the magnetic field, hence the Hall resistance must remain constant until all the impurity levels are filled. In the Hall resistance versus magnetic field curve this appears as a plateau, whenever a Landau level is filled. The discrete values

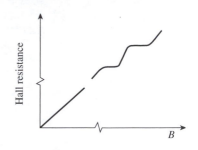

Fig. 11.22
The Hall resistance against magnetic flux density shows distinct plateaux.

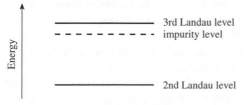

Fig. 11.23
Discrete energy levels in high magnetic fields.

of the Hall resistance at these plateau turn out to be dependent only on the fundamental constants h and e and on the number of Landau levels filled.

11.9　Some applications

Until the 1950s the only significant application of magnetic materials was for electrical machines and transformers. Modern technology brought some new applications; the most notable among them is the use of magnetism for storing information. In fact, in 1985 for the first time, the sales of magnetic information products in the US exceeded those for all other technologies. The storage densities achieved are no less remarkable. A hard disk may store information at a density of 5 billion bits per cm^2. So why are they so much outshone by semiconductors? May be because ferrite cores, one-time champions in Random Access Memories, suffered a resounding defeat at the hands of their semiconductor counterparts, and people are apt to forget the losers. Magnetic storage is certainly not on its way out, and I am not thinking of video tape recorders nor of the new digital audio tape. They still hold the market for mass storing of information in digital computers when access time is of secondary importance.

The principle of operation of magnetic memories is very simple. At 'writing' a magnetic field is applied to some area of a tape or disk, and at 'reading' this magnetic field is sensed. The actual technical solutions do need, however, some ingenuity. It is far from trivial to design a recording system in which the magnetic head may move just a mere $0.25\ \mu m$ above the surface of a disk, with a relative velocity of $160\ km\ h^{-1}$.

I shall obviously not be able to talk about all the various magnetic memories. I shall mention only two, magnetic bubbles and the newly arrived magnetic tunnel junctions.

11.9.1　Magnetic bubbles

This device, first demonstrated by Brobeck in 1967, works on the principle that small regions of magnetic materials can have differing magnetic alignments within a uniform physical shape. The technique is to grow very thin films epitaxially of either orthoferrites or garnets on a suitable substrate.* The film is only a few micrometres thick. All the domains can be aligned in a weak magnetic field normal to the film. Then by applying a stronger localized field in the opposite direction, it is possible to produce a cylindrical domain (called a 'magnetic bubble') with its magnetic axis inverted [see Fig. 11.24(a), which shows several].

An important question is, 'Does the bubble stay there when the strong field that created it is removed?' It turns out that, with suitable materials, the domain-wall coercivity is great enough to produce a stable bubble, and that the most stable bubble size results when the radius is about equal to the garnet film thickness, that is a few micrometres. The next thing is to move the bubble about in a controlled manner.

One way of achieving controlled motion is by printing a pattern of small permalloy bars on the surface. The usual manufacturing technique for this is photoengraving, using a photo-resist material similar to the process described for integrated circuits in Section 9.22.

* The usual material is garnet with the general formula $R_3Fe_5O_{12}$, where R represents yttrium or a combination of rare-earth ions. Sometimes gallium or aluminium is substituted for some of the iron, to lower the saturation magnetization. In these ways the magnetic properties are bespoke by the chemists, and a typical successful composition is $Eu_1Er_2Ga_{0.7}Fe_{4.3}O_{12}$. Chemistry was never like this when I was at school.

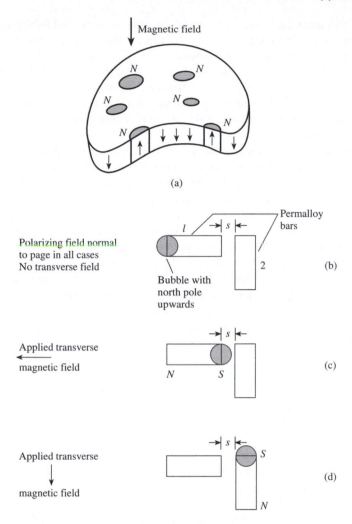

(a)

Polarizing field normal
to page in all cases
No transverse field

(b)

Applied transverse
magnetic field

(c)

Applied transverse
magnetic field

(d)

Fig. 11.24
Magnetic bubble domains. (a)
Applied field for stable bubbles.
(b)–(d) Illustrate how a bubble can be
moved.

The actual devices look fairly complicated. Our intention is just to show the basic principles of how the bubbles can be persuaded to move from one place to another, so let us consider just two typical permalloy bars on the surface and assume the presence of a bubble with its north pole upwards, as shown in Fig. 11.24(b).

In the absence of a magnetic field the permalloy bars are unmagnetized and have no effect upon the bubble. However, if a magnetic field (as shown in Fig. 11.24(c)) is applied, then bar 1 becomes magnetized, and the north pole of the bubble will move to the south pole of bar 1. How can we move the bubble to bar 2? We only need to change the direction of the magnetic field, as shown in Fig. 11.24(d). Then bar 2 becomes magnetized, and the bubble moves to the south pole of bar 2.

Note that the highest density of elements is determined by the smallest feature in the structure, that is the separation s between two bars. An increase in density has been recently made possible by the use of a contiguous structure (Fig. 11.25) produced by depositing thin layers of gold on a garnet substrate. The gold layer

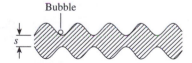

Fig. 11.25
A bubble moving along a contiguous structure.

serves as a mask for the next process of ion implantation, establishing thereby an implanted–unimplanted boundary. It turns out that under certain conditions (the garnet substrate must have suitable properties) this boundary may guide a bubble. Thus, by varying the applied magnetic field, the bubble is driven *along* the pattern and not *under* the pattern. For the same minimum feature size (s in Fig. 11.25) the bubble may be smaller. It is believed that this technique may lead to a tenfold increase in density over that achievable by the various permalloy structures.

Bubbles can be detected by making them pass under a strip of indium antimonide which has a high *magneto-resistance*, that is its electrical resistance is changed by a magnetic field. This property is closely connected with the large Hall effect in InSb, already mentioned in Section 9.20.

So you see that bubbles, by their presence or absence, may be used for storing binary information, and that the information can be read out. What is the advantage of using magnetic bubbles? Mainly density. With presently available photoengraving techniques, the contiguous structure may lead to a density of 10 million bits cm^{-2}. This is of course not a random access memory, the information must be read out serially; the achievable speed may be a few hundred $kbit s^{-1}$. It could be used at an advantage when large blocks of data need to be transferred to the main memory of a computer for processing.

11.9.2   Spintronics

This is a newly coined term presumably in analogy with electronics.* It is concerned with new phenomena and new devices based on the properties of the electron spin. In electrical and even in electronic engineering one rarely mentions spin. When we talk about the current in a wire, a diode or a transistor we do not ask any questions about the spins of the participating electrons. We just quietly assume that spin is averaged out. The new thing is that spin may very well matter.

Giant Magneto-Resistance (GMR). This is the same kind of effect, change of electrical resistivity due to a magnetic field, as mentioned in the previous section. Giant Magneto-Resistance means that the resistivity changes quite a lot. A change by a factor of two already qualifies as giant. The large (or giant) change with applied magnetic field can be measured in metallic multilayers in which ferromagnetic layers are separated by spacers of a few atoms thick made of a non-ferromagnetic transition metal. The initial lining up of the magnetic moments depends critically on the thickness of the spacers. Assume that we choose the magnetic moments in neighbouring ferromagnetic layers to line up in opposite directions. This is then the state of high resistance because the electrons have difficulties getting through the multi-layer. If their spin orientation is favourable (i.e. low scattering) for one ferromagnetic layer it must be unfavourable for the next ferromagnetic layer. The low magneto-resistance state can be reached by applying a magnetic field which will ensure that the magnetic moments of all the ferromagnetic layers line up in the same direction. Then electrons with the right spin orientation can get through with much less scattering.

Main application is in information technology. When digital information is magnetically recorded, e.g. ones and zeros are represented by opposite magnetic fields then the Giant Magneto-Resistance effect can be used for reading the information. The read head can have high or low resistance depending on the recorded information being a one or a zero. At the moment GMR is the favoured mechanism for read heads in laptop computers because they have high sensitivity and can read low magnetic fields.

Magnetic Tunnel Junctions. These devices are also based on the spin of the electrons but the spacer is an insulator rather than a non-ferromagnetic metal and the current is due to tunnelling rather than conduction. The effect is however quite similar: the resistance is low when the magnetic moments on both sides of the junction have the same orientation and the resistance is high when the magnetic moments are in opposite directions. Potential applications are not only in read heads but much more importantly in Random Access Memories where they could soon have densities comparable to those of semiconductor memories. In fact, Magnetic Tunnel Junctions may soon replace semiconductor memories in applications when the non-volatile nature of the storage (once the magnetic information is written it does not need to be continually renewed) is the main requirement.

Combination of ferromagnetic metals and semiconductors. As discussed, electrons passing through a ferromagnetic material may acquire a dominant spin orientation. The question is what happens to their spin as they enter a semiconductor. How can they be efficiently injected, once inside how far can they move before losing their spin orientation and how can that loss be influenced by external factors, for example, an applied voltage. When answers to all these question would have been obtained then it might be possible to design a FET based on spin. The source and the drain could be made of ferromagnetic metals with magnetic moments of the same orientation (so that the spin polarized current can enter the drain). Then, if the voltage applied to the gate can decohere (i.e. destroy the spin uniformity) the current, all the conditions for a three terminal device are satisfied: the current from source to drain can be controlled by a gate voltage.

11.9.3 Isolators

My next example is a device which lets an electromagnetic wave pass in one direction but heavily attenuates it in the reverse direction. It is called an *isolator*. The version I am going to discuss works at microwave frequencies and uses a ferrite rod, which is placed into a waveguide and biased by the magnetic field of a permanent magnet (Fig. 11.26). The input circularly polarized wave may propagate unattenuated, but the reflected circularly polarized wave (which is now rotating in the opposite direction) is absorbed. Thus, the operation of the device is based on the different attenuations of circularly polarized waves that rotate in opposite directions.

The usual explanation is given in classical terms. We have seen that a magnetic dipole will precess in a constant magnetic field. Now if in addition to the constant magnetic field in the z-direction there appears a magnetic vector in the x-direction (Fig. 11.27), then there is a further torque acting upon the magnetic dipole. The effect of this torque is insignificant, except when the extra magnetic field rotates with the speed of precession—and, of course, in the right direction. But this is exactly what happens for one of the circularly

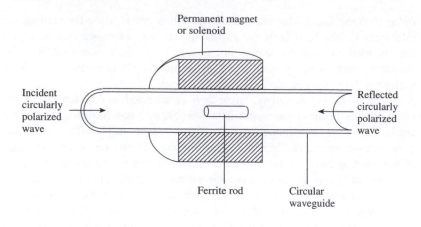

Fig. 11.26
Schematic representation of
an isolator.

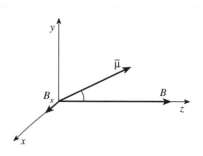

Fig. 11.27
The magnetic dipole moment
precesses around the constant
magnetic field, B. An additional
magnetic field, B_x, gives an extra
torque, trying to increase the angle of
precession.

polarized waves when its frequency is equal to the frequency of precession. The interaction is then strong, and energy is taken out of the electromagnetic wave in order to increase the angle of precession. Hence, for one given frequency (the resonance frequency) and one sense of rotation (that of the reflected wave) the electromagnetic wave is absorbed.

The quantum explanation is based on the resonance phenomena discussed in the last section. The electromagnetic wave is absorbed because its energy is used to sponsor transitions between the respective energy levels. Unfortunately, quantum mechanics provides no intuitive description of the effect of circularly polarized waves. You either believe that the result comes out of the mathematical description of the problem or, alternatively, you stick to the classical picture. This is what many quantum physicists do, but to ease their conscience, they put the offending noun between inverted commas. They do not claim that anything is really precessing, nonetheless, they talk of 'precession'.

11.9.4   Sensors

Magnets can be used to sense position, force, torque, speed, rotation, acceleration, and of course current and magnetic field. Since the advent of the light and powerful neo-magnets, their use has been rapidly expanding, as for example in anti-lock brakes and in activating airbags.

11.9.5   Medical imaging

The traditional medical imaging technique, X-rays, are used less and less, owing to their harmful biological effects. The imaging technique that is becoming more widespread is based on Nuclear Magnetic Resonance. It is usually referred to as Magnetic Resonance Imaging or MRI (the word 'nuclear' has been omitted so as not to be associated with anything dangerous and warlike). It is used primarily to measure the concentration of protons in tissues. There is also important information in the decay time of the absorption, as tested by short rf (radio frequency) pulses. Apparently, cancer tumours have a longer decay, by a factor of two, than normal tissues.

Image clarity improves with increasing magnetic field. Hence, the usual choice is to generate the magnetic field by currents in superconductors (see Chapter 14). The disadvantage of that is the necessity of cooling and the concomitant high running costs. A cheaper, although not quite that satisfactory, solution is to use permanent magnets. The structures have to be pretty big because the magnetic field is required within a large volume. A flux density of 0.2 T may be achieved by a mere 2.6 tons of neo-magnet. If this seems excessive, it is worth noting that the alternative ferrite magnets would weigh 21 tons.

11.9.6 Electric motors

This is one of the oldest applications of magnetism, the conversion of electrical into mechanical power. Have there been any major advances? Well, our illustration of what happened to the magnetic circuit of moving-coil meters (Fig. 11.13) applies just as well to electric motors. They have come down in size, so much so that in a modern motor car there is room for as many as two dozen permanent magnet motors, which drive practically everything that moves (apart from the car, of course).

Some thirty years ago all the electric motors that we moved about the laboratory or our homes were 'fractional horse power'. In fact, a $\frac{1}{2}$ h.p. motor was rather large and heavy. Anything greater than 1 h.p. was classed as 'industrial' and had a built-in fan or water-cooling. Now our domestic motors are smaller, cooler, and quieter; and where power is needed, such as in portable drills, lawn-mowers, and shredders ratings of up to 1.6 kW (i.e. more than 2 h.p.) are quite common and reasonably portable. What has happened besides the discovery of better magnetic alloys? This is something we should have mentioned in the last chapter in the section about insulators. The makers of motors woke up to the fact that new polymeric insulations were available that were more effective than the brown paper soaked in transformer oil which they had been using for the previous century. I am telling you this story because it illustrates that some improvements in technology, which the public is hardly aware of, can have a significant impact upon how we live.

Exercises

11.1. Check whether eqn (11.12) is dimensionally correct. Take reasonable values for N_a, Z and r and calculate the order of magnitude of the diamagnetic susceptibility in solids.

11.2. The magnetic moment of an electron in the ground state of the hydrogen atom is 1 Bohr magneton. Calculate the induced magnetic moment in a field of 1 T. Compare the two.

11.3. A magnetic flux density, B is applied at an angle θ to the normal of the plane of a rectangular current loop (Fig. 11.28).

(i) Determine the energy of the loop by finding the work done by the magnetic field when lining up (bringing to a stable equilibrium) the loop.

(ii) By defining the energy of a magnetic dipole as

$$E = \mu_m \cdot \mathbf{B}$$

and by identifying the loop with a magnetic dipole determine the magnetic moment vector of the loop.

(iii) Confirm that in the stable equilibrium position the magnetic field of the loop augments the applied field.

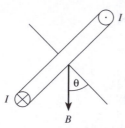

Fig. 11.28

11.4. How have domains in ferromagnetic materials been observed?

11.5. Check the calculation leading to the values of the Weiss constant, magnetic moment and saturation magnetization for iron given in eqn (11.28).

11.6. Show that the data for the magnetic susceptibility of nickel given below is consistent with the Curie law [eqn (11.26)] and evaluate the Curie constant and temperature. Hence find the effective number of Bohr magnetons per atom. Atomic weight 58.7, density $8850 \, \mathrm{kg \, m^{-3}}$.

$T(^\circ\mathrm{C})$	500	600	700	800	900
$\chi_m \, 10^5$	38.4	19.5	15	10.6	9.73

11.7. An alloy of copper and cobalt consists of spherical precipitates, averaging 10 nm diameter, of pure cobalt in a matrix of pure copper. The precipitates form 2 per cent by volume of the alloy. Cobalt is ferromagnetic, with saturation magnetization of $1.4 \, \mathrm{MA \, m^{-1}}$. Each cobalt precipitate is a single domain, and acts as a strong dipole, which responds to any external field as a paramagnetic dipole. The effect is called 'superparamagnetism'. Calculate the susceptibility of the alloy at 300 K.

[Hint: The total magnetic moment of each precipitate is equal to the product of magnetic moment density (saturation magnetization) with the volume of the precipitate.]

11.8. A system of electron spins is placed in a magnetic field $B = 2$ T at a temperature T. The number of spins parallel to the magnetic field is twice as large as the number of antiparallel spins. Determine T.

11.9. In a magnetic flux density of 0.1 T at about what frequencies would you expect to observe (i) electron spin resonance, (ii) proton spin resonance?

11.10. The energy levels of a free electron gas in the presence of an applied magnetic field are shown in Fig. 11.29 for absolute zero temperature. The relative numbers of electrons with spins 'up' and 'down' will adjust so that the energies are equal at the Fermi level. Show that the paramagnetic susceptibility is given by the approximate expression

$$\chi_m = \mu_m^2 \mu_0 Z(E_F)$$

where μ_m is the magnetic moment of a free electron, μ_0 the free space permeability, and $Z(E_F)$ the density of states at the Fermi level. Assume that $\mu_m \mu_0 H \ll E_F$.

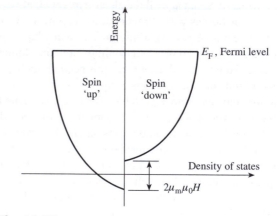

Fig. 11.29

What goes up must come down.
It's all done with mirrors.
              19th century aphorisms

12.1 Equilibrium

We have several times arrived at useful results by using the concept of equilibrium. It is a pretty basic tenet of science and like a similar idea, conservation of energy, it is always coming in handy. When we say that the electrons in a solid have a Fermi–Dirac distribution of energies, we are really saying two things: first, that the system is in equilibrium; second, that it has a particular temperature. Temperature is a statistical concept and is bound up with the idea of equilibrium. On the one hand, we cannot meaningfully speak of the temperature of a single particle; on the other, if we have a system of particles that is perturbed from equilibrium, say by accelerating *some* of them, then for a transient period the temperature cannot be specified, since there is no value of T that will make the Fermi function describe the actual distribution. Of course, for electrons in a solid, or atoms in the gaseous state, the effect of collisions rapidly flattens out the perturbation, the whole system returns to its equilibrium state, and the *idea* of temperature becomes valid again, although its actual value may have changed.

We have on one or two occasions considered perturbed equilibrium. We saw, for example, in Chapter 1 that large currents may flow in a conductor with a very slight change in the energy distribution. Thus, we could describe low field conduction in metals and semiconductors without departing from the equilibrium picture.

Lasers are different. They have massively perturbed population distributions that are nevertheless in some kind of equilibrium. But when we come to consider what temperature corresponds to that equilibrium, it turns out to be negative. Now you know that 0 K is a temperature that can never quite be obtained by the most elaborate refrigerator; so how can we get a negative temperature? It is not inconsistent really because, as we shall show, a negative temperature is hotter than the greatest positive temperature. But before going further into Erewhon*let us return to earth and start from the beginning.

* Erewhon (approximately 'nowhere' backwards) was a country in the book of the same name by Samuel Butler, where all habits and beliefs were the opposite of ours and were justified with impeccable logic and reasonableness.

12.2 Two-state systems

Let us consider a material in which atoms have only two narrow allowed energy levels, as illustrated in Fig. 12.1. Provided that the whole system containing the material is in thermal equilibrium, the two allowed levels will be populated corresponding to a dynamic energy equilibrium between the atoms. The

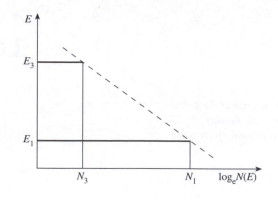

Fig. 12.1
Number of atoms in a natural two-state system as a function of energy. The dotted line shows the Boltzmann function, decaying exponentially with increasing energy.

population of the energy levels is, therefore, accurately described by the temperature, T, of the system and its appropriate statistics, which we shall take as Boltzmann statistics.

The two levels we are considering are labelled E_1 and E_2 in Fig. 12.1. Later on we shall see what happens in a three-level system, with the third level called E_2; but for the moment do not be put off by this notation; we are still talking of only two levels. The numbers of electrons N_1, N_3 in the levels E_1, E_3 are related by the Boltzmann function, so that they will be of the general form

$$N = N_0 \exp(-E/kT), \tag{12.1}$$

N_0 is a constant.

Therefore,

$$N_3 = N_1 \exp\left(-\frac{E_3 - E_1}{kT}\right). \tag{12.2}$$

As I said above, the atoms are in dynamic equilibrium, which means that the number of atoms descending from E_3 to E_1 is the same as the number leaping from E_1 to E_3. An atom at E_3 can lose the energy $E_3 - E_1$ either by radiative or by non-radiative processes. I shall consider only the former case here. When a radiative transition between E_3 and E_1 occurs during the thermal equilibrium process, it is called *spontaneous* emission for the 'down' process and photon absorption for the 'up' process. In each case, the photon energy is given by

We shall follow here the custom adopted in laser theory of using the frequency, ν, instead of the angular frequency, ω.

$$h\nu_{31} = E_3 - E_1. \tag{12.3}$$

What do we mean by talking about photons being present? It is a very basic law of physics that every body having a finite temperature will radiate thermal or 'black body' radiation. This radiation comes from the sort of internal transitions that I have just mentioned. As we saw in Chapter 2, the whole business of quantum theory historically started at this point. In order to derive a radiation law that agreed with experiments, Planck found it necessary to say that atomic radiation was quantized. This famous radiation equation is

$\rho(\nu)$ is the radiation density emanating from a body at temperature, T, in a band of the frequency spectrum of width, $d\nu$, and at a frequency, ν.

$$\rho(\nu)d\nu = \frac{8\pi n^3 h\nu^3}{c^3} \frac{d\nu}{\exp(h\nu/kT) - 1}. \tag{12.4}$$

The derivation can be found in many textbooks.

So far we have talked about photons generated within the material. Now if photons of energy $h\nu_{31}$ are shone on to the system from outside, a process called *stimulated* emission occurs. Either the photon gets together with an atom in a lower (E_1) state and pushes it up to E_3; or, less obviously, it *stimulates* the emission by an E_3-state atom of a photon ($h\nu_{31}$). In the latter case one photon enters the system, and two photons leave it. It was one of Einstein's many remarkable contributions to physics to recognize, as early as 1917, that both these events must be occurring in a thermodynamical equilibrium; he then went on to prove that the probabilities of a photon stimulating an 'up' or a 'down' transition were exactly equal. The proof is simple and elegant.

Consider our system, remembering that we have two states in equilibrium. The rate of stimulated transitions ($R_{1\to3}$) from the lower to the upper state will be proportional to both the number of atoms in the lower state and the number of photons that can cause the transition. So we can write

$$R_{1\to3} = N_1 B_{13}\rho(\nu_{31})d\nu, \tag{12.5}$$

The constant of proportionality B_{13} is the probability of absorbing a photon, often referred to as the Einstein B-coefficient.

For the reverse transition, from E_3 to E_1 we have a similar expression for stimulated emission, except that we will write the Einstein B-coefficient as B_{31}. There is also spontaneous emission. The rate for this to occur will be proportional only to the number of atoms in the upper state, since the spontaneous effect is not dependent on external stimuli. The constant of proportionality or the probability of each atom in the upper state spontaneously emitting is called the Einstein A-coefficient, denoted by A_{31}. Hence,

$$R_{3\to1} = N_3\{A_{31} + B_{31}\rho(\nu_{31})\}d\nu. \tag{12.6}$$

In equilibrium the rates are equal:

$$R_{1\to3} = R_{3\to1}, \tag{12.7}$$

that is

$$N_1 B_{13}\rho(\nu_{31})d\nu = N_3\{A_{31} + B_{31}\rho(\nu_{31})\}d\nu. \tag{12.8}$$

After a little algebra, using eqn (12.2) to relate N_3 to N_1, we get

$$\rho(\nu_{31})d\nu = \frac{A_{31}d\nu}{B_{13}\exp(h\nu_{31}/kT) - B_{31}}. \tag{12.9}$$

Comparing this with eqn (12.4), which is a universal truth as far as we can tell, we find that our (or rather Einstein's) B-coefficients must be equal

$$B_{13} = B_{31}, \tag{12.10}$$

that is stimulated emission and absorption are equally likely. Also,

$$A_{31} = B_{31}\frac{8\pi n^3 h\nu_{31}^3}{c^3}, \tag{12.11}$$

that is the coefficient of spontaneous emission is related to the coefficient of stimulated emission.

What is the physical significance of A_{31}? It is a measure of the spontaneous depopulation of state 3. Assuming, as usual, an exponential decay, the rate of change of population is

$$-\frac{dN_3}{dt} = A_{31}N_3, \tag{12.12}$$

which leads to a decay time constant, called spontaneous lifetime, by defining

$$t_{\text{spont}} = \frac{1}{A_{31}}. \tag{12.13}$$

We should, by now, have quite a good picture of what happens when light of frequency v_{31} shines on the two-state system. In the presence of an input light spontaneous decay is usually negligible, and although the probabilities of upward and downward transitions are exactly equal, there will be more transitions from E_1 to E_3 because there are many more atoms in the lower state. In other words, the result is a net absorption of photons. This we often see in nature. For example, many crystalline copper salts have two energy bands, separated by photon energy corresponding to yellow light. Thus, when viewed in white light, the yellow part is absorbed, and the crystal transmits and reflects the complementary colour, blue. Ruby (chromium ions in crystalline alumina) has an absorption band in the green by this mechanism, and hence looks red in white light.

When light is absorbed, the population of the upper level is increased. Normally this perturbation from the equilibrium condition is small. But if we have an increasingly intense 'pump' light source, the number in level 3 will go on increasing, by the same amount as those in level 1 decrease. Fairly obviously, there is a limit, when the levels are equally populated, and the pump is infinitely strong. This is illustrated in Fig. 12.2. For the case of intense pumping, the non-equilibrium level populations (denoted by an asterisk) become almost equal:

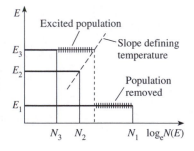

Fig. 12.2
The three-level system. The strong 'pump' signal has equalized levels E_1 and E_3, so that E_3 now has a greater population than E_2. The dotted line shows how population is changing with energy, as in Fig. 12.1., but now it has a positive slope.

$$N_1^* \simeq N_3^* \simeq \frac{N_1 + N_3}{2}. \tag{12.14}$$

Now let us consider a three-level system, with the third level E_2 between E_1 and E_3, also shown in Fig. 12.2. The pumping will have no effect on its population, which is the equilibrium value N_2. So with the three-level system strongly pumped, the number of electrons in the three states are N_1^*, N_2, and N_3^*. Suppose that some photons come along with energy

$$hv_{32} = E_3 - E_2. \tag{12.15}$$

They will clearly interact with the system, causing stimulated emission by transitions from E_3 to E_2 and absorption by transitions from E_2 to E_3. But now we have an unnatural occurrence: there are more electrons in the upper state (E_3) than in the lower (E_2). So instead of there being a net *absorption* of photons of energy, hv_{32}, there will be a net *emission*. The three-level system will amplify a photon of frequency, v_{32}, which is called the *signal frequency*. The whole thing is called a *laser*, which stands for *l*ight *a*mplification by *s*timulated *e*mission of *r*adiation.

When there are more atoms in an upper than a lower level, as in the case of E_3 and E_2 in Fig. 12.2, it is justifiable jargon to speak of an 'inverted population'.

The other point we should clear up, before describing some real system with inverted populations, is the one concerned with temperature. From eqn (12.1) the locus of the line representing the populations of the various energy levels

$$\frac{\mathrm{d}E}{\mathrm{d}N} = -\frac{kT}{N} \qquad (12.16)$$

(shown as a dotted curve in Fig. 12.1) has a negative slope proportional to T/N. Now look at Fig. 12.2. First consider the populations N_3^* and N_1^*. They are in a steady state in the sense that as long as the pump continues steadily, they do not change with time. But for these two levels with a finite energy difference there is virtually no difference in population. Therefore, if we regard eqn (12.16) as a way of defining temperature, for a well-pumped two-level system the temperature is infinite. If we now consider the energy-level populations at E_3 and E_2 in Fig. 12.2, we see that

$$N_3^* > N_2 \qquad (12.17)$$

and the $\mathrm{d}E/\mathrm{d}N$ locus has a positive slope, which by eqn (12.16) corresponds to a *negative temperature*.

Again, this is a fairly reasonable shorthand description of there being more atoms in an upper state than in a lower one. Now if you imagine a natural-state system pumped increasingly until it attains an infinite temperature and then eventually an inverted population, you will see there is *some* sense in the statement that a negative temperature is hotter than a positive one.

12.3 Lineshape function

So far we have assumed that energy levels are infinitely narrow. In practice they are not, and they cannot be as we have already discussed it in Section 3.10. All states have a finite lifetime, and one can use the uncertainty relationship in the form

$$\Delta E \Delta t = \hbar. \qquad (12.18)$$

We may now identify Δt with t_{spont}.

Since $E = h\nu$, we shall find for the uncertainty in frequency (which we identify with the frequency range between half power points, called also the linewidth)

$$\Delta \nu = \frac{1}{2\pi t_{\text{spont}}}. \qquad (12.19)$$

Unfortunately, the uncertainty relationship will not yield the shape of the line function. To find that we need to use other kind of physical arguments. But before trying to do that, let us define the lineshape, $g(\nu)$. We define it so that $g(\nu)\mathrm{d}\nu$ is the probability that spontaneous emission from an upper to a lower level will yield a photon between ν and $\nu + \mathrm{d}\nu$. The total probability must then be unity, which imposes the normalization condition

$$\int_{-\infty}^{\infty} g(\nu)\mathrm{d}\nu = 1. \qquad (12.20)$$

Let us stick for the moment to spontaneous decay (or natural decay) and derive the linewidth by a circuit analogy to which we have already appealed in Section 5.11.

A lossless resonant circuit has a well defined resonant frequency. However, in the presence of losses the resonance broadens. In what form will the voltage decay in a lossy resonant circuit? If the losses are relatively small, then circuit theory provides the equation

$$U(t) = U_0 \exp(-t/\tau) \cos 2\pi \nu_0 t. \tag{12.21}$$

What is the corresponding frequency spectrum? If the oscillations decay, then they can no longer be built up from a single frequency. The range of necessary frequencies, that is the spectrum, is given by the Fourier transform

$$f(\nu) = \int_0^\infty U(t) \exp(i2\pi \nu t) \mathrm{d}t. \tag{12.22}$$

Restricting ourselves to the region in the vicinity of ν_0 and after proper normalization, we obtain

$g(\nu)$ is known as a Lorentzian lineshape.

$$g(\nu) = \frac{(1/2)\pi \tau}{2\pi[(\nu - \nu_0)^2 + (1/2(\pi \tau))^2]}, \tag{12.23}$$

If we work out now the frequency range between the half-power points, we obtain

$$\Delta \nu = \frac{1}{2\pi \tau}, \tag{12.24}$$

which is the same as eqn (12.19) provided we identify the decay constant of the circuit with the spontaneous lifetime of the quantum mechanical state. So again, a simple argument based on the uncertainty relationship agrees with that based on a quite different set of assumptions.

In a practical case spontaneous emission is not the only reason why a state has finite lifetime. Interaction with acoustic waves could be another reason (electron–phonon collision in quantum mechanical parlance) or collisions with other atoms. The latter becomes important when lots of atoms are present in a gas, leading to so-called pressure broadening.

All those mentioned so far belong to the category of homogeneous broadening, where homogeneous means that conditions are the same everywhere in the material. When conditions differ (say strain varies in a solid) then we talk of inhomogeneous broadening.

The best example of inhomogeneous broadening is the so-called Doppler broadening, owing to the fact that an atom moving with velocity, v, will emit at a frequency,

$$\nu = \nu_0 \left(1 + \frac{v}{c}\right). \tag{12.25}$$

In thermal equilibrium the atomic gas has a Maxwellian velocity distribution, hence the corresponding broadening may be calculated. The result (see example 12.8) for the normalized lineshape is

$$g(\nu) = C_1 \exp[-C_2(\nu - \nu_0)^2], \tag{12.26}$$

where

M is the atomic mass.

$$C_1 = \frac{c}{\nu_0} \left(\frac{M}{2\pi kT}\right)^{1/2} \quad \text{and} \quad C_2 = \frac{M}{2kT} \left(\frac{c}{\nu_0}\right)^2. \tag{12.27}$$

12.4 Absorption and amplification

Let us look now at energy levels 2 and 3 and consider the induced transition rate between them. It is

$$W_{32} = B_{32}\rho(v) = \frac{c^3\rho(v)}{8\pi n^3 h v^3 t_{\text{spont}}}, \tag{12.28}$$

where eqns (12.11) and (12.13) have been used. The transition rate will of course depend on the lineshape function, so we need to multiply eqn (12.28) by $g(v)$. We shall also introduce the power density (measured in $W\,m^{-2}$) instead of the radiation density (measured in $J\,m^{-3}$) with the relation

$$I = \frac{c}{n}\rho, \tag{12.29}$$

leading to the form

$$W_{32} = \frac{c^2 I g(v)}{8\pi n^2 h v^3 t_{\text{spont}}}. \tag{12.30}$$

Now the number of induced transition per second is $N_3 W_{32}$ per unit volume, and the corresponding energy density per second is $N_3 W_{32} hv$. For upward transitions, we obtain similarly $N_2 W_{32} hv$, and hence the power lost in a dz thickness of the material is $(N_3 - N_2)W_{32}hv\,dz$. Denoting the change in power density across the dz element by dI, we obtain the differential equation,

$$\frac{dI}{dz} = \gamma(v)I, \quad \gamma(v) = (N_3 - N_2)\frac{c^2 g(v)}{8\pi n^2 v^2 t_{\text{spont}}}, \tag{12.31}$$

which has the solution,

$$I(Z) = I(0)\exp\gamma(v)z. \tag{12.32}$$

Under thermal equilibrium conditions $N_3 < N_2$, and consequently, the input light suffers absorption. However, when $N_3 > N_2$, that is there is a population inversion, the input light is amplified.

12.5 Resonators and conditions of oscillation

As we have said before, the energy levels are not infinitely narrow, hence emission occurs in a finite frequency band. For single-frequency emission (by single-frequency, we mean here a single narrow frequency range) all the excited states should decay in unison. But how would an atom in one corner of the material know when its mate in the other corner decides to take the plunge? They need some kind of coordinating agent or—in the parlance of the electronic engineer—a feedback mechanism. What could give the required feedback? The photons themselves. They stimulate the emission of further photons as discussed in the previous section and also ensure that the emissions occur at the right time. If we want to form a somewhat better physical picture of this feedback mechanism, it is advisable to return to the language of classical physics and talk of waves and relative phases. Thus, instead of a photon being emitted, we may say that an electromagnetic wave propagates in a way in which any two points bear strict phase relationships relative to each other. This phase information will be retained if we put perfect reflectors in the path of the waves on both sides, constructing thereby a resonator. The electromagnetic wave will then bounce to and fro between the two reflectors establishing standing waves, which also implies that the region between the two reflectors must be an integer multiple of half wavelengths. Thus, in a practical case, we have a relatively wide frequency band in which population inversion is achieved, and the actual frequencies of oscillation within this band are determined by the possible resonant frequencies of the resonator.

A resonator consisting of two parallel plate mirrors is known as a Fabry–Perot resonator after two professors of the Ecole Polytechnique, who followed each other (mind you, in the wrong order, Perot preceded Fabry).

What will determine the condition of oscillation* in a resonator? Obviously, the loop gain must be unity. If we denote the attenuation coefficient by α (discussed previously in Chapter 1, Section 5 and Chapter 10, Section 13— talking about lossy waves) the intensity in a resonator of length, l, changes by a factor, $\exp[(\gamma - \alpha)2l]$. Denoting further the two mirror's reflectivity by R_1 and R_2, respectively, we find that the condition for unity loop gain is

$$R_1 R_2 \exp[(\gamma - \alpha)2l] = 1. \qquad (12.33)$$

We know what determines γ. How can we find α? It represents all the losses in the system except mirror losses, which may be summarized as ohmic losses in the material, diffraction losses in the cavity, and losses due to spontaneous emission.[†]

Just one more word on diffraction losses. If the resonator consists of two parallel mirrors, then it is quite obvious that some of the electromagnetic power will leak out. In any open resonator there is bound to be some diffraction loss. Then why don't we use a closed resonator, something akin to a microwave cavity? The answer is that we would indeed eliminate diffraction losses, but on the whole we would lose out because ohmic losses would significantly increase.

12.6   Some practical laser systems

How can we build a practical laser? We need a material with suitable energy levels, a pump, and a resonator. Is it easy to find a combination of these three factors which will result in laser oscillation? It is like many other things; it seems prohibitively difficult before you've done it and exceedingly easy afterwards. By now thousands of 'lasing' materials have been reported, and there must be millions in which laser oscillations are possible.

There are all kinds of lasers in existence; they can be organic or inorganic, crystalline or non-crystalline, insulator or semiconductor, gas or liquid, they can be of fixed frequency or tunable, high power or low power, CW or pulsed. They may be pumped by another laser, by fluorescent lamps, by electric arcs, by electron irradiation, by injected electrons, or by entirely non-electrical means, as in a chemical laser. You can see that a mere enumeration of the various

* In fact, lasers are nearly always used as oscillators rather than amplifiers. So the phenomenon should be referred to as light oscillation by stimulated emission of radiation but, somehow, the corresponding acronym never caught on.

† This is one of the reasons why it is more difficult to obtain laser action in the ultra-violet and soft X-ray region. According to eqn (12.11) the coefficient of spontaneous emission increases by the third power of frequency.

realizations could easily take up all our time. I shall be able to do no more than describe a few of the better known lasers.

12.6.1   Solid state lasers

The first laser constructed in 1960 was a ruby laser. The energy-level diagram for the transitions in ruby (Cr ions in an Al_2O_3 lattice) is given in Fig. 12.3. I remarked above that ruby owed its characteristic red colour, to absorption bands of the complementary colour, green. This absorption is used in the pumping process. A typical arrangement is sketched in Fig. 12.4, which shows how the light from a xenon discharge flash tube 'pumps' the ruby to an excited state. Now the emission process is somewhat different here from that which I sketched previously for three-level systems. The atoms go from level 3 into level 2 by giving up their energy to the lattice in the form of heat. They spend a long enough time* in level 2 to permit the population there to become greater than that of level 1. So laser action may now take place between levels 2 and 1, giving out red light.

The ruby itself is an artificially grown single crystal that is usually a cylinder, with its ends polished optically flat. The ends have dielectric (or metal) mirrors evaporated on to them. Thus, as envisaged in the previous section, the resonator comprises two reflectors. Some power is certainly lost by diffraction, but these losses are small provided the dimensions of the mirror are much larger than the wavelength. It needs to be noted that one of the mirrors must be imperfect in order to get the power out.

Another notable representative of solid-state crystalline lasers is Nd^{3+}: YAG, that is neodymium ions in an yttrium–aluminium–garnet. It is a four-level laser radiating at a wavelength of $1.06\,\mu m$ pumped by a tungsten or mercury lamp.

Laser operation at the same frequency may be achieved by putting the neodymium ions into a *glass* host material. Glasses have several advantages in comparison with crystals: they are isotropic, they can be doped at high concentrations with excellent uniformity, they can be fabricated by a variety of processes (drilling, drawing, fusion, cladding), they can have indices of refraction in a fairly wide range, and last but not least, they are considerably cheaper than crystalline materials. Their disadvantage is low thermal conductivity, which makes glass lasers unsuitable for high average power applications.

* Energy levels in which atoms can pause for a fairly long time (a few milliseconds in the present case) are called 'metastable'.

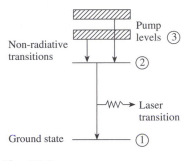

Fig. 12.3
Energy levels of the Cr^{3+} ion in ruby. The pump levels are broad bands in the green and blue, which efficiently absorb the flash tube light. Level 2 is really a doublet (two lines very close to each other) so that the laser light consists of the two red lines of wavelengths 694.3 and 692.9 nm.

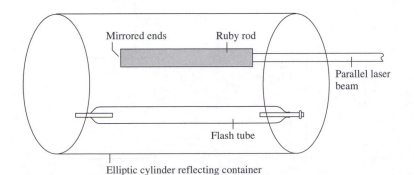

Fig. 12.4
General arrangement of a ruby laser. The ruby and the flash tube are mounted along the foci of the elliptic cylinder reflector for maximum transference of pump light.

12.6.2   The gaseous discharge laser

When a current is passed through a gas, as happens in a fluorescent lamp or a neon sign, most of the charged particles making up the current come from gas atoms that have been ionized by collision. But as well as completely dispossessing atoms of their electrons, the collisional process causes some bound electrons to gain extra energy and go into a higher state, that is, a state described by higher quantum numbers. You will remember that we had a formula for the simplest gas, hydrogen, in Chapter 4:

$$E_n = -\frac{13.6}{n^2}. \tag{12.34}$$

This shows that there is an infinite number of excited states above the ground state at $-13.6\,\text{eV}$, getting closer together as the ionization level $(0\,\text{eV})$ is approached.

In the helium–neon laser the active 'lasing' gas is neon, but there is about 7–10 times as much helium as neon present. Consequently, there are quite a lot of helium atoms excited to states about 20 eV above the ground state (Fig. 12.5). Now helium atoms in these particular states can get rid of their energy in one favourable way   by collision with other atoms that also have levels at the same energies. Since neon happens to have suitably placed energy levels, it can take over the extra energy making the population of the upper levels ($3a'$, $3b'$) more numerous than that of the lower level ($2'$), and thus laser action may occur. It is, of course, necessary to adjust gas pressures, discharge tube dimensions, and current quite critically to get the inverted population; in particular it is obtained only in a fairly narrow range of gas pressures around 1 Torr.

The reflectors are external to the tube, as shown in Fig. 12.6. Note that the windows are optical flats, oriented at the Brewster angle, θ_B, in order to

Fig. 12.5

The energy levels of interest for a helium–neon laser. Helium atoms get excited to levels $3a$ and $3b$ due to the impact of accelerated electrons. Neon atoms, which happen to have the same energy levels ($3a'$, $3b'$) collide with helium atoms and take over the extra energy. Laser action may now occur at two distinct wavelengths, corresponding to radiative transitions from levels $3a'$ and $3b'$ to a lower level $2'$.

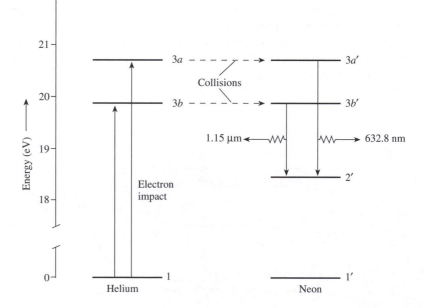

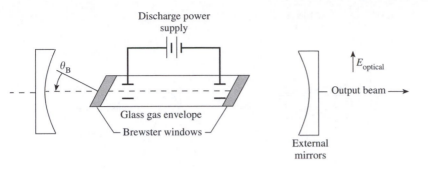

minimize reflections for the desired polarization. The advantage of spherical mirrors is that their adjustment is not critical, and they also improve efficiency. Dielectric mirrors are also used, not only because they give better reflections than metal mirrors but because they can also select the required wavelength from the two possible transitions shown in Fig. 12.5.

A close, though more powerful, relative of the He–Ne laser is the *argon ion* laser, operating in a pure Ar discharge. The pumping into the upper level is achieved by multiple collisions between electrons and argon ions. It can deliver CW power up to about 40 W at 488 and 514 nm wavelengths. It is in the company of the He–Ne laser, the one most often seen on laboratory benches.

The CO_2 laser is capable of delivering even higher power (tens of kW) at the wavelength of 10.6 μm. It is still a discharge laser, but the energy levels of interest are different from those discussed up to now. They are due to the internal vibrations of the CO_2 molecule. All such molecular lasers oscillate in the infrared; some of them (e.g. the HCN laser working at 537 μm) approach the microwave range.

12.6.3 Dye lasers

This is an interesting class of lasers, employing fluorescent organic dyes as the active material. Their distinguishing feature is the broad emission spectrum, which permits the tuning of the laser oscillations.

The energy levels of interest are shown in Fig. 12.7(a). The heavy lines represent vibrational states, and the lighter lines represent the rotational fine structure, which provides a near continuum of states. The pump (flashlamp or another laser) will excite states in the S_1 band ($A \rightarrow b$ transition) which will decay non-radiatively to B and will then make a radiative transition ($B \rightarrow a$) to an energy level in the S_0 band. Depending on the endpoint, a, a wide range of frequencies may be emitted. Finally, the cycle is closed by the non-radiative $a \rightarrow A$ transition. Unfortunately, at any given frequency of operation, there are some other competing non-radiative processes indicated by the dotted lines. A photon may be absorbed by exciting some state in the higher S_2 band, or there might be a non-radiative decay to the ground state via some other energy levels. There is net gain (meaning the gain of the wave during a single transit between the reflectors) if the absorptive processes are weaker than the fluorescent processes.

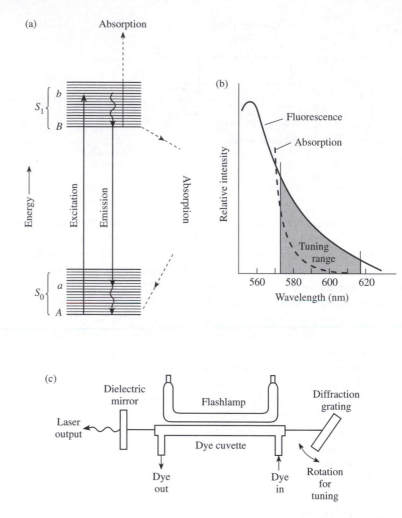

Fig. 12.7
(a) The relevant energy levels of a dye molecule. The wavy arrows from *b* to *B* and from *a* to *A* represent non-radiative transitions. The broken lines leading to the right also represent non-radiative transitions in which some other states are involved. (b) The tuning range of rhodamine 6G as a function of wavelength. (c) Schematic representation of a tuneable dye laser.

Note that this range is not the end of the dye laser's tuneability. By choosing the appropriate dyes any frequency within the visible range may be obtained.

The tuning range of a specific dye laser (rhodamine 6G) is shown in Fig. 12.7(b) by the shaded area, where the fluorescent and absorption curves are also plotted as a function of wavelength. Laser action becomes possible when the absorption curve intersects the fluorescence curve. At the long wavelength extreme, the gain of the laser (meaning the gain of the wave during a single transit between the reflectors) becomes too small for oscillation, as a result of the decrease in fluorescence efficiency.

How can we tune the laser? An ingenious solution is shown in Fig. 12.7(c), where one of the mirrors is a rotatable diffraction grating. The oscillation frequency of the laser will be determined by the angular position of the grating, which will reflect a different frequency at each position. The tuneable range is a respectable 7%.

12.6.4 Gas-dynamic lasers

The essential difference between these lasers and all the others discussed so far is that no electric input is needed. One starts with a hot gas (e.g. CO_2)

in the so called stagnation region. Then most of the energy is associated with the random translation and rotation of the gas molecules and only about one-tenth of the energy is associated with vibration. Next, the gas is expanded through a supersonic nozzle, causing the translational and rotational energies to change into the directed kinetic energy of the flow. The vibrational energy would disappear entirely if it remained in equilibrium with the decreasing gas temperature. But the vibrational relaxation times are long in comparison with the expansion time, hence the population of the vibrational levels remains practically unchanged. At the same time, the lower level population diminishes rapidly with the expansion, leading to significant population inversion after a few centimetres downstream. For CO_2 gas the emission wavelength is again $10.6\,\mu m$, using other gases the typical range is from 8 to $14\,\mu m$, although oscillations may be achieved at much shorter wavelengths, as well.

The advantage of gas-dynamic lasers is the potential for high average powers because waste energy can be removed quickly by high-speed flow.

12.6.5 Excimer lasers

Excimers are molecules which happen to be bound in an excited state and not in the ground state, so their operation differs somewhat from the general scheme. Their main representatives are the rare gas halides like KrF or XeCl. They need powerful pumps in the form of discharges, optical excitation or high-current, high-voltage electron beams. Their advantages are high efficiency and high pulse energy in a part of the spectrum (in the ultraviolet down to wavelengths of about 100 nm) which was inaccessible before. Most of them are inherently broadband and offer the further advantage of tuneability.

12.6.6 Chemical lasers

As the name suggests, the population inversion comes about as a result of chemical reactions. The classification is not quite clear. Some of the excimer lasers relying on chemical reactive collisional processes could also be included into this category. The clearest examples are those when two commercially available bottled gases are let together, and monochromatic light emission is brought about by the chemical reaction.*

12.7 Semiconductor lasers

We shall dwell on semiconductor lasers a little longer because they are in a quite special category. For us they are important for the reason that we have already invested much effort in understanding semiconductors, so that any return on that investment is welcome. There are, though, some more compelling reasons as well.

1. They are of interest because the technology and properties of semiconductors are better known than those of practically any other family of materials.
2. Laser action is due to injection of charged carriers, so semiconductor lasers are eminently suitable for electronic control.
3. They have high efficiency.

* The advent of chemical lasers raises an intriguing problem I have often asked myself. What path would technology have followed if electricity had never been discovered? The question may be posed because electricity and technology developed separately, the former being a purely scientific pastime until the fourth decade of the last century. Had scientists been less interested in electricity or had they been just a bit lazier, it is quite conceivable that the social need for fast communications (following the invention of the locomotive) would have been satisfied by systems based on modulated light. In the search for better light sources, the chemical laser could then have been invented by the joint efforts of chemists and communication engineers a century ago.

4. They operate at low voltage.
5. They are small.
6. They are robust.
7. They have long life.
8. The technique of their production is suitable for mass manufacture, so they are potentially inexpensive.
9. They can be produced in arrays.
10. They may be made to work in the wavelength range in which optical fibres have favourable loss (near to minimum) and dispersion properties.

I am sure if I tried hard, I could come up with a few more advantages but, I think, ten are enough to show that semiconductor lasers merit special attention.

How does a semiconductor laser work? The basic idea is very simple. It is radiative recombination in a direct gap semiconductor which leads eventually to laser action. Why a direct gap? Because we want the probability of a transition from the bottom of the conduction band to the top of the valence band to be high. What else do we need? We need a piece of material in which there are lots of electrons in the conduction band eager to descend, and in which there are lots of empty states at the top of the valence band eager to receive the electrons. A homogeneous piece of semiconductor is obviously not suitable because we cannot achieve both conditions simultaneously, only one at a time. But that gives an idea. We can have lots of electrons in a degenerate (discussed in Section 9.10 when talking of tunnel diodes) n-type semiconductor and, similarly, we can have lots of holes in a degenerate p-type semiconductor. So let us put them together, that is produce a p–n junction, and then in the middle of it both conditions may be expected to be satisfied, provided the forward bias, U_1, is close to the energy gap.

The energy band structure and the distribution of electrons and holes for this case are shown in Fig. 12.8(a) and (b) for thermal equilibrium and for forward bias, respectively. The overlap region in the middle of the junction, where both electrons and holes are present with high density, is called the active region, and that is where radiative recombination takes place. In order to keep up the process, whenever an electron–hole pair disappears by emitting a photon, it must be replaced by injecting new carriers.

(a)

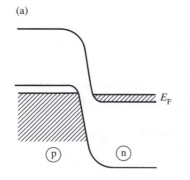

(b)

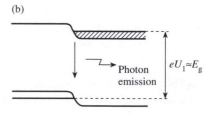

Fig. 12.8

A degenerate p–n junction at (a) thermal equilibrium, (b) forward bias.

If we examine the simple model shown in Fig. 12.9, the total number of electrons in the active region is $N_e\,lwd$, where N_e, as usual, denotes the density of electrons. The rate of change of the number of electrons, due to spontaneous recombination, is $N_e\,lwd/t_{rec}$, and this loss should be replenished by injection of electrons, that is the number required is $(I_i/e)\eta$, where I_i is the injected current and η, the quantum efficiency, is the fraction of injected electrons which recombine radiatively, leading to the formula

$$\frac{N_e lwd}{t_{rec}} = \frac{I_i \eta}{e}. \tag{12.35}$$

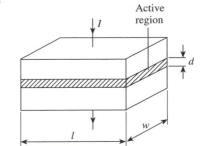

Fig. 12.9
Schematic representation of a laser diode.

Recognizing now that the recombination time in the above equation corresponds to t_{spont} discussed before, we may now use eqn (12.31) to find the amplification of the optical wave. For simplicity, we may take the population of the lower level as zero, and obtain

$$\gamma(\nu) = \frac{c^2 g(\nu)\eta}{8\pi n^2 \nu^2 elwd} I_i. \tag{12.36}$$

For laser oscillations we need the loop gain to be unity. When both mirrors have the same reflectivity, the condition of oscillation is

$$R \exp(\gamma - \alpha)l = 1, \tag{12.37}$$

whence the threshold current density is

$$\frac{I_i}{lw} = \frac{8\pi n^2 \nu^2 ed}{c^2 \eta g(\nu)} \left(\alpha - \frac{1}{l}\ln R \right). \tag{12.38}$$

As we have said before α represents the losses in the material. But are there any losses at all? The optical wave propagating in the active region will surely grow and not decay. True, but there is no reason why the optical wave should be confined to the active region. A well calculated plunge (one we shall not take here) into the mysteries of electromagnetic theory would show that a not inconsiderable portion of the electric field propagates outside the active region, where there is no population inversion. The losses there are mainly caused by the so called free-carrier absorption, which comes about by electrons and holes excited to higher energies within their own bands.

What is the value of R? In the simplest laser diode the mirror consists of the cleaved end of the semiconductor crystal, that is one relies on the difference in refractive index between semiconductor and air. A typical* refractive index is 3.35, which yields for the reflection coefficient, $R = 0.292$. For a practical case (see example 12.10) the threshold current comes to a value of about $820\,\mathrm{A\,cm^{-2}}$. This is quite a large value. Can we reduce it by some clever trick? Yes, we can, and the trick is to use a heterojunction instead of a homojunction. A schematic drawing of the device is shown in Fig. 12.10.

What is the role of the various layers? The insulating SiO_2 layers are there in order to steer the current toward the middle of the device and thus increase the current density. The heavily doped GaAs layers next to the metal electrodes are there to provide ohmic contacts. The p-type and n-type AlGaAs layers serve to provide the p–n junction, and then we come to the star of the show, the thin

* We are concerned here with power. R is obtained by squaring the amplitude reflection coefficient given by eqn (10.17).

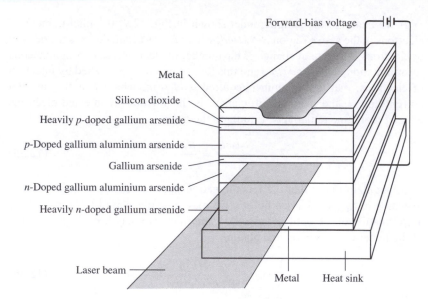

Fig. 12.10

Schematic drawing of a double
heterojunction diode laser.

layer ($\approx 100\,$nm) of GaAs in between. Very remarkably, we can kill two birds
with that one layer. It will serve both to confine the wave and to confine the
carriers.

It is very easy to see why the wave is confined. GaAs has a higher index of
refraction than AlGaAs, hence the mechanism of confinement is simply that
of total internal reflection, as mentioned in Section 10.13 when talking about
optical fibres.

Why are the carriers confined? Well, we have been through this before,
haven't we? We discussed this type of heterojunction in Section 9.15 and came
clearly to the conclusion that the electrons are confined to a very narrow potential well. So why do I ask this question again? The reason is that the confinement
of carriers is due now to a different mechanism. The crucial thing is still the
lower energy gap of GaAs relative to AlGaAs, but we no longer rely on the
triangular potential well for confinement.

To see in detail what happens in the p-type AlGaAs–undoped GaAs–n-
type AlGaAs heterojunction, I shall first show the energy diagram at thermal
equilibrium [Fig. 12.11(a)]. This is drawn by exactly the same technique which
led to Fig. 9.41(c). The triangular well we have seen in Section 9.15 is there at
the right-hand junction. A new kind of triangular well, in which the holes are
confined, may be seen at the left-hand junction.

What happens when we apply a forward bias? The barriers decline
[Fig. 12.11(b)], but in contrast to those in homojunction, the remaining barriers (for electrons towards the left and for holes towards the right) are still high
enough to prevent carriers spilling over into the oppositely doped region and
disappearing by the wrong kind of non-radiative recombination. The electrons
injected from the left have little other choice but to take the plunge into the
empty states in the valence band and emit a photon, meanwhile.

The threshold current density of our heterojunction will be much smaller
because the fraction of electrons which recombine radiatively will be much

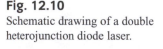

Note, however, that our aim is
now to confine the carriers to the
narrow GaAs region and not to
the extremely narrow triangular
wells. These blips in the energy
diagram are now embarrassments
rather than assets. In fact, by gradually increasing the proportion of
Al in the junction, the blips can be
removed (we no longer show them
in Fig. 12.11(b)).

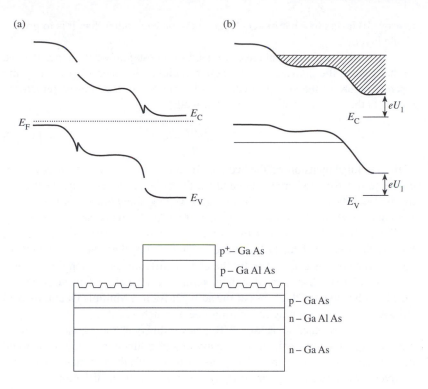

Fig. 12.11
An AlGaAs–GaAs heterojunction (a) at thermal equilibrium, (b) at a forward bias of eU_1.

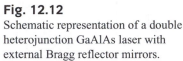

Fig. 12.12
Schematic representation of a double heterojunction GaAlAs laser with external Bragg reflector mirrors.

higher. The threshold current will also be small on account of the small thickness, d, of the active region [see eqn (12.38)] which is 100 nm in our example.

Can we further decrease the threshold current? Yes, both common sense and eqn (12.38) tell us that we need higher reflectivity mirrors. One way of doing it is to use an external Bragg reflector, as shown in Fig. 12.12. Each slight corrugation will cause a small reflection which all add up in phase at the right wavelength.* By these means threshold currents as small as 0.5 mA have been achieved.

It is very nice, indeed, to reduce the threshold current because that will reduce the power consumption of devices (e.g. compact disk players) using semiconductor lasers. But those lasers have to deliver a certain amount of power. There is no way of getting out a fair amount of power without putting in a fair amount of power, so it is also of crucial importance how the output power increases as the current exceeds its threshold value.

Let us now come back to the role of d, the thickness of the active region. As we reduce it, the threshold current decreases simply because fewer electrons need to be supplied to make up for spontaneous emission. Fewer electrons being available will also reduce the achievable power output. For this reason d cannot be usefully reduced to a value smaller than about 100 nm. Actually, if our aim is to reduce the thickness further without losing output power, we could simply increase the number of wells, say by a factor of 10, and make each of them a thickness of $d/10$. The advantages of doing so are far from obvious. On the other hand, it is easy to see a major disadvantage. It must be more complicated

* The corrugations (or slight bumps) can actually be inside the laser, in which case we talk about a Distributed Bragg Reflector laser or DBR.

to grow 10 layers of AlGaAs–GaAs layers upon each other than it is to grow a single layer.

In order to see the advantages, we need to investigate what happens as we further reduce the thickness of the active region. The main effect is that the discrete nature of the energy levels will be more manifest. Let us remember (eqn 6.1) the energy levels in a potential well:

$$E = \frac{h^2 n^2}{8m L^2}. \tag{12.39}$$

If the lateral dimension of the well is 10 nm, then the lowest energy level comes to 0.056 eV (where we have taken the effective mass of the electron at $m^* \, m^{-1} = 0.067$). In terms of the energies we talk about this is not negligible. It comprises about 4% of the energy gap of GaAs. If this is the lowest energy available above the bottom of the conduction band, and similarly, there is a highest discrete level for holes in the valence band, then the wavelength of emitted radiation is determined by the energy difference between these levels (see Fig. 12.13). Thus, one advantage should now be clear. Our laser can be tuned by choosing the thickness of the active layer in a Multiple Quantum Well (MQW) device. The tuning range might be as much as 20%.

Are there any other advantages? To answer this question, we need to make a digression and look again at the density of states function which we worked out in Chapter 6. Let us start with the energy levels of a three-dimensional well, as given by eqn (6.2) but permitting well dimensions to be different:

$$E = \frac{h^2}{8m} \left(\frac{n_x^2}{L_x^2} + \frac{n_y^2}{L_y^2} + \frac{n_z^2}{L_z^2} \right). \tag{12.40}$$

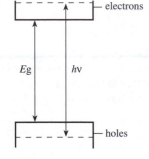

Fig. 12.13

The frequency of the emitted radiation, v, depends on the positions of the quantized energy levels in the conduction and valence bands.

In MQW lasers the dimensions L_y and L_z are much larger than $L_x = d$, the thickness of the active region. We may just as well take $L_y = L_z = l$, with which eqn (12.40) modifies to

$$E = E_0 \left[n_x^2 + \left(\frac{d}{l} \right)^2 \left(n_y^2 + n_z^2 \right) \right], \tag{12.41}$$

where

$$E_0 = \frac{h^2}{8m \, d^2}. \tag{12.42}$$

It is clear from eqn (12.41) that n_x has a much higher influence on the allowed energies then n_y and n_z. There will be big steps at $n_x = 1$, 2, 3, etc. Our primary interest is in the density of states because that will tell us that how many electrons within an energy range dE can make the plunge downwards.

Next, let us determine the density of states in the region between $n_x = 1$ and $n_x = 2$. This is then determined by n_y and n_z. Within a radius of $n(\gg 1)$ the number of possible states are πn^2, since there is a state for each integer value of n_y and n_z. Solving for $n^2 = n_y^2 + n_z^2$ from eqn (12.41) we obtain

$$n^2 \pi = \pi \left(\frac{l}{d} \right)^2 \left(\frac{E}{E_0} - 1 \right). \tag{12.43}$$

* This is exactly analogous to the calculation of the three-dimensional density of states we performed in Section 6.2.

Thus, eqn (12.43) gives the number of states* having energies less than E.

Similarly, the number of states having energies less than $E + dE$ is

$$\pi \left(\frac{l}{d}\right)^2 \left(\frac{E + dE}{E_0} - 1\right). \tag{12.44}$$

Consequently, the states having energies between E and $E + dE$ (and that defines the density of states function $Z(E)$) may be found as

$$Z(E)dE = \pi \left(\frac{l}{d}\right)^2 \frac{dE}{E_0}. \tag{12.45}$$

Remember that only positive integers count, so we need to divide by 4. On the other hand, there is spin as well, which is taken into account by multiplying by 2. Thus, the density of states in eqn (12.45) needs to be divided by 2. That is actually a minor detail. The important thing is that the density of states is independent of energy in the range $n_x = 1$ to $n_x = 2$. Taking $l = 300\,\mu m$ and $d = 10\,nm$, we get $(l/d)^2 = 9 \times 10^8$. Thus, when $n^2 = n_y^2 + n_z^2 = 27 \times 10^8$, then E reaches the value of $4E_0$. We may, however, alternatively obtain an energy $4E_0$ with n_y and n_z being very small and $n_x = 2$. It is clear that above $4E_0$, the same energy level may be reached in two different ways: with n_y and n_z relatively small and $n_x = 2$ or with n_y and n_z large and $n_x = 1$. Thus, the number of available states suddenly double at $E = 4E_0$. Between $E = 4E_0$ and $E = 9E_0$, the density of states remains constant again, and there is a new contribution at $E = 9E_0$, which leads to trebling of the initial density of states.

The fruit of our calculations, $Z(E)$ as a function of energy for a two-dimensional potential well, is shown in Fig. 12.14. The density of states increases stepwise at the discrete points $E_n = n_x^2 E_0$, where it reaches the value

$$Z(E_n) = \frac{n_x}{2E_0} \left(\frac{l}{d}\right)^2. \tag{12.46}$$

Eliminating n_x, we may obtain the envelope function (dotted lines) as

$$Z(E_n) = \frac{1}{2} E_n^{1/2} (E_0)^{-3/2} \left(\frac{l}{d}\right)^2, \tag{12.47}$$

which gives the same functional relationship as that found earlier (eqn 6.10) for the three-dimensional density of states function.

The truth is that it is the two-dimensional density of states function which is responsible for the superior performance of quantum wells but to provide a quantitative proof is beyond the scope of the present book. We shall, instead, provide a qualitative argument:

In lasers without two-dimensional confinement, the low energy states near to the bottom of the band play no role. When the probability of occupation is taken into account, maximum inversion (which defines the centre frequency of the laser) occurs at an energy higher than the gap energy. These low lying states are wasted, hence their elimination in MQW lasers is beneficial. Consequently, when the injected current is increased above its threshold, better use is made of the available electrons. There is, therefore, a much higher increase of output power with current. This also implies a faster reaction to the increase or decrease

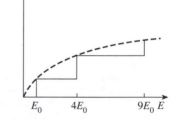

Fig. 12.14
The two-dimensional density of states as a stepwise function of energy.

of current, hence MQW lasers may be electronically modulated up to higher frequencies.

O.K., you might argue, MQW lasers are superior in performance, but surely they are much more expensive. Oddly enough they are not. When they are produced by one of the new techniques (MBE or MOCVD) they hardly cost more than ordinary semiconductor lasers. With GaAs comprising the active region, it is possible to produce lasers in the wavelength region 650–850 nm by varying the thickness of the quantum wells, although they are commercially available only at a few of these wavelengths. If it is possible to confine carriers in one dimension then, surely, it is possible to confine them in two dimensions. The resulting structures are called quantum wires. We can determine the density of states for that configuration by following the arguments used for quantum wells. Assuming a wire of square cross section with side d which is of the order of 100 nm, we can take in Eqn (12.40)

$$L_x = L_y = d \quad \text{and} \quad L_z = \ell \quad \text{where } d \ll \ell \quad (12.48)$$

Now the variation in n_x and n_y lead to sudden discrete changes in energy whereas the variation in n_z can be regarded as smooth continuous change so we can still talk about the density of states. The calculation is left to the reader. The result for the $n_x = n_y = 1$ case is

$$Z(E) = \frac{1}{2}\frac{\ell}{d}\big[E_0(E - 2E_0)\big]^{-1/2} \quad (12.49)$$

There is a singularity at $E = 2E_0$ which means a discrete state. And there are of course singularities for all integral values of n_x and n_y.

Is there any interest in producing devices using quantum wires? There are a few laboratories interested but, on the whole, it has not been a success. It is a kind of half-way house. If we want to do more carrier confinement, why not go the whole hog and confine them in all three dimensions? This leads us to the quantum dot in which all dimensions* are small. All the energy states, both for electrons and holes, are now discrete and the levels depend on size. What are the advantages of quantum dot lasers? Higher spectral purity, lower threshold current and ability to work at high temperatures, all because the energy levels are discrete and there is a much more efficient use of the available electrons. It is also easy to design a quantum dot laser to work at a given wavelength. The energy levels depend only on the size of the dot. The problems are technological, how to make them of the same size, how to control their spatial distribution and how to incorporate them in the active layer. Some quantum dot lasers have already appeared on the market at the infrared end. They have high potential but even if they win eventually it will take quite some time. Meanwhile we have to be satisfied with more traditional techniques, with confinement in one dimension that can nevertheless produce diodes capable of lasing at practically any required wavelength from the infrared to the ultraviolet. It is done by a technique called bandgap engineering based on the properties of semiconductor compounds. Each compound has a lattice spacing and an energy gap as shown in Fig. 12.15, where the lattice spacing is plotted against energy gap. It may for example be seen that the line connecting GaAs with AlAs is nearly horizontal, that is by adding judicious amount of Al to GaAs, we can realize compounds

* That raises the question that how many dimensions quantum dots have. If quantum wells are two-dimensional and quantum wires one-dimensional then quantum dots, which confine the electrons in one fewer dimension, cannot be anything but zero-dimensional. It is an odd terminology but one can get used to it.

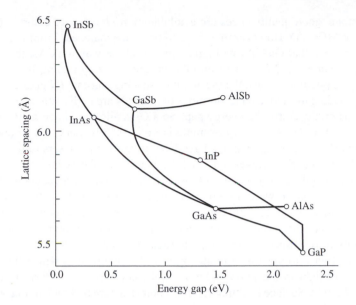

Fig. 12.15
Relationship between energy gap and lattice spacing for some mixed III–V semiconductors.

which have a fairly large range of energy gaps with roughly the same lattice constant.

The general problem may be stated as follows. Once the wavelength is chosen (say 1.55 μm, desirable for fibre communications) 3 compounds with approximately the same lattice spacing* must be found: compound 1 to serve as the highly doped substrate, compound 2 to provide the active region, and compound 3 to provide the material with the higher energy gap and lower refractive index.

Bandgap engineering has become a sophisticated science. A guide to materials and laser wavelength ranges is given in Table 12.1 and 'road map' of how to construct ternary compounds for particular band gaps and lattice spacing is given in Fig. 12.15, where we give most of the III–V compounds having a cubic lattice (zinc blende types). The hexagonal lattice of the nitrides does not fit this pattern, neither do they fit comfortably on any substrate. However, considerable alloying is possible within the InN, GaN and AlN materials to get practically any energy gap between 1.9 and 6.2 eV, even below 1.9 eV because of band bowing. For the cubic structures, alloys even stretch to quaternary compounds, which gives a possibility of a range of energy gaps for a fixed lattice, or vice versa.

The nitrides have added another facet to bandgap engineering. Generally among the III–Vs we find that large lattice spacing corresponds to small energy gap—see the extremes of InSb and GaP in Fig. 12.15; nitrogen is the smallest atom we consider in the semiconductor components of Table 8.2, even smaller than carbon and boron which have lower atomic numbers and weights. So adding N as isoelectronic replacement for As or P will decrease the mean lattice spacing, but the highly electronegative and piezoelectric nature of N means that the band gap is reduced by the large bowing factor.[†] There is an interesting group called the 'Gina' alloys (Ga In N As). Gallium arsenide has the most advanced technology of all the III–Vs so it is much in demand as

* Strain introduced by having somewhat different lattice constants can actually be beneficial as pointed out in Section 9.25 on heterostructures.

Table 12.1 *Compounds for laser diodes*

Wavelength range (nm)	Laser diodes based on
375–700	InGaN
600–900	AlGaAs
630–750	GaInP
870–1040	InGaAs
1040–1600	InGaAsP
1100–1670	GaInNAs

[†] When Ga and In are mixed in the proportion of $1 - x$ and x, one would expect the resulting energy gap to be $E_g(\mathrm{Ga}_{1-x}\mathrm{In}_x\mathrm{N}) = (1 - x)E_g(\mathrm{GaN}) + xE_g(\mathrm{InN})$ but the actual energy gap turns out to be smaller. An empirical formula gives the reduction in the form $bx(1 - x)$ where b is called the bowing factor. In any case it is difficult to predict the exact energy gap owing to the fact that nitrides do not form large single crystals or uniform alloy. There are compositional variations and strains as well as piezoelectric effects throughout the MOCVD layers.

a substrate, good quality slices are available to grow on other compounds by MBE or MOCVD. However they have to be lattice matched. From Fig. 12.15 it can be seen that GaAlAs will lattice match all the way to AlAs, that is the bandgap can go to 2.2 eV well into the visible. Now by adding N and In to GaAs it is possible to keep the mean lattice spacing constant (N goes down, In up), providing the added concentrations of In and N are in the ratio of 3:1. Both these materials reduce the energy gap. So a Gina alloy to lattice match GaAs is $Ga_{1-3x} In_{3x} N_x As_{1-x}$. An example is $3x = 0.53$ which gives a band gap of 0.74 eV. These infrared alloys have been used to make lasers to match the desirable optical fibre wavelength of 1.3 μm and for solar cells.

What should we do if we wish to have a high power semiconductor laser? Instead of one laser, we can produce an array of lasers (Fig. 12.16) grown on the same substrate and lightly coupled to each other. There may be as many as 40 diodes in an array capable of producing several watts of output power. The difficulty is to persuade all the lasers to radiate in phase.

Next, I wish to mention a relatively new development in which the diode lasers emit light in the same direction as the current flows. They are called Vertical Cavity Surface Emitting Lasers. Their structure is shown in Fig. 12.17. The active layer still consists of multiple quantum wells. The main difference is that the Bragg reflectors are at the top and bottom. They can be produced by the same techniques as the wells, and they can be made highly reflective. In the realization of Fig. 12.17, the reflector at the top has a reflection coefficient very near to unity, whereas the reflection coefficient of the bottom reflector is somewhat smaller, allowing the radiation to come through the transparent substrate. The area of the laser can now be made very small leading to even smaller threshold currents ($\ll 0.1$ mA). A further advantage is the ease with which arrays can be made. A two-dimensional array is shown in Fig. 12.18, where each microlaser may work at the same wavelength (to produce a high output) or may be tuned to different wavelengths.

Fig. 12.16
An array of lasers.

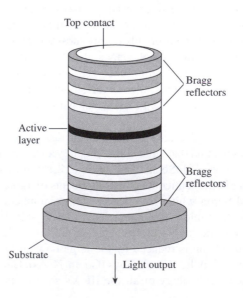

Fig. 12.17
Schematic representation of a Vertical Cavity Surface Emitting Laser.

Top contact

Bragg reflectors

Active layer

Bragg reflectors

Substrate

Light output

Before concluding the story of semiconductor lasers, it may be worth mentioning a relative, the Quantum Cascade laser, that does not quite belong to the family. The family trait, as repeated many times, is the descent of the electron from the conduction band to the valence band and the subsequent emission of a photon, of one single photon. The Quantum Cascade laser, conceived in the early 1970s, is an exception. All the things that matter happen in the conduction band.

The basic principle of operation of the Quantum Cascade laser is shown in Fig. 12.19. There are two semiconductor materials, A and B, which are alternately deposited upon each other (say, one hundred of them) by Molecular Beam Epitaxy (Fig. 12.19). A is the active material which has a conduction band edge much below that of semiconductor B. Lasing action takes place between energy levels 1 and 2. The wavelength of the emitted light depends on ΔE, the difference between the two energy levels. There is also a voltage applied across the whole sandwich. For simplicity let us assume that there is a voltage drop, V_B across each piece of semiconductor B but none across semiconductor A and choose this voltage so that $eV_B = \Delta E$. This ensures that energy level 1 on the left-hand side of B coincides with energy level 2 on the right-hand side of B.

What happens next? If semiconductor B is thin enough (just a few atomic layers), then electrons will tunnel from energy level 1 on the left-hand-side to energy level 2 on the right-hand-side. Once the electrons find themselves in the next piece of semiconductor A they are ready to descend again from level 2 to level 1 and produce a photon. If there are 50 layers of semiconductor A, then a single electron will produce 50 photons. From the point of view of the electron, this is like a cascaded ornamental waterfall. By the end the electron will have lost all its energy. From the point of view of the photon, this is an exercise in gathering strength.

The energy difference between levels 1 and 2 depends on the thickness of semiconductor A. Hence, the laser wavelength can be changed by choosing the appropriate material thickness. The wavelength range Quantum Cascade lasers can cover is large, from about 3 to 17 μm.

The principles upon which Quantum Cascade lasers work were enunciated in the 1970s but they have only very recently entered the market place. Why? You can appreciate the reasons: it is the extreme accuracy required. Each layer must have a certain number of atoms, not an atom more not an atom less.

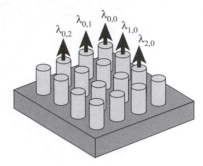

Fig. 12.18
An array of VCSEL lasers.

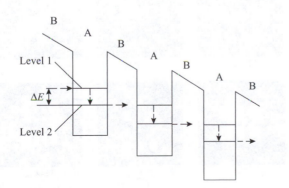

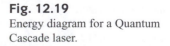

Fig. 12.19
Energy diagram for a Quantum Cascade laser.

As you can see, semiconductor lasers of all kinds have made and are making great leaps forward. They are poised to acquire the same dominance in lasers as other semiconductor devices enjoy in generating and amplifying lower frequency signals and in the field of switching.

12.8 Laser modes and control techniques

Having discussed the principles of operation of a large number of lasers, let us see now in a little more detail how the electric field varies inside a laser resonator and describe a few methods of controlling the mode purity and the duration of laser oscillations.

12.8.1 Transverse modes

What will be the amplitude distribution of the electromagnetic wave in the laser resonator? Will it be more or less uniform, or will it vary violently over the cross-section? These questions were answered in a classical paper by Kogelnik and Li in 1966, showing both theoretically and experimentally the possible modes in a laser resonator. The experiments were performed in a He–Ne laser, producing the mode patterns of Fig. 12.20. For most applications we would

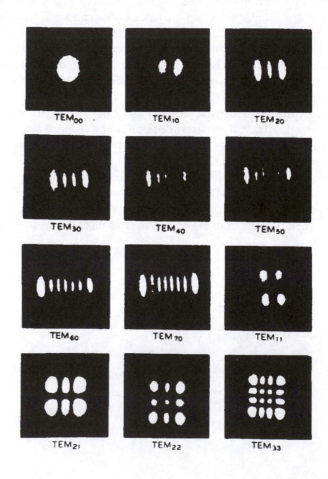

Fig. 12.20

Experimentally measured transverse mode patterns in a He–Ne laser having a resonator of rectangular symmetry (H. Kogelnik and T. Li, *Proc. IEEE 54*, pp. 1312–1329, Oct. 1966).

like a nice, clean beam as shown in the upper left-hand corner. How can we eliminate the others? By introducing losses for the higher order modes. This may be done, for example, by reducing the size of the reflector. Since the higher order modes have higher diffraction losses (they radiate out more), this will distinguish them in favour of the fundamental mode. However, this will influence laser operation in the fundamental mode as well; thus a more effective method is to place an iris diaphragm into the resonator, which lets through the fundamental mode but 'intercepts' the higher order modes.

12.8.2 Axial modes

As mentioned in Section 12.5 and shown in Fig. 12.21, laser oscillations are possible at a number of axial modes, each having an integral number of half wavelengths in the resonator. The frequency difference between the nearest modes is $c_m/2L$ (see Exercise 12.7) where L is the length of the resonator, and c_m is the velocity of light in the medium. How can we have a single frequency output? One way is to reduce the length of the resonator so that only one mode exists within the inversion range of the laser. Another technique is to use the good offices of another resonator. This is shown in Fig. 12.22, where a so-called Fabry–Perot etalon, a piece of dielectric slab with two partially reflecting mirror, is inserted into the laser resonator. It turns out that the resonances of this composite structure follow those of the etalon, that is the frequency spacing is $c_m/2d$, where d is the etalon thickness. Since $d \ll L$, single frequency operation becomes possible.

Are we not losing too much power by eliminating that many axial modes? No, we lose very little power because the modes are not independent of each other. The best explanation is a kind of optical Darwinism or the survival of the fittest. Imagine a pack of young animals (modes) competing for a certain amount of food (inverted population). If the growth of some of the animals is prevented, the others grow fatter. This is called *mode competition*.

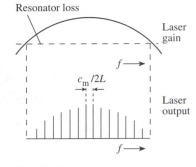

Fig. 12.21
The inversion curve of a laser and the possible axial modes as a function of frequency.

12.8.3 Q switching

This is a method for concentrating a large amount of power into a short time period. It is based on the fact that for the build-up of oscillations a feedback

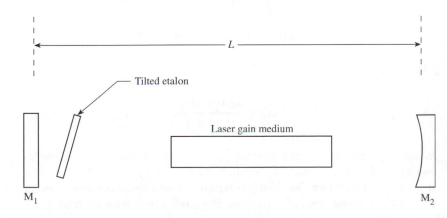

Fig. 12.22
Schematic representation of a laser oscillator in which single mode operation is achieved with the aid of an etalon.

mechanism is needed, usually provided by mirrors. If pumping goes on, but we spoil the reflectivity of one of the mirrors (i.e. spoil the Q of the resonator) by some means, then there will be a lot of population inversion without any output. If the reflectivity is restored (i.e. the Q is switched) for a short period to its normal value, the laser oscillations can suddenly build up, resulting in a giant pulse output. The pulse duration might be as short as a few nanoseconds, the power as much as 10^{10} W, and the repetition frequency may be up to 100 kHz. The easiest, though not the most practical, way of spoiling the Q is by rotating the mirror. The Q is then high only for the short period the mirrors are nearly parallel.

12.8.4 Cavity dumping

This is another, very similar method for obtaining short pulses also based on manipulating the Q of the resonator (called also 'cavity'; that's where the name comes from). We let the pump work and make the reflectivity 100% for a certain period, so the oscillations can build up but cannot get out. If we now lower the reflectivity to zero, all the accumulated energy will be dumped in a time equal to twice the transit time across the resonator. The method may be used up to about a repetition rate of 30 MHz.

12.8.5 Mode locking

We have implied earlier that it is undesirable to have a number of axial modes in a laser. This is not always so. The large number of modes may come useful if we wish to produce very short pulses of the order of picoseconds. The trick is to bring the various axial modes into definite relationships with each other. How will that help in producing short pulses? It is possible to get a rough idea by doing a little mathematics. Let us assume that there are $N + 1$ modes oscillating at frequencies $\omega_0 + l\omega$, where $l = (-N/2, \ldots, 0, \ldots, N/2)$, that they all have the same phase and amplitude, and they all travel in the positive z-direction (the set travelling in the opposite direction will make a similar contribution). The electric field may then be written in the form

$$\mathcal{E}(z,t) = \mathcal{E}_0 \sum_{l=-N/2}^{N/2} \exp[-\mathrm{i}(\omega_0 + l\omega)(t - z/c_\mathrm{m})]$$

$$= \mathcal{E}_0 \exp[-\mathrm{i}\omega_0(t - z/c_m)] F(t - z/c_\mathrm{m}), \qquad (12.50)$$

where $\mathcal{E}_0$ is a constant, and

$$F(x) = \frac{\sin(\frac{1}{2}N\omega x)}{\sin(\frac{1}{2}\omega x)}. \qquad (12.51)$$

Equation (12.50) represents a travelling wave, whose frequency is ω_0, and its shape (envelope) is given by the function, F. If $N \gg 1$, F is of the form of a sharp pulse of width $4\pi/N\omega$, and it is repeated with a frequency of ω. Taking $N = 100$, a resonator length of 10 cm, and a refractive index of 2, we get a pulsewidth of 27 ps and a repetition frequency of 750 MHz. The situation is, of course, a lot more complicated in a practical laser, but the above figures give good guidance. The shortest pulses to date have been obtained in dye lasers with pulsewidths well below 1 ps.

How can we lock the modes? The most popular method is to put a saturable absorber in the resonator which attenuates at low fields but not at high fields. Why would a saturable absorber lock the modes? A rough answer may be produced by the following argument: when the modes are randomly phased relative to each other, the sum of the amplitudes at any given moment is small, hence they will be adversely affected by the saturable absorber. However, if they all add up in phase, their amplitude becomes large, and they will *not* be affected by the saturable absorber. Thus, the only mode of operation that has a chance of building up is the one where the modes are locked, consequently, that will be the only one to survive in the long run (where long means a few nanoseconds).

12.9 Parametric oscillators

In principle, this is the same thing as already explained in connection with varactor diodes in Section 9.13. The main differences are that in the present case (i) the non-linear capacitance is replaced by a non-linear optical medium, (ii) the dimensions are now large in comparison with the wavelength; hence wave propagation effects need to be taken into account, and (iii) instead of amplifiers, we are concerned here with oscillators (although optical parametric amplifiers also exist). What is the advantage of parametric oscillators? Why should we worry about three separate frequencies, when we can easily build oscillators at single frequencies? The reason is that we can have tuneable outputs.

A schematic diagram of the optical parametric oscillator is shown in Fig. 12.23. The parametric pump (not to be confused with the pump needed to make the laser work) is a laser oscillating at ω_3. There is also a resonator which may resonate at ω_1 and ω_2. If the waves at these frequencies satisfy both the

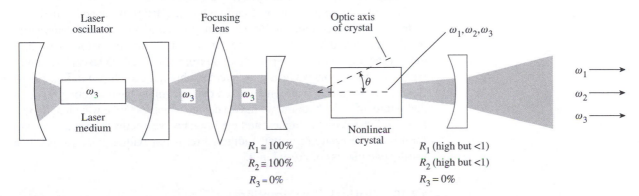

Fig. 12.23

Schematic representation of a tuneable parametric oscillator. R_1, R_2, and R_3 are the reflectivities of the mirrors at frequencies ω_1, ω_2, ω_3 respectively.

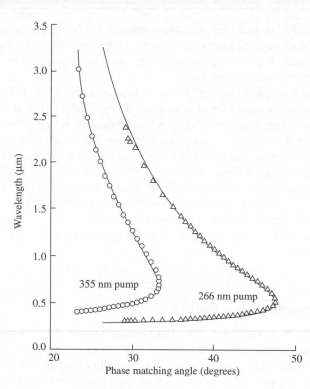

Fig. 12.24

Theoretical (solid lines) and experimental ($\circ$ and $\triangle$) tuning curves using BBO as the nonlinear medium.

$\omega_3 = \omega_1 + \omega_2$ and the $k_3 = k_1 + k_2$ (k = propagation coefficient as in our classical studies of Chapter 1) conditions, then there is a parametric interaction between the waves. The power at ω_1 and ω_2 builds up from the general noise background at the expense of pump power. Thus, we have output at all three frequencies. But why is this set-up tuneable? Because of the particular properties of the chosen non-linear medium. It is a crystal in which the dielectric constant is dependent on the direction of propagation. By rotating the crystal, the matching condition for the propagation constants is satisfied at another set of frequencies, ω_1' and ω_2', still obeying $\omega_3 = \omega_1' + \omega_2'$.

The crystal used most often is barium borate (BBO), produced abundantly in the People's Republic of China. It may be used in a pulsed parametric oscillator, pumped by either the third harmonic (355 nm) or by the fourth harmonic (266 nm) of the 1.066 μm radiation from a Nd^{3+}:YAG laser. The tuning range for either pump wavelengths is remarkably large as shown in Fig. 12.24. Remember, for a dye laser with a given dye, we might have a tuning range in the vicinity of 10%, but now we have a device which can tune wavelength by a factor of 7 between the highest and the lowest wavelengths. The price we pay for it is the necessity to use an additional resonator with a piece (in practice usually two pieces) of crystal in it.

12.10   Optical fibre amplifiers

It is quite obvious that amplifiers can be built on the same principles as oscillators, but usually there is less need for them. A field, however, in which amplifiers have crucial importance is long-distance communications. One might be able to span the oceans of the world by optical fibres without the need to regenerate the signal in repeaters, if the signals propagating in the fibres could be amplified.

The idea of using fibre amplifiers is just about as old as the oldest laser. There were experiments in the early 1960s with fibres doped with Nd. Population inversion could be achieved by pumping it with a flash-lamp, which then served to amplify a signal. This idea was resurrected in the middle 1980s, using another rare-earth element, erbium, as the dopant. Today, erbium doped fibre amplifiers (acronym EDFA) pumped by diode lasers are standard components in an optical fibre communication system. Without them the World Wide Web could have hardly come into existence.

12.11   Masers

As you probably know, the original acronym was 'maser', standing for microwave amplification by stimulated emission of radiation. Masers used to be employed as low noise amplifiers, but I believe they are out of favour now. They lost out to cooled parametric amplifiers. Nonetheless, I shall say a few words about them, partly to show an example of magnetic tuning and partly out of historical interest.

When discussing paramagnetism in Section 11.7, we came across the splitting of energy levels in an applied magnetic field. The possible energies are given by eqn (11.41):

$$E = -M_J g \mu_{\mathrm{mB}} B, \quad M_J = J, J - 1, \ldots, -J. \qquad (12.52)$$

The material used is ruby, which happens to be good both for lasers and masers. For trebly ionized chromium, the outer 3d-shell has three electrons of identical spin. Hence its total spin contribution is 3/2. The contribution from orbital angular momentum is taken as zero,* thus $j = 3/2$, leading to the values

$$M_J = \frac{3}{2}, \frac{1}{2}, -\frac{1}{2}, -\frac{3}{2}. \qquad (12.53)$$

Taking further $g = 2$ (corresponding to pure spin), we find the energy levels shown in Fig. 12.25(a). The energy levels found experimentally are illustrated in Fig. 12.25(b). The dependence on magnetic field may be seen to be well predicted by the simple theory, but not the split at zero magnetic field. As far as the maser is concerned, what matters is that its frequency of operation may be changed by varying the magnetic field. In other words, we have a tuneable maser. The magnetic fields required are reasonable and can be realized in practice.

* There is some theoretical justification for doing so, but the real reason is that unless orbital momentum is disregarded, there is no resemblance at all between theory and experiment.

12.12   Noise

Why can masers be used as low noise amplifiers? Mainly, because their operation is not dependent on the motion of charge carriers, whose density and velocity are subject to fluctuations. So, we managed to get rid of one source of noise, but we have now another kind of noise, namely that due to spontaneous emission.

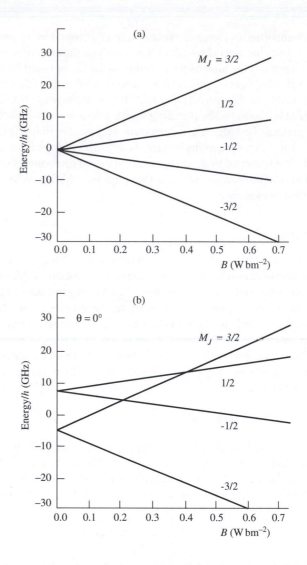

Fig. 12.25

The splitting of energy levels of Cr^{3+} ions in ruby as a function of magnetic field. (a) Plot of eqn (12.52) for the case when the orbital momentum is quenched and the angular momentum is due to spin only. (b) Experimental curves for B in the direction of the symmetry axis of the ruby crystal, $\theta = 0$.

The amount of noise generated in an amplifier may be characterized by a parameter called 'noise temperature'; a low 'noise temperature' means a small amount of noise. For a maser it can be shown that under ideal conditions* this noise temperature is numerically equal to the negative temperature of the emission mechanism. Hence, the aim is to have a low negative temperature, that is, large population inversion.

* Ideal conditions mean high gain, no ohmic losses, and no reflections from a noisy load.

How can we achieve large population inversion? With reference to our three-level maser scheme, we have to do two things: (i) pump hard so that the population of levels 3 and 1 become roughly equal; (ii) keep the device at a low temperature†, so that the relative number of atoms is higher in level 1.

† Low temperatures help incidentally in reducing the ohmic losses as well.

Be careful, we are talking now of three different 'temperatures'. The maser has to work at a low (ordinary) temperature to get a low negative (inversion) temperature, which happens to be equal to the noise temperature of the amplifier. Now what is the minimum noise temperature one can achieve? Can we approach the zero negative temperature and thus the zero noise temperature? We can certainly approach the zero negative temperature by cooling the amplifier towards 0 K. As the actual temperature approaches absolute zero, the ratio

$$\frac{\text{number of atoms in level 1}}{\text{number of atoms in level 2}}$$

tends to infinity. Hence, after pumping, the negative temperature tends to zero. But spontaneous emission does not disappear, since it is proportional to the number of atoms in level 3. Thus, the noise temperature cannot reach zero.

It turns out that, as the inversion temperature tends to zero, the noise temperature tends to the finite value of $h\nu/k$, where ν is the frequency of operation. When $\nu = 5 \times 10^9$ Hz, the limiting noise temperature comes to about 0.25 K. Experimental results on masers cooled to liquid helium temperatures are not far from this value. Noise temperatures around 2 K have actually been measured.

All I have said so far about noise applies to lasers as well, though the numerical values will be radically different. For the argon laser mentioned before, $\nu = 6.6 \times 10^{14}$ Hz, giving $T_{\text{noise}} \sim 30\,000$ K as the theoretically available minimum.

12.13 Applications

Finally I should like to say a few words about applications. What are lasers good for? Surprisingly, an answer to this question was not expected when the first lasers were put on the bench. It is true to say that never has so much effort been expended on a device with so little regard to its ultimate usefulness. Lasers were developed for their own sake.

I suppose that in the trade most people's reaction was that sooner or later something useful was bound to come out of it. Radio waves have provided some service (even allowing for the fact that radio brought upon us the plague of pop music); microwaves have been useful (how else could you see the Olympic finals in some far away country from your armchair in Tunbridge Wells); so coherent light should be useful for something.

The military who remembered that radar was useful also hoped that laser would be good for something, and they gave their blessing (and their money too!).

Another powerful contributing factor was the human urge to achieve new records. I could never understand why a man should be happier if he managed to run faster by one-tenth of a second than anyone else in the world. But that is how it is. If once a number is attached to some performance, there will be no shortage of men trying to reduce or increase that number (whatever the case may be). And so it is with coherent radiation. Man feels his duty to explore the electromagnetic spectrum and produce coherent waves of higher and higher frequencies.

There may have been some other motives too, but there was no unbridled optimism concerning immediate applications. What can we say some 40 years later? Well, the military were apparently right. They got a guidance system out of it which can direct a bomb dropped by an aeroplane into the middle of a plate of lentils, and there are, very likely, lots of other applications in the pipeline. The ray-gun, that favourite dream of boys, science fiction writers, and generals may not be very far from realization. What about civilian applications? There are many of them in the medical field; there is optical radar, but of course the most important applications to date have been the compact disc and optical communications. There are many scientific applications too. We shall start with them.

12.13.1   Nonlinear optics

The whole subject, the study of non-linear phenomena at optical frequencies, was practically born with the laser.

12.13.2   Spectroscopy

An old subject has been given a new lease of life by the invention of tunable lasers. Spectroscopists have now both power and spectral purity previously unattainable.

12.13.3   Photochemistry

Carefully selected high-energy states may be excited in certain substances, and their chemical properties may be studied.

12.13.4   Study of rapid events

With the aid of picosecond and sub-picosecond light pulses, a large number of rapidly occurring phenomena may be studied in physics, chemistry, and biology. The usual technique is to generate a phenomenon by a strong pulse and probe it by another time-delayed pulse. A field in which these techniques have been successfully utilized is the creation and decay of excitons in a semiconductor crystal.

12.13.5   Plasma diagnostics

Many interesting properties of plasmas may be deduced by their scatter of laser light.

12.13.6   Plasma heating

A plasma may be heated to high temperatures by absorbing energy from powerful lasers.

12.13.7   Acoustics

Properties of high-frequency (in the GHz range) acoustic waves in solids may be studied by interacting them with laser light.

12.13.8   Genetics

Chromosomes may be destroyed selectively by illuminating single cells with focused laser beams.

12.13.9 Metrology

The velocity of light may be determined from the relationship, $c = \nu\lambda$, by measuring the frequency and wavelength of certain laser oscillations. The laser is stabilized by locking it to a molecular absorption line, and its frequency is measured by comparing it with an accurately known frequency, which is multiplied up from the microwave into the optical range. The wavelength is measured independently by interferometric methods. The accuracy with which we know the velocity of light was improved this way by a factor of hundred.

I am sure there are hundreds of other scientific applications. A laser is a wonderful device for any research worker who has (temporarily, of course) run out of ideas. Whatever he is investigating, he may always ask the question: 'And how will this phenomenon be influenced by laser light?'

Let us leave now the scientific applications and look at some others. Perhaps one of the most obvious applications is the optical radar. If it can be done with microwaves, why not with lasers? The wavelength is much smaller, so we may end up with higher accuracy in a smaller package. This is indeed the case; some of the optical radars may weigh less than 20 kg and can recognize a moving car as a car and not a blotch on a screen. They can also determine the position of objects (e.g. clouds, layers of air turbulence, agents of pollution) which do not give sufficient reflection at microwave frequencies. The two lasers used most often are YAG lasers at 1.06 μm and CO_2 lasers at 10.6 μm, the latter has the merit of being able to penetrate fog, haze, and smoke. Optical radars are best known under the name of Lidar (light detection and ranging) but also as Ladar (laser detection and ranging) and Oadar (optical aids to detection and ranging).

A less obvious application is the compact disc known as the CD. I mean it is obvious in hindsight, but it was not easy to come to the idea and put it in practice. It is certainly the most successful consumer product that uses lasers. Since its introduction in 1982, about half a billion players have been sold, each one containing a semiconductor laser (probably a MQW laser). It stores information in digital form, regardless of its origin. The first application was for sound, and that is of course the main application at the moment, but CD-ROMS have recently become very popular too. The principle of operation is quite simple: recording is by a focused laser beam which burns holes in a disc; replay is by another laser beam which gets modulated by the holes.

The standard CD format will soon be abandoned for a new one called Digital Versatile Disk (DVD) which will be the same size but will be able to store about 15 times that much information. The increase is only partly due to the fact that the laser wavelength is reduced from 780 nm to about 640 nm (yielding a factor of $(780/640)^2 = 1.5$ increase in capacity), the rest is achieved by using more sophisticated modulation and error-correcting techniques. Note that DVDs will be capable of storing a full length film. A further big increase in capacity has come now when blue semiconductor lasers are commercially available. GaN lasers working at 410 nm enables a capacity of 9.4 Gbytes on a standard DVD.

Medical applications are growing fast, particularly in the United States, where the medical profession is much less conservative than in Europe. The essential property of lasers that comes useful is that the radiated energy can be concentrated on a small spot and that different tissues have different absorptions. A uniquely useful application is, for example, the reattachment of the human retina by providing the right amount of heat at the right place. Surgeons may use higher laser energy to vaporize tissue (a useful way to get rid of malignant tumours) or lower laser energy to coagulate tissue, that is stop bleeding. It is actually possible to make bloodless cuts without causing pain. The number of various medical applications is high (I understand in ophthalmology alone, as many as forty different problems are treated by lasers) but not very widespread as yet. I want to finish the list by mentioning one more, rather bizarre, application practised by some gynaecologists; to open up the Fallopian tube.

A laser beam can of course be focused not only upon human tissue but upon inanimate matter as well, making possible laser *machining* and *welding*. The same technique may also be useful for writing patterns on high-resolution photographic plates (possibly to be further reduced) used in integrated circuit technology.

An interesting application, which has recently been introduced to service, is for navigation, which necessitates the measurement of rotation. Lasers can detect rotational movement as low as a thousandth degree per hour. The basic principles may be understood from Fig. 12.26. This is a so-called ring laser, in which resonance is achieved by a ray biting its own tail. The condition of resonance is now that the total length around the ring should be an integral multiple of the wavelength. When the system is at rest (or moving with uniform velocity) the clockwise and anticlockwise paths are equal, and consequently the resonant wavelengths are equal too. However, angular rotation of the whole system makes one path shorter than the other one, leading to different frequencies of oscillation. The two beams are then incident upon a photodetector, which produces a current at the difference frequency. The rate of rotation may be deduced by measuring this difference frequency.

A simpler variant, aiming to do the same thing, uses a cylinder upon which hundreds of metres of optical fibres are wound. If the cylinder rotates, then the light path going clockwise is different from the light path going

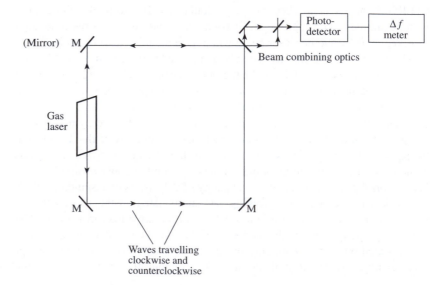

Fig. 12.26
Schematic representation of a laser
rotation sensor.

counterclockwise. The rotation rate may then be determined by measuring the path difference.

Talking of fibres, I must mention that rotation rate is just one of the numerous physical quantities which can be measured with the aid of light propagating in fibres. Sensors have already been built for measuring angular position, temperature, pressure, strain, acceleration, magnetic field, etc. The availability of fibres and semiconductor lasers has made it fashionable to convert all kind of input variables into light signals. The reason is low attenuation, flexibility (in both senses of the word), high information capacity, compactness, light weight and, last but not least, the potentially low price of the device.

I have already mentioned Communications several times. There is no doubt that lasers, combined with the advent of optical fibres, are responsible for the enormous increase in volume, and for the drastic decline in the cost of Transatlantic and Transcontinental calls. In the second edition of this book, published in 1979 the following prediction was made about optical communications: 'It is bound to come, and bound to be followed by cheap intercontinental communications. In 10 years time you will probably be able to call Uncle Billy in New York for ten pence.' Well, ten years was not quite the right prediction but if we talk about the present time, 2003, any point in America can be reached for three pence a minute. The future? Increased telephone traffic, increased data transmission traffic.

While on Communications I must mention a new development that may very well come. The problem to solve is known as that of the Last Mile concerned with transmitting information from the fibre terminal to each home. Since installing fibres into every house is rather expensive the present solution is to change to coaxial cable at the fibre terminal. The new solution envisaged is to put at the terminal the information on infrared lasers and radiate it directly, without any cables, into one's sitting room.

Let us turn now to some potential applications which may acquire high importance in the future. Take *laser fusion* for example. The chances of success seem fairly small, but the possible rewards are so high that we just cannot afford to ignore the subject. The principles are simple. As I have already mentioned, a plasma may be heated by absorbing energy supplied by a number of high-power pulsed lasers. The fusion fuel (deuterium and tritium) is injected into the reactor in the form of a solid pellet, evaporated, ionized, and heated instantly by a laser pulse, and the energy of the liberated neutrons is converted into heat by (in one of the preferred solutions) a lithium blanket, which also provides the much needed tritium.

Next in importance is another nuclear application, namely *isotope separation*. With the change from fossil to fissile energy sources, we shall need more and more enriched uranium. The cost of uranium enrichment in the USA for the next 20 years has been estimated at over 100 000 million dollars. Thus, the motivation for cheaper methods of separation is strong.

The laser-driven process, estimated to be cheaper by a factor of 20, is based on the fact that there is an optical isotope shift in atomic and molecular spectra. Hence, the atoms or molecules containing the desired isotope can be selectively excited by laser radiation. The separation of excited atoms may, for example, be achieved by a second excitation in which they become ionized and can be collected by an electric field.

A disadvantage of the process is that, once perfected, it will enable do-it-yourself enthusiasts (with possibly a sprinkling of terrorists among them) to make their own atomic bombs.

As the last application, I would like to mention *holography*, a method of image reconstruction invented by Dennis Gabor in 1948. It is difficult to estimate at this stage how important it will eventually turn out to be. It may remain for ever a scientific curiosity with some limited applications in the testing of materials. On the other hand it might really take off and might have as much influence on life in the 21st century, as the nineteenth century invention of photography has upon our lives. The technique is by no means limited to the optical region; it could in principle be used at any frequency in the electromagnetic spectrum, and indeed, holography can be produced by all kinds of waves including acoustic and electron waves. Nevertheless, holography and laser became strongly related to each other, mainly because holographic image reconstruction can most easily be done with lasers at optical frequencies.

The basic set-up is shown schematically in Fig. 12.27. The laser beam is split into two, and the object is illuminated by one of the beams. The so-called 'hologram' is obtained by letting the light scattered from the object interfere with the other beam. The pattern that appears depends both on the phase and on the amplitude of the scattered light, storing this information on a photographic plate, we have our 'picture', which bears no resemblance to the object at all. However, when the hologram is illuminated by a laser (Fig. 12.28) the original object will dutifully spring into life. The reconstructed wave forms appear to diverge from an image of the object. Moving the eye from *A* to *B* means

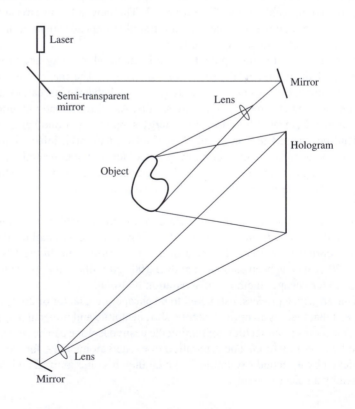

Fig. 12.27
Schematic representation of taking a hologram.

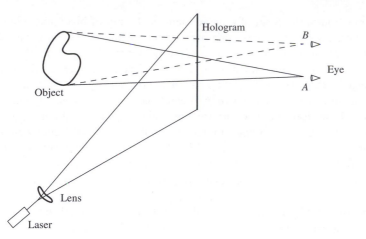

Fig. 12.28
Schematic representation of viewing a
hologram.

viewing the object from a different angle, and it looks different indeed just as
in reality. So the picture we obtain is as good as the object itself, if not better.
For examining small biological specimens, for example, the picture may be
better than the original because the original will not sit motionless under the
microscope. A hologram can be investigated at leisure without losing any of
the details, and one can actually focus the microscope to various depths in the
three-dimensional picture.

A variation on the same theme offering some advantages is volume holography, to which we shall return in more detail in Section 13.5. It uses a certain
volume in a photosensitive medium, in which a three-dimensional interference pattern is recorded in the form of refractive index variation. One of the
advantages of volume holograms is that the holographic reproduction is strongly
wavelength dependent—it is the Bragg effect again, contributions must be
added in the correct phase. Hence, the image may be viewed in white light,
with the hologram selecting the wavelength it can respond to from the broad
spectrum available. The Bragg effect is of course strongly dependent on incident angle as well. The wavelength and angular dependence together make it
possible to record multiple holograms in the same material.

The entertainment provided by holography has so far been confined to the
laboratory with the exception of a few exhibitions and a few art shops. To
produce a holographic image is expensive, and the rich have decided to spend
their money on other objects of luxury. So are there any commercial applications? Probably, only two. There are, as everyone knows, holograms on credit
cards (to make forgery more difficult), and there are, as some people know,
holograms in every supermarket scanner that reads bar-codes.

A scanner works in the following way. A series of holograms are recorded on
a disc. As the disc rotates and is illuminated by a laser (the original reference
beam) each small part of the disc gives rise to an object beam moving in a
different direction. The total effect is a continuously moving light beam that
scans the bar-code.

Finally, I wish to mention a potential application that has been talked about
for at least a quarter of a century. It is holographic storage. It is based on
the recording of multiple holograms. Each object beam may carry digital

information in the form of a two-dimensional array of black and white spots and there can be a very large number of object beams. When illuminated by the original reference beams, the object beams are reconstructed and the white or black spots are read by an array of detectors. The advantage of this method of storage is the massive parallel operation and the density of storage achievable. In principle, it has been shown by theoreticians, each wavelength cube of the material can store one bit of information. Taking a wavelength of 0.5 μm and a disc of 30 mm diameter and 3 mm thickness the theoretical figure would come to about 20 terabit, i.e. 20×10^{12} bits. In laboratories, well over 1 terabit has already been achieved, about 20 times as much as available from magnetic memories. Will holographic memories ever be a commercial success? I am sceptical for two reasons: the technical problems for mass production are still numerous and by the time they will be solved conventional memories might not be too far behind in capacity and would be a lot cheaper.

12.14   The atom laser

It would be quite legitimate to ask why we need another section on atom lasers, when so much has already been said about the various energy states of atoms and how leaping from one energy state to another one may lead to laser action.

The atom laser is only called a laser. It is not a proper laser in the sense that it has nothing to do with Light Amplification by Stimulated Emission of Radiation. A less often used alternative name, matter wave laser, however, gives away the secret. It is concerned with coherent matter waves in much the same way as ordinary lasers are concerned with coherent electromagnetic waves.

What do we need to produce an ordinary laser? We need to confine the photons by a resonator and ensure that they all have the same energy. In a matter wave laser the atoms need to be confined to a finite space, and all of them must be in the same state. If many atoms are to be collected in the same state, they must be bosons, as we briefly mentioned in Chapter 6.

How can we confine the atoms? If they have a magnetic moment, they can be trapped by magnetic fields. The simplest example of a trap is a magnetic field produced by two parallel coils carrying opposite current, which yield zero magnetic field in the centre. An atom moving away from the centre will be turned back.

How can we have a sufficient number of atoms in the ground state? By cooling the assembly of atoms, we can make more of them remain in the ground state. The lower the temperature, the larger the number of particles in the ground state. When the density is sufficiently large and the temperature is sufficiently low, we have a so-called Bose–Einstein condensation, which means that most of the atoms are in their ground state.

How do we know if we achieved a Bose–Einstein condensation? In the same manner as we know whether we have coherent electromagnetic radiation, we derive the two beams from a laser and make them interfere with each other. Coherence is indicated by the appearance of an interference pattern. Can we do the same thing with an atom laser? We can.

In a particular experiment, sodium atoms were trapped in a double well: there were two separate condensates, each one containing about five million atoms. The trap was then suddenly removed, and the atom clouds were let to fall for 40 ms. They were then illuminated by a probe beam from an ordinary laser. The absorption of the light as a function of space showed an interference pattern in which the fringes were about 15 μm apart.

The subject, you have to realize, is still in its infancy. Will it be useful when it reaches adulthood? Nobody can tell. Remember that nobody knew what to do with ordinary lasers when they first appeared on the scene.

Exercises

12.1. Calculate (a) the ratio of the Einstein coefficients A/B and (b) the ratio of spontaneous transitions to stimulated transitions for

(i) $\lambda = 693$ nm, $T = 300$ K
(ii) $\lambda = 1.5$ cm, $T = 4$ K

Take the index of refraction to be equal to 1.

At what frequency will the rate of spontaneous transitions be equal with the rate of stimulated transitions at room temperature?

12.2. What causes the laser beam on a screen to appear as if it consisted of a large number of bright points, and why do these points appear to change their brightness as the eye is moved?

12.3. An atomic hydrogen flame is at an average temperature of 3500 K. Assuming that all the gas within the flame is in thermal equilibrium, determine the relative number of electrons excited into the state $n = 2$.

If the flame contains 10^{21} atoms with a mean lifetime of 10^{-8} s, what is the total radiated power from transitions to the ground state? Is the radiation in the visible range?

12.4. The gain constant γ is found to be equal to 0.04 cm^{-1} for a ruby crystal lasing at $\lambda = 693$ nm. How large is the inverted population if the linewidth is 2×10^{11} Hz, $t_{spont} = 3 \times 10^{-2}$ s and $n = 1.77$.

12.5. In a laser material of 3 cm length the absorption coefficient is 0.14 cm^{-1}. Applying a certain amount of pump power the achieved gain coefficient is 0.148 cm^{-1}. One of the mirrors at the end of the laser rod has 100% efficiency.

Determine the minimum reflectivity of the second mirror for laser action to be possible.

12.6. A typical argon ion laser has an output of 5 W and an input current of 50 A at 500 V. What is the efficiency of the laser?

12.7. Assuming that the inversion curve of a laser is wide enough to permit several axial modes, determine the frequency difference between nearest modes.

12.8. The Doppler broadened lineshape function is given by eqn (12.26).

(i) Show that the half-power bandwidth is given by the expression

$$\Delta \nu = 2\nu_0 \left(\frac{2\,kT\,\mathrm{ln}2}{Mc^2} \right)^{1/2}$$

(ii) Work out the half-power bandwidth for an argon laser emitting at 514.5 nm at a temperature of 5000 K.

(iii) How many longitudinal modes are possible at this line if the length of the cavity is 1.5 m?

12.9. Can the inverted population be saturated in a semiconductor laser? Can one make a Si or Ge p–n junction 'laser'?

12.10. Determine the threshold current density of a GaAs junction laser which has a cleaved edge. Assume the following values: linewidth $= 10^{13}$ Hz, attenuation coefficient, $\alpha = 10^3$ m^{-1}, $l = 0.2$ mm, $d = 2$ μm, $n = 3.35$, $\eta = 1$.

12.11. A microwave cavity of resonant frequency ω is filled with a material having a two-level system of electron spins, of energy difference $\hbar\omega$. The microwave magnetic field strength H can be considered uniform throughout the cavity volume V. Show that the rate of energy loss to the cavity walls is

$$\frac{\omega\mu_0 H^2 V}{Q},$$

where Q is the quality factor of the cavity.

If the probability for induced transitions between levels per unit time is αH^2 and spontaneous emission is negligible, show that the condition for maser oscillation is

$$\Delta N > \frac{\mu_0}{\alpha\hbar Q},$$

where ΔN is the excess of upper level population density over lower. [Hint: $Q = \omega \times$ energy stored/energy lost per second.]

13 Optoelectronics

Mehr Licht
Attributed dying words of Johann Wolfgang von Goethe

I saw Eternity the other night,
Like a great ring of pure and endless light
Henry Vaughan *The World*

13.1 Introduction

What are optoelectronic devices? I do not think it is easy to answer this question; as far as I know, there is no accepted definition. There was certainly a time when people talked about photoelectronic devices. I believe they were devices which had something to do both with photons and electrons, in particular with the interaction of light with electrons. Then there was (and still is) electro-optics, concerned with the effect of electric fields upon the propagation of light. So, maybe we should define optoelectronics as a broader discipline which covers both photo-electronics and electro-optics. But what about the interaction of light and acoustic waves, or nonlinear optics, are they part of optoelectronics? We seem to be driven towards the definition that any modern way of manipulating light (interpreted generously, so as to include infrared and ultraviolet) will qualify as optoelectronics.

Some cynics have an alternative definition. They maintain that this new subject emerged when the relevant grant-giving authorities (on both sides of the Atlantic) came to the conclusion that they had very little money to spare for research in the traditional subjects of photoelectronics, electro-optics, acousto-optics, etc. So those compelled by necessity to spend a considerable part of their time writing grant applications invented a revolutionary new subject with a brilliant future . . .

Having decided that optoelectronics is a very broad subject, the next thing I have to do is to say that I can treat only a small fraction of it. The choice is bound to be subjective. I shall, of course, include important devices like junction type photodetectors, and I shall certainly talk about new topics like integrated optics or nonlinear optics. Where I might deviate a little from the general consensus is in choosing illustrations like phase conjugation in photorefractive materials or electroabsorption in quantum well structures, which I find fascinating. I do hope though that I shall be able to give a 'feel' of what is happening in this important field.

13.2 Light detectors

Let me start with the conceptually simplest method of light detection, photoconduction. As described before, photons incident upon a piece of semiconductor may generate extra carriers (the density of generated carriers is usually proportional to the input light). For a given applied voltage, this increase in mobile carrier density leads to an increase of current, which can be easily measured. This is how the CdS cells that are used extensively in exposure meters and in automatic shuttering devices in cameras work. The advantage of using such photoconductors is that they are cheap because they are easy to construct and can be made from polycrystalline material. On the other hand, they are relatively slow and require an external voltage source.

Next, we shall consider a p–n junction. We choose reverse bias because for sensitive detection we require a large fractional change—it is very noticeable if one microamp current doubles, but it is quite difficult to see a one microamp change in one milliamp. If photons of the right wavelength shine on the p-side, they create electrons that are minority carriers, and these will be driven across the reverse bias junction. This is the basis of a sensitive photodetector that is made by producing a shallow layer of p-type material on an n-type substrate, so that the junction is very close to the illuminated surface (Fig. 13.1).

The photodetection properties of p–n junctions may be improved by turning them into p–i–n junctions, that is by adding an extra intrinsic layer, as shown in Fig. 13.2. Since the number of carriers in the intrinsic layer is small, we need only a small reverse bias (a few volts) to extend the depletion region all the way through the i region. A large depletion region gives a large volume in which carriers can be usefully generated in a background of small carrier concentration. In practice the reverse bias is maintained at a value considerably higher than the minimum, so that the intrinsic region remains depleted of carriers, even under high illumination. A typical p–i–n diode would withstand 100 V reverse bias and would have a current of about 2 nA at a voltage of −20 V at a temperature of 25°C.

The response time of p–i–n detectors is related to the transit time of the carriers across the intrinsic region. In a high field this is small, therefore p–i–n detectors are fast; fast enough, in fact, to be used in optical communications systems.

A further possibility is to use a metal–semiconductor junction [as shown in Fig. 9.16(b)] for the detection of carriers. There is then again a depletion region in which the carriers can be generated and which are driven through an external resistance by an applied voltage. Its main advantage is that it can work in the blue and near-ultraviolet region, since the metallic layer (usually gold) can be made thin enough to be transparent.

We can now ask the following question: can we improve the efficiency of the detection method by amplifying the photocurrent? The answer is yes. I shall mention two variants, the avalanche photodiode and the phototransistor.

In the avalanche photodiode the reverse bias is so high that the generated carriers traversing the depletion region have sufficient energy to create further carriers by impact ionization; the additional carriers create ever more carriers by the same mechanism, leading to an avalanche, as discussed in Section 9.12.

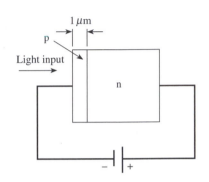

Fig. 13.1
The p–n junction as a light detector.

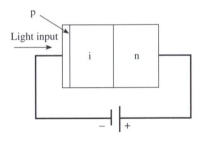

Fig. 13.2
A p–i–n junction used as a light detector.

In a phototransistor the base is not connected. Instead, it is exposed to the input light, which creates the carriers providing the base current. The base current is then amplified in the usual manner.

Perhaps I should add that amplification is a good thing but not the only thing to consider. Under certain conditions noise performance may be a more important criterion when choosing a particular detector.

All the photodetectors mentioned so far needed an applied voltage. It should be noted, however, that it is not necessary in this case. Light may be detected in a p–n junction simply by short-circuiting it via a microammeter of very low resistance, which serves as the load. The electrons and holes generated by light in the junction will move under the effect of the built-in voltage and drive a current through the ammeter. The measured current turns out to be proportional to the input light intensity. This arrangement is usually referred to as the photovoltaic operation of the junction.

A photo-voltaic cell energized by the sun becomes a solar battery. When we used to worry about the 'energy crisis' it was hoped that acres of solar cells would replace nasty, dirty power stations. This has not quite come off, mainly for economic reasons. A simple calculation shows that, even with a generous estimate for the lifetime of a solar cell, the total energy it will generate is less than that required to purify and fabricate the single crystal slice from which it is made. So single crystals are definitely out, except for applications when money is no object—as in space vehicles. Amorphous materials, however, which we briefly discussed in Section 8.9, can be used, as the economics are more favourable. The amorphous semiconductor with the most advanced technology is silicon. It is possible to process it in a wide variety of ways, so that its texture absorbs light well, and the actual absorption edge can be shifted to give a better match to the sun's output than is obtained by the much more clearly defined single crystal. Amorphous silicon is usually deposited in a vacuum or reduced gas pressure as a thin film. This makes it possible to optimize the film thickness—thick enough to absorb light, but not so massive that the much shorter carrier lifetimes and diffusion lengths lead to loss of carriers before they participate in useful current. A typical solar cell would consist of successive n- and p-layers sputtered on a metallized substrate and superposed by a transparent metal top electrode. There are many varieties and, in fact, the variables are so numerous that solar cells have made many PhD theses but have not yet solved the energy problem. However, one commercial realization, now commonplace, is the solar battery driven calculator.

Other materials that have seemed promising include CdTe and $CuInSe_2$ as components in multistage solar converters which absorb some of the light photovoltaically and reflect or transmit other wavelengths to different energy gap devices.

One advantage of working as a photovoltaic engineer is that you have two levels of costing and seeking efficiency. On the one hand, there is the space vehicle market where photovoltaics are looked upon as essential but relatively cheap accessories to a vehicle whose cost is astronomical, so a bit more efficiency is worth paying for. On the other hand there is the power station replacement market where you need acres of photo cells and your competitor is coal which comes out of the ground with low cost.

Figure 13.3 is a schematic drawing of a four level solar cell. It uses a thin top layer of GaInP with an energy gap of 1.85 eV, then successive layers of materials to utilize the longer wavelengths in sunlight, including a GINA alloy mentioned in Section 12.7. The expected conversion efficiency is 41% compared with about 30% for a single junction device. The four stage device is made by MOCVD so is more expensve than coal, but it can be used with mirrors or lenses to concentrate light up to 500 suns, where its expected efficiency is over 50%. Possibly some hope for the terrestrial market as the technology advances.

So far we have talked about light detectors but since the principles are more or less the same, this is also the place to treat infrared*detectors. The primary form of detection in that region too is the change in the conductivity of the material induced by incident radiation. The materials are again semiconductors. The devices rely on band-to-band transition up to 10 μm, requiring small energy gap semiconductors. Semiconductors of even smaller energy gap can be found, so in principle, band-to-band transition could be used for even longer wavelengths, but in practice this range from 10–100 μm is covered by impurity semiconductors in which the increased conductivity is obtained by exciting an electron from a donor level into the conduction band (or from the valence band into an acceptor level).

Infrared radiation between† 100 μm and 1 mm is usually detected with the aid of the so called 'free carrier absorption'. This is concerned with the excitation of electrons from lower to higher energy levels in the conduction band. The number of electrons available for conduction does not change, but the mobility does, owing to the perturbed energy distribution of the electrons. the change (not necessarily increase) in conductivity may then be related to the strength of the incident infrared radiation. Since electrons may be excited to higher energy levels by lattice vibrations as well (thus masking the effect of the input infrared radiation) the crystal is usually cooled.

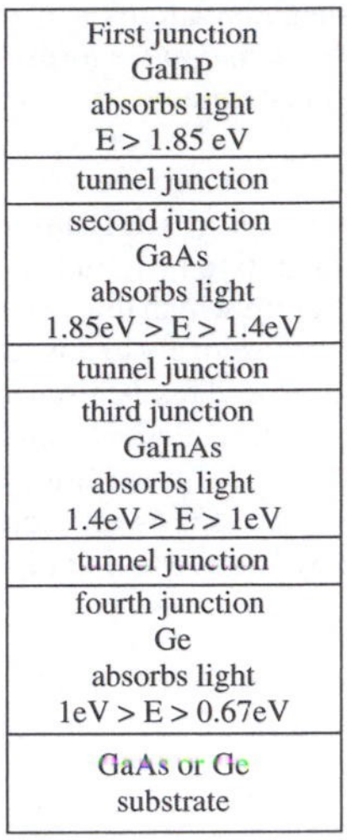

13.3   Light emitting diodes (LEDs)

Semiconductor lasers have been discussed in quite detail in Section 12.7. We came to the conclusion that it is a good thing to have a good semiconductor laser: it has many applications. Are bad lasers good for anything? Yes, certainly, as long as they can emit light. How do we make a bad laser? By depriving a good laser of its cavity. The injected carriers may still recombine and emit light but in the absence of the cavity there is no agent to take care of the phase relationships and consequently the emitted light will not be coherent and will not emerge as a narrow beam. But that is exactly what we want if the aim is to use the semiconductor diode for lighting: we want a diffuse light coming out at all directions. This light source does indeed exists. It is called a light emitting diode, abbreviated as an LED.

In fact, LEDs were invented in 1960, a few years before semiconductor lasers. The first specimen emitted red light. They were made initially of GaAsP alloys (GaAs has an infrared energy gap, which is increased by adding the lighter element P). Too much P induces an indirect gap, so the other alloy GaAlAs has been used to get well into the visible. These energy gaps are shown in Fig. 12.15

| First junction GaInP absorbs light E > 1.85 eV |
| tunnel junction |
| second junction GaAs absorbs light 1.85eV > E > 1.4eV |
| tunnel junction |
| third junction GaInAs absorbs light 1.4eV > E > 1eV |
| tunnel junction |
| fourth junction Ge absorbs light 1eV > E > 0.67eV |
| GaAs or Ge substrate |

Fig. 13.3
Schematic drawing of proposed 4 junction solar cell, parts of which have been made and tested.

* Purists might object saying that the command 'let there be light' was restricted to the visible region, but we think that is a too narrow interpretation.

† Some care needs to be exercised with terminology. This region used to be called the submillimetre region. The latest fashion is to talk about frequency rather than wavelength. 300 GHz would correspond to 1 mm wavelength, 1 THz to 0.3 mm and 100 THz to 3 μm.

* Electronegative atoms

When a covalent bond is formed between two different elements such as the III–Vs or II–VIs, the electron distribution is not symmetrical between the elements because it is energetically favourable for the electron pair to be found closer to one atom. The atom which draws the bonding electrons more closely to it is called more electronegative. This is a polar bond, in other words the covalent bond is partly ionic, as we have shown for some cases in Table 8.3. The following table has been compiled by Linus Pauling (Table 13.1).

Table 13.1 Electronegativities of elements

Be	B	C	N	O
1.5	2.0	2.5	3.0	3.5
Mg	Al	Si	P	S
1.2	1.5	1.8	2.1	2.5
Ca	Ga	Ge	As	Se
1.0	1.6	1.8	2.0	2.4
Sr	In	Sn	Sb	Te
1.0	1.7	1.8	1.9	2.1

The above table appears to have been produced by intuition and genius, a powerful combination. Somewhat similar results can be obtained by equating the electronegativity to the average of the ionisation energy and electron affinity of the element, namely,

$$E = \tfrac{1}{2}(I + E_{\text{aff}})$$

This is common sense, since high E_{aff} means the atom wants to collect spare electrons, and high I means it will hold on to them. From the table, it can be seen that the nitrides of Ga, In and Al have a marked imbalance of electrons, giving a large ionic or polar component to the covalent bond.

† A crucial factor in the success of the InGaN story was that the Nichia Chemical Industries outfit in Japan had alone and presciently worked on making and processing GaN during most of the 1980s. Their principal scientist Shuji Nakamura and his colleagues published the crucial paper in 1992, and many further results since.

which also shows that the lattice parameters of GaAs and AlAs are very similar, so AlAs layers can be grown on GaAs substrate. LEDs in the blue and green were intially made from ZnS and ZnSe, using MBE, or even ion implantation, to get both n and p types.

But the whole scenario of applications and usefulness of LEDs has changed in the past decade by the advent of InGaN devices. The nitrides were regarded by most people making LEDs in 1990 as 'scientifically interesting' which to commercially minded scientists, meant 'black hole for time and money'. However, the basic idea was attractive; if alloys of InN, GaN and AlN were available, LEDs could be made from direct gap semiconductors from the red (1.89 eV) into the UV (6.2 eV). The main problem had been that there was no way of getting a suitable substrate that could be used for epitaxial growth. The nitrides, typically GaN, have the wurtzite structure, with strong ionicity due to the small, electronegative N atoms.* So, the ionic as well as covalent nature of the bond gives a very high melting point. Attempts to grow single crystals by the method outlined in Section 8.11, and even more modern and sophisticated versions of this apparatus failed to produce anything better than fractured crystals a few millimetres across. The first breakthrough came when it proved possible to grow a good mirror finish GaN layer on a sapphire substrate. This used a MOCVD process. First a 'nucleating' layer was put down at a relatively low substrate temperature 700–800 K, then a second faster deposition at 1100–1300 K, this coalesced the first layer and surprisingly good LEDs were made. Surprising, because the large lattice mismatch (about 16%) caused threading defects to reach through to the active layer even though different buffer layers and thicker active layers were tried. The dislocation density could not be reduced below 10^8–10^{10} cm^{-2}. But the LEDs shone brilliantly, particularly in the blue and green where the II–VI opposition devices immediately became obsolete. LED law, based on GaAsP and GaP(ZnO), the earlier red diodes which some of us still have in our pocket calculators, was that dislocation density greater than 10^4 cm^{-2} killed light emission. InGaN was outside this law. The In is important because pure GaN does not luminesce very well. Around 1 part in 100 of In makes all the difference, and of course you need more In than that to get through the visible spectrum. A second problem that had to be solved before these diodes could be made was that initially it was difficult to make p-type material. This is a snag that cropped up many years ago when CdS seemed a good optical device material. It was never completely solved although some p-type CdS was made by MBE. The crucial thing was to avoid thermodynamic equilibrium that caused p-type centres to be swamped. The same sort of problem with GaN was occasioned largely by the fact that heat treatment (faster growth) of a p-type impurity led to compensating reactions (electrons cluster round holes). It was solved by using Mg impurity in the MOCVD process and annealing in N$_2$ which stopped the compensation. Thus, by 1992†good quality InGaN films were available to make the first successful blue DH LEDs. A further refinement was to add a single quantum well (SQW) to the structure and soon after, a multiple quantum well (MQW). A typical LED is sketched in Fig. 13.4. Quite complicated compared to the simple pn junction of earlier days. In particular, there is a special thin InGaN emitting layer, where the recombination of electrons and holes from the pn layers occurs. This controls the operating

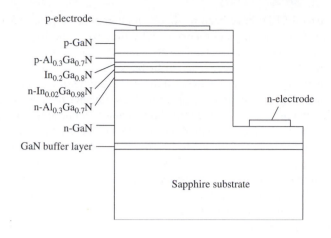

Fig. 13.4
Typical layer structure of a QW based nitride LED.

wavelength*. Forming and consolidating the buffer layer is almost an art form. A black art many say. However, an elaborate process called epitaxial lateral overgrowth (ELOG) involving chemical vapour deposition (CVD) of a layer of SiO_2, subsequently penetrated by windows a few microns wide by conventional masking and etching. This structure is overlaid by further InGaN MOCVD grading and growth layers. The distortion causes lots of extra strain, but in some regions of the resulting overlayer the dislocation density is below 10^6 cm^{-2}. Unfortunately, there is very little change in device performance across such a layer, except that for pure GaN, which is not very luminescent, there is a marked improvement at low defect density. So the practical answer is to keep some In and no ELOG. This is borne out by Fig. 13.5 which shows how the quantum efficiency varies with In content, from work done a few years ago.

Recently the quantum efficiency of InGaN LEDs reached 30% at violet-blue wavelengths at 20 mA current, a power of 21 mW emitted in the blue, and 7 mW in the green, corresponding to about 60–30 lumens W^{-1}. High brightness red LEDs are up to 50% efficient. This has led the way to commercial usage of LEDs. There are large colour displays, signal and traffic lights, automotive uses including stop and rear lights, also for cycles.

The major interest is now room lighting. Blue LEDs coated with clever phosphors have produced white light, fairly closely matched to sunlight with 15 lumens W^{-1}, about 20 lumens per diode. The snag for room lighting is that each diode is quite expensive and that a lot are needed to produce as many lumens as a 100 W bulb. However if advances in performance in the next 5 years match the previous 5, we could all have nitride lighting, at increased capital costs, but everlasting in the sense that our grandchildren, not us will have to replace the bulbs. The estimated life of InGaN LEDs at present, from accelerated and projected life testing is about 10 years continuous use. For an average room light duty cycle this gives 60 years! The long life, low maintenance, is also the main selling point for traffic lights. It makes some local authorities, even in the U.K., look further forward than the next financial year, normally their absolute limit to thinking ahead.

So how can we assess the state of InGaN LED technology in early 2003? The nitrides are still difficult to grow as reasonable epitaxial layers, and the

*Not only composition defines the wavelength. The nitrides are very non-centro symmetric and have a large piezo-electric constant (Section 10.11), so the strains and dislocations cause a big internal field, this gives wavelength shift by the Stark effect, particularly in the quantum well structure (Section 13.10.3 and 4).

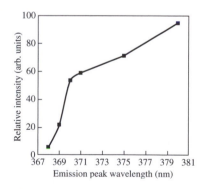

Fig. 13.5
Relative output power of uv InGaN SQW LEDs as a function of emission peak wavelength. The dots show increasing In mole fractions from practically 0% on the left to 4% on the right.

very successful LEDs have been made on small crystals, typically 5 nm thick active layer. There is no good theory of why they work. So we are in a situation of technological success leading to a rapidly expanding semiconductor lighting industry without a proper scientific backing. This has happened before, maybe steam engines were somewhat like this before thermodynamics caught up. But following the precise science/technology advance of solid state circuit electronics, it a bit of a shock. We must hope that all will become clearer when better material is available. Meanwhile, we can conjecture that the necessary small quantity of In in GaN leads to localised deep levels that produce relatively stable excitons; and as their constituents move in the strong fields the recombination is rapid. It has been estimated that the diffusion length of minority carriers (before recombination) is 50–60 nm. Thus radiative recombination would occur before a non radiative encounter at a threading dislocation, even with dislocation densities of 10^{10} cm^{-2}.

As we have said several times, the variable quality, internal strains and piezo electric fields mask the correlation between band gap and emitted light. An interesting summary based on experimental results of numerous InGaN LEDs has been made by P. K. O'Donnell[*] who gives the following equation in terms of x, the molecular fraction of In in InGaN (i.e. $In_x Ga_{1-x} N$) for the peak of the emitted light E_p (in eV).

[*] In the book edited by B. Gil listed in further reading Appendix V.

$$E_p = 3.41 - 4.3x$$

There is a spread of about 10% in these coefficients, and it applies only for x between 0–0.5.

Finally, a word about the environment. Electricity is mainly produced by fossil fuel which causes the undesirable CO_2 emission. Over one fifth of the electricity consumption in developed countries is due to lighting and the proportion is even higher in developing countries. If the use of LED lamps will halve that consumption that, would save, it has been calculated, 300 Megatons of CO_2 emission in the US alone. Will that happen? As I said I feel quite certain that white light LEDs will come in due course but I do not quite believe in the saving. If we have to pay less for our lighting, we may use more of it, but hopefully not twice as much.

13.4   Electro-optic, photorefractive, and nonlinear materials

Before talking more about applications, I shall first review some properties of materials which make them suitable components for devices.

In electro-optic materials, the application of an electric field will affect the index of refraction that an optical wave 'sees'. Note that waves with different electric polarizations are differently affected. The exact relationships are given by tensors, a subject I am reluctant to enter, but if you are interested you can attempt Exercise 13.6. Let me just say that the dielectric tensor (which relates the three components of the electric field to the three components of the dielectric displacement) has nine terms, and each of these terms may depend on the three components of the electric field. Thus, altogether, the electro-optic tensor has twenty-seven components (only eighteen if the symmetry of the

Table 13.2 *Properties of electro-optic materials*

Substance	Wavelength (μm)	Electro-optic coefficient (10^{-12} m V^{-1})	Index of refraction	Static dielectric constant
$Bi_{12}SiO_{20}$	0.514	2.3	2.22	56
$BaTiO_3$	0.514	820	2.49	4300
CdTe	1.0	4.5	2.84	9.4
GaAs	1.15	1.43	3.43	12.3
$KNbO_3$	0.633	380	2.33	50
$LiNbO_3$	0.633	32.6	2.29	78
ZnO	0.633	2.6	2.01	8.15

dielectric tensor is taken into account). There is no need to worry. In practice, usually, only one of the numerous components is needed, and the effect may be presented in the form

$$\Delta\left(\frac{1}{\epsilon_r}\right) = r\mathcal{E}. \tag{13.1}$$

r is the electro-optic coefficient.

Since $\epsilon_r = n^2$, the change in refractive index may be written as

$$\Delta n = -\tfrac{1}{2}n^3 r\mathcal{E}. \tag{13.2}$$

Note that *r* may be positive or negative depending on crystal orientation.

Taking $LiNbO_3$ as an example, $n = 2.29$, and for a certain orientation of the crystal we have $r = 3.08 \times 10^{-11}$ m V^{-1}. With reasonable voltages, one may obtain an electric field of about 10^6 V m^{-1}, causing a change in the refractive index of $\Delta n = 1.86 \times 10^{-4}$. It does not seem a lot, but it is more than enough for a number of applications. The indices of refraction and the electro-optic coefficients for some often used materials are given in Table 13.2.

The essential thing to remember is that in electro-optic crystals the refractive index, and consequently the propagation of the wave, may be changed by applying an electric field.

The properties of electro-optic materials are dependent, of course, on direction, but since the aim is no more than to give a general idea of the ranges involved, only the largest components are listed for each material. The wavelengths at which these values were measured are also indicated.

Photorefractive materials represent a rather special class of crystals which are both electro-optic and photoconductive. Some representatives of these materials are $LiNbO_3$, $Bi_{12}SiO_{20}$, $BaTiO_3$. I shall return to them in the next section.

Non-linear materials are usually characterized by relating the dielectric polarization, P, to the electric field. The linear relationship given by eqn (10.5) may be generalized and written in the form:

$$P = \epsilon_0[\chi^{(1)}\mathcal{E} + \chi^{(2)}\mathcal{E}^2 + \chi^{(3)}\mathcal{E}^3]. \tag{13.3}$$

$\chi^{(1)}$ is the linear susceptibility (what we called before simply susceptibility and denoted by χ), and $\chi^{(2)}$ and $\chi^{(3)}$ are known as the quadratic and cubic susceptibilities.

In some materials the nonlinearity may be more conveniently expressed with the aid of the index of refraction as

$$n = n_0 + n_2 I. \tag{13.4}$$

Intensity dependent absorption is also possible. In fact one of the very interesting devices to be presented in Section 13.11 operates on that basis.

13.5 Volume holography and phase conjugation

I have already mentioned (Section 12.13) some of the interesting optical phenomena holography can produce. I shall now briefly talk about one particular branch of holography, known as volume holography, and discuss what happens in the simplest possible case, when both the reference beam and the object beam are plane waves (Fig. 13.6). The distinguishing feature of volume holography is that the recording process takes place in the volume of the photosensitive material.

Let us now do a little mathematics. The amplitudes of the two waves may be written in the form:

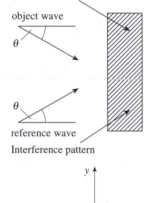

$$A_{\mathrm{ref}} = A_{10} \exp[ik(x \cos\theta + y \sin\theta)] \tag{13.5}$$
$$A_{\mathrm{obj}} = A_{20} \exp[ik(x \cos\theta - y \sin\theta)], \tag{13.6}$$

leading to the interference pattern (note that the intensity, I, is proportional to the square of the amplitude):

$$I = |A_{\mathrm{ref}} + A_{\mathrm{obj}}|^2 = A_{10}^2 + A_{20}^2 + 2A_{10}A_{20} \cos(2ky \sin\theta). \tag{13.7}$$

It may now be seen from the above equation that the intensity varies periodically in the y-direction with a period,

$$\Lambda = \frac{2\pi}{2k \sin\theta} = \frac{\lambda}{2n \sin\theta}, \tag{13.8}$$

Fig. 13.6
Two plane waves incident upon a photosensitive medium.

which is nothing else but the Bragg relation once more. The symbol, Λ, is usually referred to as the grating spacing.

After recording comes the processing (a black art for all known photosensitive materials) with a result that the interference pattern is turned into a modulation of the dielectric constant, that is the end product is a dielectric constant varying as

$$\epsilon_r = \epsilon_{r0} + \epsilon_{r1} \cos(2ky \sin\theta). \tag{13.9}$$

What happens when we illuminate the hologram with the reference wave? According to the rules of holography, the object wave springs into existence. Interestingly, we could reach the same conclusion, considering Bragg diffraction. If a wave is incident upon a material with a periodic structure at an angle and wavelength that satisfies eqn (13.8), then a diffracted beam of appreciable amplitude will emerge. In fact, under certain circumstances it is possible to transfer all the power of the incident reference wave into the diffracted object wave.

What happens when the object wave is not a plane wave but carries some pictorial information in the form of some complicated amplitude and phase distribution? Since any wave can be represented by a set of plane waves, and since the modulation of the dielectric constant is small, one can apply the principle of superposition, leading to the result that each constituent plane wave, and thus the whole picture, will be properly reconstituted.

The photosensitive media most often used are silver halide emulsion (basically the same as that used for photography but with smaller grain size) and dichromated gelatin. In the former case, the refractive index modulation is due to the density variation of silver halide in the gelatin matrix. In the latter case, the mechanism has still not been reliably identified. It is quite likely that the refractive index modulation is caused again by density variations mediated by chromium, but explanations claiming the presence of voids cannot be discounted. Unfortunately, it is rather difficult to know what goes on inside a material when chemical processing takes place. We can, however, trust physics. It involves much less witchcraft. So I shall make an attempt to give an explanation of the origin of dielectric constant modulation in photorefractive materials.

As I mentioned previously, a photorefractive material is both photoconductive and electro-optic. Let us assume again that two plane waves are incident upon such a material, but now a voltage is applied as well as shown in Fig. 13.7. The light intensity distribution [given by eqn (13.7)] is plotted in Fig. 13.8(a). How will the material react? The energy gap is usually large, so there will be no band-to-band transitions but, nevertheless, charge carriers (say electrons) will be excited from donor atoms, the number of excited carriers being proportional to the incident light intensity. Thus, initially, the distribution of electrons (N_e) and ionized donor atoms (N_D^+) is as shown in Fig. 13.8(b) and (c). Note, however, that the electrons are mobile, so under the forces of diffusion (due to a gradient in carrier density) and electric field (due to the applied voltage) they will move in the crystal. Some of them will recombine with the donor atoms while some fresh electrons will be elevated into the conduction band by the light still incident. At the end an equilibrium will be established when, at every point in space, the rate of generation will be equal to the rate of recombination. A sketch of the resulting electron and donor densities shown in Fig. 13.8(d). Since the spatial distributions of electrons and ionized donors no longer coincide, there is now a net space charge, as shown in Fig. 13.8(e). Now remember Poisson's equation. A net space charge will necessarily lead to the appearance of an electric field [Fig. 13.8(f)]. So weh ave got an electric field which is constant in time and periodic in space. Next, we invoke the electro-optic property of the crystal which causes the dielectric constant to vary in proportion with the electric field. We take r, the electro-optic coefficient, as positive, so the dielectric constant is in anti-phase with the electric field. We have come to the end of the process. The input interference pattern has now been turned into the dielectric constant variation shown in Fig. 13.8(g).

We may now argue again that the process works for more complicated waves as well, so we have got the means to record holograms in photorefractive materials. Noting that the process by which the dielectric grating is produced may occur quickly (it is in a range extending from nanoseconds to seconds) we have here a real-time holographic material.

In fact, the major application of photorefractive materials is not for real-time holography but for wave interaction. The phenomenon by which the incident

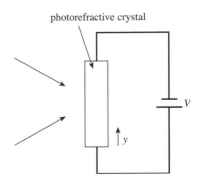

Fig. 13.7

Two plane waves incident upon a photorefractive crystal across which a voltage is applied.

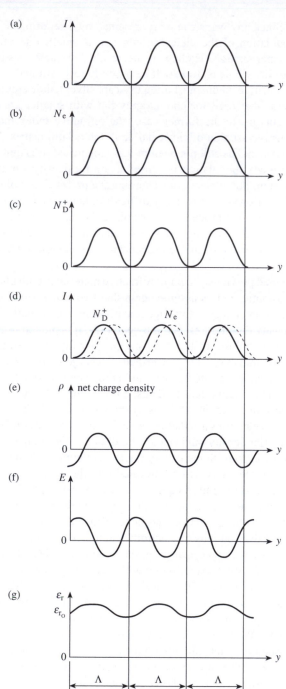

Fig. 13.8
The variation of a number of physical
quantities in a photorefractive crystal
in the y-direction: (a) intensity,
(b) electron density, (c) ionized donor
density, (d) electron and ionized
donor density in the stationary case,
(e) the net charge density, (f) the
resulting electric field, (g) the
resulting dielectric constant.

light brings forth a dielectric constant modulation and the way this modulation
reacts back (by diffracting the waves) upon the light beams, leads to all sorts
of interesting effects. I shall mention only the most notable one among them,
phase conjugation.

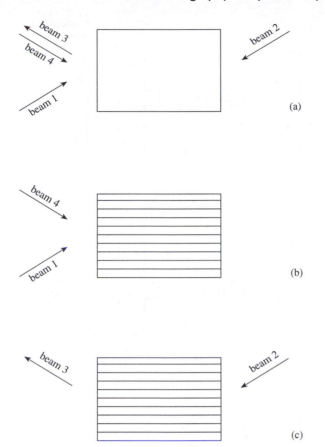

Fig. 13.9
Schematic representation of phase conjugation. (a) Incident beams 1, 2 and 4 produce the phase conjugate beam 3. (b) Beams 1 and 4 produce a grating. (c) Beam 2 is diffracted into beam 3 by the grating recorded.

The physical configuration is shown in Fig. 13.9. There are three beams incident upon the material, and a fourth beam is generated. For this reason the phenomenon is often referred to as four wave mixing. Beams 1 and 2 are known as the pump beams and beam 4 as the probe beam (usually much weaker than the pump beams). As a result of the interaction, beam 3, the so-called phase conjugate beam, is generated.

The physical mechanism is fairly easy to explain. Beams 1 and 4 create a dielectric grating, as shown in Fig. 13.9(b). Beam 2, incident upon the grating, is then diffracted to produce beam 3, as shown in Fig. 13.9(c).

What is so interesting about beam 3? Well, it is in a direction opposite to beam 4, but there is a lot more to it. If beam 4 consists of a range of plane waves, each separate plane wave is reversed to create, in the official jargon, a phase conjugate beam. The whole device is called a phase conjugate mirror.

In what respect is a phase conjugate mirror different from an ordinary mirror? We shall give two examples. In Fig. 13.10(a) a piece of dielectric is in the way of an incident plane wave. The wavefront of the plane wave moving to the right is illustrated by continuous lines: 1 is that of the incident wave, and 2 is the wavefront after passing partially through the dielectric. After reflection by an ordinary mirror, the retarded part of the wavefront is still retarded as given by 3 (dotted lines). After passing through the dielectric once more, there is a further retardation of the wavefront, as indicated by 4.

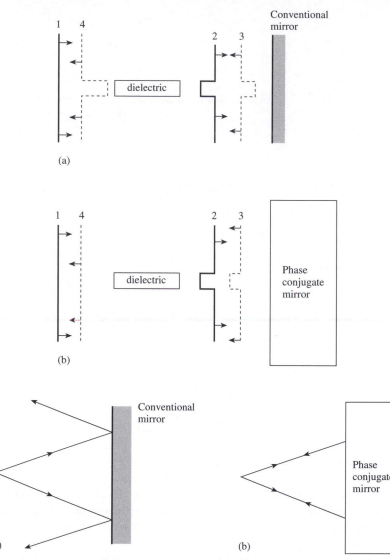

Fig. 13.10

Plane waves passing through a dielectric and reflected by (a) a conventional mirror, (b) a phase conjugate mirror.

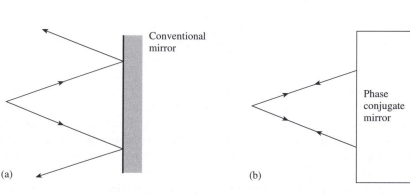

Fig. 13.11

A divergent beam reflected by (a) a conventional mirror, (b) a phase conjugate mirror.

In Fig. 13.10(b) wavefronts 1 and 2 are the same as previously. The phase conjugate mirror, however, 'reverses' the input wave. The wavefront that was retarded will now be promoted to the front as shown by 3. After passing through the dielectric for the second time, the wavefront 4 will again be smooth. The conclusion is that the phase conjugate mirror corrected the wavefront distortion introduced by the dielectric. And this would actually be true for other kinds of disturbances as well. The phase conjugate mirror reflects the incident wave with an opposite phase and direction.

My second example is a beam diverging towards the mirror. After reflection the conventional mirror will make the beam diverge further [Fig. 13.11(a)] whereas the phase conjugate mirror will produce a convergent wave [Fig. 13.11(b)]. A fascinating phenomenon you must agree but, I am afraid, still at the laboratory stage.

13.6 Acousto-optic interaction

We have seen that periodic variation of the dielectric constant within a volume of material may help to produce a diffracted beam by the mechanism of Bragg interaction. The periodic variation may be achieved by using photosensitive and photorefractive materials, but there is one more obvious possibility which we shall now discuss. An acoustic wave propagating in a material will cause a strain, and the strain may cause a change in the dielectric constant (refractive index). The relationship between the change in the dielectric constant and the strain is given by the so-called strain–optic tensor. In the simplest case, when only one coefficient needs to be considered, this may be written in the form:

$$\Delta \left(\frac{1}{\epsilon_r} \right) = pS, \tag{13.10}$$

where p is the photoelastic coefficient, and S is the strain. Thus, we can produce a volume hologram simply by launching an acoustic wave. But is a volume hologram much good if it moves? Well, it is all relative. For us anything moving with the speed of sound appears to be fast, but for an electromagnetic wave which propagates by nearly five orders of magnitude faster than a sound wave, the hologram appears to be practically stationary. There is, however, an effect characteristic to moving gratings that I must mention, and that is the Doppler shift. The frequency of the electromagnetic wave is shifted by the frequency of the acoustic wave, an effect that comes occasionally useful in signal processing.

Let us now work out the frequency of an acoustic wave needed to deflect an optical wave of 633 nm wavelength (the most popular line of a He–Ne laser) by, say, 2°. The Bragg angle is then 1°. Taking further LiNbO$_3$ as the material in which the waves interact, we find for the grating spacing,

$$\Lambda = \frac{\lambda}{2n \sin \theta} = \frac{633 \times 10^{-9}}{2 \times 2.29 \times \sin 1°} = 7.92 \,\mu\text{m}, \tag{13.11a}$$

Λ is the required wavelength of the acoustic wave.

Noting that the velocity of a longitudinal wave in LiNbO$_3$ is $6.57 \times 10^3 \text{ m s}^{-1}$ [see Table 13.3] we find for the frequency of the acoustic wave $f = 8.30 \times 10^8$ Hz.

This is regarded as quite high frequency, which cannot be excited without exercising due care, but nevertheless such an acoustic wave can be produced in bulk LiNbO$_3$ and can be duly used for deflecting an optical beam. The device is known as a Bragg cell.

Table 13.3 *Properties of some materials used for acousto-optic interaction*

Substance	Wavelength (μm)	Density (10^3 kg m^{-3})	Index of refraction	Sound velocity (10^3 m s^{-1})
Water	0.633	1	1.33	1.5
Fused quartz	0.633	2.2	1.46	5.95
GaAs	1.15	5.34	3.43	5.15
LiNbO$_3$	0.633	4.7	2.29	6.57
LiTaO$_3$	0.633	7.45	2.18	6.19
PbMoO$_4$	0.633	6.95	2.4	3.75
ZnS	0.633	4.10	2.35	5.51

A device which can deflect an optical beam can, of course, be used for modulation as well. When the acoustic wave is on, the power in the transmitted beam decreases, and a diffracted beam appears. Thus, by varying the amplitude of the acoustic wave, both output beams are modulated. It may be an advantage to use the diffracted beam as the modulated beam because the power in it is completely cut off when the acoustic wave is absent, whereas it is less straightforward to extinguish the transmitted beam.

Could we use an acoustic wave for scanning the optical beam within a certain angular region? It can be done in more than one way. I shall just show the arrangement which makes the best sense in principle.

Let us say we have an optical beam incident at an angle θ to the horizontal direction. If we want to deflect it by 2θ, then we need an acoustic wave propagating in the vertical direction [see Fig. 13.12(a)] and having a wavelength of $\lambda_{ac} = \lambda_{op}/2n \sin \theta$. How could we deflect the beam by a further $\Delta\theta$? In order to have the Bragg interaction, we need to change the acoustic

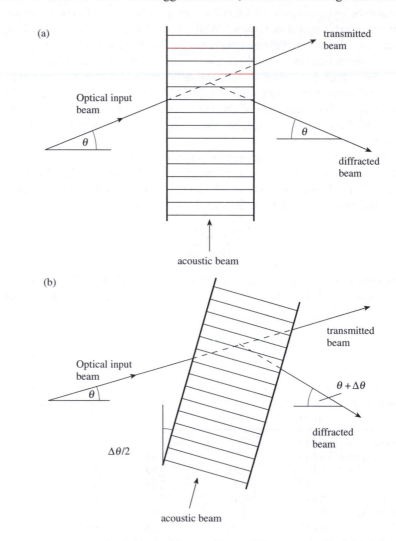

Fig. 13.12
Bragg reflection of a light beam by an acoustic wave. (a) Deflection angle of 2θ at an acoustic wavelength of λ_c, (b) deflection angle of $2\theta + \Delta\theta$ at an acoustic wavelength of $\lambda_c + \Delta\lambda_c$.

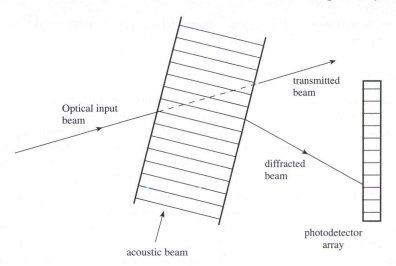

Fig. 13.13
A spectrometer relying on acousto-optic interaction.

wavelength to

$$\lambda_{\text{ac}} + \Delta\lambda_{\text{ac}} = \frac{\lambda_{\text{op}}}{2n\sin(\theta + \Delta\theta)} \qquad (13.11b)$$

and, in addition, tilt the acoustic wave (it can be done by using an appropriate launching array) by $\Delta\theta/2$, as shown in Fig. 13.12(b).

Our final example is a spectrum analyser. It is essentially the same device as the beam scanner but used the other way round. The unknown input frequency to be determined is fed into the device in the form of an acoustic wave via an acoustic transducer. It will deflect the input optical beam by an amount which depends on the frequency of the acoustic wave. The deflected optical beam is then detected by an array of photodetectors (Fig. 13.13). The position of the photodetector upon which the beam is incident will then determine the unknown frequency.

13.7 Integrated optics

This is not unlike integrated circuits, a subject we have discussed in detail when talking about semiconductor devices. The basic idea was to 'integrate', that is to put everything on a single chip and by doing so achieve compactness, ruggedness, economy of scale, etc. The same idea of integration (advanced towards the end of the 1960s) can also be applied to optical circuits with all the corresponding advantages. In principle one could have lasers, waveguides, optical processing circuits, all on the same chip. In practice, the results have been rather limited due to technological difficulties and, probably, inadequacy of the scale of effort. The economic imperative which was the driving force behind the integrated circuit revolution has simply not been there for their optical counterparts. It looks as though optical communications can come about on a wide scale, without the benefit of integrated optics, so unless there is some new and urgent impetus provided by the need to develop optical computers or some other forms of optical processing, further progress is likely to remain slow. Nevertheless, it is a very promising technique, so I must give at least an introduction to its basic precepts.

13.7.1   Waveguides

The principle is very simple. If a material exhibiting a certain index of refraction is surrounded by a material of lower index of refraction, then a wave may be guided in the former material by successive total internal reflections. Optical fibres (mentioned before) represent one such possibility for guiding waves, but that is not suitable for integrated optics. We can however rely on the fact that the refractive index of GaAs is higher than that of AlGaAs and, consequently, a GaAs layer grown on the top of AlGaAs will serve as a waveguide. As may be seen in Table 13.2, GaAs is an electro-optic crystal, it is also suitable for producing junction lasers, microwave oscillators, and transistors. Thus, altogether, GaAs seems to be the ideal material for integrated optics. Well, it is indeed the ideal material, but the problems of integration have not as yet been solved. It is still very much at the laboratory stage.

Nearer to commercial application are the LiNbO$_3$ devices, which I shall describe in more detail. In these devices the waveguides are produced by indiffusing Ti into a LiNbO$_3$ substrate through appropriately patterned masks (the same kind of photolithography we met in Section 9.22 when discussing integrated circuits). Where Ti is indiffused the refractive index increases sufficiently to form a waveguide.

13.7.2   Phase shifter

Considering that LiNbO$_3$ is electro-optic, we may construct a simple device, using two electrodes on the surface of the crystal on either side of the waveguide, and apply a voltage to it as shown in Fig. 13.14. With a voltage V_0, we may create an electric field roughly equal to V_0/d, where d is the distance between the electrodes. Hence, the total phase difference that can be created is

L is the length of the electrodes.

$$\Delta\phi = \frac{2\pi L \Delta n}{\lambda}. \tag{13.12}$$

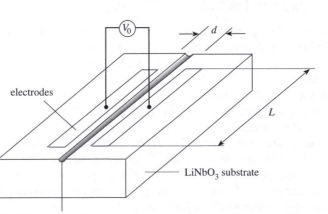

Fig. 13.14

A phase shifter relying on the change of dielectric constant caused by the applied voltage.

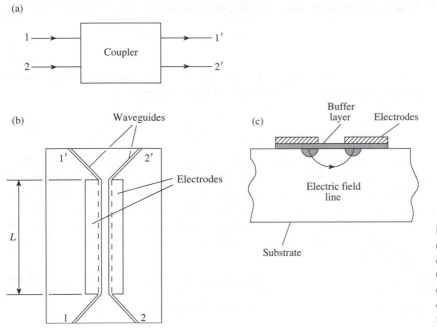

Fig. 13.15
(a) Schematic representation of a directional coupler. (b) Integrated Optics realization of the directional coupler. (c) Cross-section of the device showing also an electric field line.

A voltage of 5 V with a distance of 5 μm between the electrodes gives an electric field of 10^6 V m^{-1} for which we found previously (Section 13.4) $\Delta n = 1.86 \times 10^{-4}$. A little algebra will then tell us that in order to produce a phase difference of π at a wavelength of about 1.5 μm (good for optical communications) we need electrode lengths of about 4 mm. So we have now a phase shifter, or if we keep on varying the voltage between 0 and 5 V, we have a phase modulator.

13.7.3 Directional coupler

One of the elementary requirements of signal processing is the facility to direct the signals into different locations. In its simplest form it means [see Fig. 13.15(a)] that a signal coming in at port 1 should be divided between output ports 1′ and 2′ in any desired proportion, including the possibility that all the input power should appear at one single output port. And, similarly, power coming in at input 2 should be divided between the same output ports. The realization in integrated optics form is shown in Fig. 13.15(b). For a length of L, the two waveguides are so close to each other that there is leakage of power from one to the other one. In addition, it is possible to change the relative velocity of wave propagation in the two waveguides by applying a voltage between the two electrodes. As shown in Fig. 13.15(c), the vertical component of the electric field is in the opposite direction for the two waveguides. Hence, according to eqn (13.1) the indices of refraction will vary in the opposite direction.

Let us now formulate this problem mathematically. A wave propagating in the positive z-direction with a wavenumber k_1 is of the form (recall Chapter 1)

$$A_1 = A_{10} \exp ik_1 z. \tag{13.13}$$

A_{10} is the amplitude of the wave at $z = 0$.

This may be described by the differential equation,

$$\frac{dA_1}{dz} = ik_1 A_1. \tag{13.14}$$

Another wave propagating in the same direction with a wavenumber, k_2, would then analogously be described by the differential equation,

$$\frac{dA_2}{dz} = ik_2 A_2. \tag{13.15}$$

Let us identify now waves 1 and 2 with those propagating in waveguides 1 and 2. Next, we shall have to take into account coupling between the waveguides. In order to do so, we may advance the following argument. If there is coupling between the two waveguides, then the rate of change of the amplitude of the wave in waveguide 1 will also depend on the amplitude of the wave in waveguide 2, and the higher the coupling, the larger is the effect of wave 2. In mathematical form,

$$\frac{dA_1}{dz} = ik_1 A_1 + i\kappa A_2. \tag{13.16}$$

And, similarly, the rate of change of the amplitude of wave 2 is

The coupling coefficient has been taken as $i\kappa$.

$$\frac{dA_2}{dz} = i\kappa A_1 + ik_2 A_2. \tag{13.17}$$

Note that we have met this type of coupled differential equation before when discussing quantum mechanical problems in Chapters 5 and 7. The solution is not particularly difficult; I shall leave it as an exercise (13.7) for the reader. I shall give here only the solution for the case when $k_1 = k_2 = k$ and when all the input power appears at port 1 with an amplitude A. We obtain then for the two outputs:

$$A_1 = A_{10} \exp(ikz) \cos \kappa z \tag{13.18}$$

and

$$A_2 = iA_{10} \exp(ikz) \sin \kappa z. \tag{13.19}$$

It may be clearly seen that the amount of power transfer depends on the length of the coupler section. When $z = L = \pi/2\kappa$, all the power from waveguide 1 can be transferred to waveguide 2. With $L = \pi/\kappa$, the power launched in waveguide 1 will first cross over into waveguide 2, but it will then duly return. At the output, all the power is in waveguide 1.

This exchange of power may take place when $k_1 = k_2$, that is when the velocities are identical. If we apply a voltage, the velocity of propagation increases in one waveguide and decreases in the other one. In the absence of synchronism the amount of power transferable may be shown to decrease. When the velocities in the two waveguides are radically different, then they simply ignore each other; there is no power transfer from one to the other irrespective of the amount of coupling.

What can we use such a coupler for? Well, it is obviously a switch. In the absence of a voltage, all the power can be transferred from waveguide 1 to waveguide 2. Destroying then the synchronism by applying a voltage, we can switch the power to waveguide 1 or vice versa.

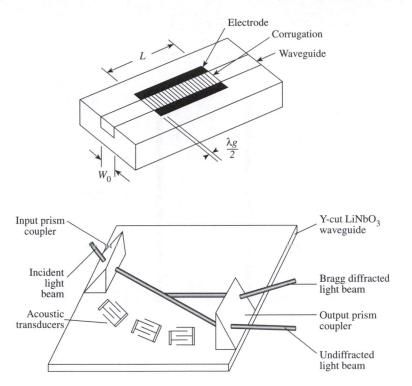

Fig. 13.16
A Bragg type filter employing grooves as the reflecting elements.

Fig. 13.17
A beam deflector in Integrated Optics form.

13.7.4 Filters

One type of filter which reflects the signal in a certain wavelength band and transmits the rest may be realized by relying once more on Bragg reflection. Cumulative reflection may be obtained by placing reflecting elements at the right period into the waveguide. This is shown in Fig. 13.16, where the reflecting elements are grooves at a distance of $\lambda_g/2$ from each other, with λ_g being the wavelength in the waveguide.

Obviously, a large number of other devices exist which I cannot possibly include in this course, but let me just briefly mention one more, namely the integrated optics realization of the acousto-optic beam deflector. In this case, the steerable acoustic column is provided by interdigital surface acoustic wave transducers (see Section 10.11) and the optical beam is confined to the vicinity of the surface by a so-called planar waveguide. The optical beam will then sense the periodic perturbation caused by the surface acoustic wave and will be duly diffracted, as shown in Fig. 13.17.

13.8 Spatial light modulators

We have several times mentioned light modulators which modulate the intensity of the incident light beam. Note that in those devices there is only one light beam, and it is affected everywhere in the same way by the modulation. Spatial light modulators do the same thing, but different parts of the beam are differently affected. A simple definition would be that a spatial light modulator is a device

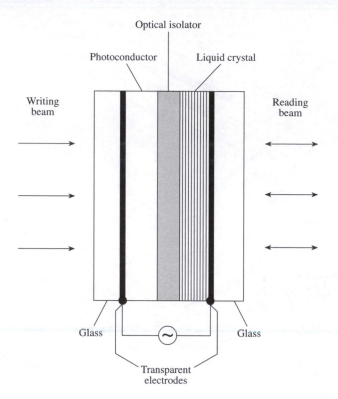

Fig. 13.18

An incoherent to coherent light
converter.

which gives a desired light distribution over a certain area. Thus, in principle,
all programmable display devices may be regarded as spatial light modulators,
including possibly a display at a railway station which announces the departure
of trains. Other examples are a cathode-ray tube in a television set or a liquid
crystal display in a calculator.

I shall discuss here the operation of only one of the modern spatial light
modulators, which may also be called an incoherent-to-coherent light converter.
Such a device is needed because coherent light is usually more suitable for
further processing than incoherent light.

A schematic diagram of such a converter is shown in Fig. 13.18. In the
absence of input incoherent light from the left (the writing beam) the photoconductor does not conduct, and consequently there is a high voltage drop across
the photoconductor and a low voltage across the liquid crystal (in practice the
liquid crystal layer is much thinner than the photoconductor). The role of the
liquid crystal is to transmit or to absorb the coherent light (reading light) coming
from the right, depending on whether there is a voltage across it or not. Thus,
the intensity modulation of the writing beam is converted into the intensity
modulation of the reading beam. The optical isolator is usually in the form of
a wide band dielectric mirror, which separates the writing and reading beams
from each other.

We have already mentioned the quest for a liquid crystal display in
Section 10.15. The ideal device that still needs to be produced is a spatial
light modulator which is voltage addressable with a resolution of the order of a

wavelength and a speed (say nanoseconds) comparable with that of fast digital computers.

13.9 Nonlinear Fabry–Perot cavities

A number of interesting effects occur when either the absorption coefficient or the dielectric constant depend on the intensity of light. In this section we shall discuss one particular manifestation of this non-linear effect when a dielectric, whose index of refraction obeys eqn (13.4), forms a Fabry–Perot cavity.

It is fairly easy to show that the relationship between I_t, and I_i, the output and input intensity of the cavity, is

$$\frac{I_t}{I_i} = \left[1 + \frac{4R}{(1-R)^2}\sin^2 kl\right]^{-1}, \qquad (13.20)$$

where l is the length of the cavity, k is the wave number, and R is the power reflection coefficient. Equation (13.20) makes good sense. When $kl = m\pi$ or $l = m\lambda/2$, the transmission is maximum (all power is transmitted) whereas minimum transmission occurs when $kl = 2(m+1)(\pi/2)$, or $l = (2m+1)(\lambda/4)$.

For $R = 0.36$ and 0.7, eqn (13.20) is plotted in Fig. 13.19 for one period as a function of kl. Obviously, the greater is R the sharper is the resonance. Large values of R at a given frequency can be easily achieved by multiple element dielectric mirrors. But, even in the simplest case when we rely upon reflection at a dielectric–air interface, we can get quite high values. For InSb, for example, which has been used in bistability experiments, $\epsilon_r = 15.9$, and $R = 0.36$.

If the dielectric is linear, then an increase in input intensity would lead to a proportional increase in output intensity. Consider now the case when the index of refraction obeys the equation $n = n_0 + n_2 I$ (assuming that the intensity inside the cavity is the same as that leaving) and take a point on the I_t/I_i curve, where the function is increasing (A in Fig. 13.19). What happens now if I_i is increased? Then first, we may argue, I_t will increase as in the linear regime. But an increase in I_t will lead to an increase in n and k, and consequently, we

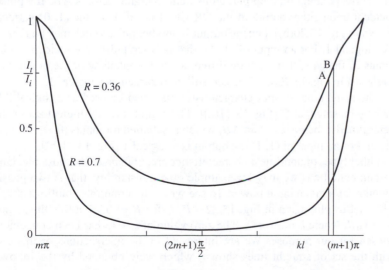

Fig. 13.19
A plot of eqn (13.20) for $R = 0.36$ and 0.7.

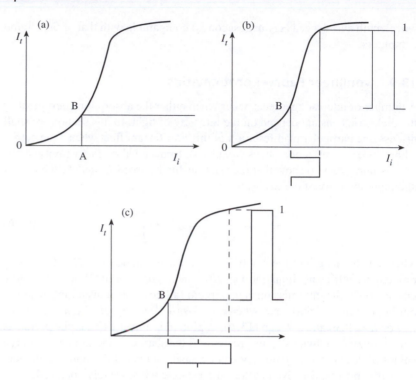

Fig. 13.20

Non-linear, I_t versus I_i characteristics
used as an OR gate.

need to move up to a point higher on the curve, say B. But this means that, for a given value of I_i, we have an even higher value of I_t, which increases n even further, and in turn makes us move higher up on the curve, etc. Due to this positive feedback, a small increase in I_i may lead to a large increase in I_t. So the I_t versus I_i curve may turn out to be highly non-linear.

What could we use this high non-linearity for? An input–output characteristic shown in Fig. 13.20(a) can be used as an OR gate. The bias input light is OA. The corresponding light output, equivalent to digital zero, is AB. If a pulse is incident upon either output of the OR gate [the (0, 1) or the (1, 0) variety as shown in Fig. 13.20(b)] then the output is another pulse which may be regarded as a logical 1. For an input of (1, 1), that is when pulses are incident on both inputs of the gate, there is little difference in the intensity of the output beam as shown in Fig. 13.20(c). This can still be regarded as a logical 1.

The same input–output characteristic may also be used as a logical AND gate by biasing it at C [Fig. 13.21(a)]. The output level for logical zero is now determined as being less than AB, which is satisfied for inputs (0, 0), (0, 1) and (1, 0). For an input of (1, 1) the output is a logical 1 [Fig. 13.21(b)].

Other types of non-linear characteristics are, of course, also possible. Under certain conditions we may for example obtain bistability, that is two possible outputs for a given input power. To see when such multiple solutions are possible, let us plot again in Fig. 13.22—I_t/I_i for $R = 0.7$ but this time against $(k - k_0)l$, where $k = 2\pi n/\lambda$. The zero value of $k - k_0$ has been conveniently chosen for our purpose. We are interested in the intersections of this curve with the set of straight lines shown, which were obtained by the following

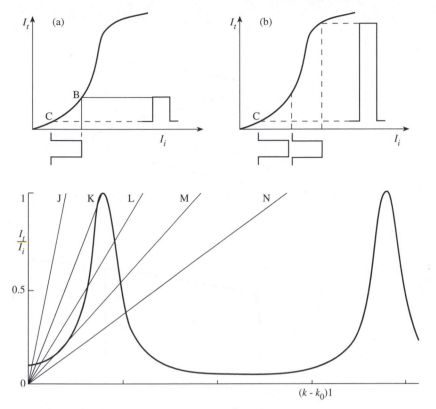

Fig. 13.21
A non-linear I_t versus I_i
characteristics used as an AND gate.

Fig. 13.22
A graphical construction finding the
intersection of the curve given by eqn
(13.20) with the straight lines given
by eqn (13.21).

considerations. From eqn (13.4)

$$I_t = \frac{n - n_0}{n_2} = \frac{\lambda}{2\pi n_2}(k - k_0), \tag{13.21}$$

whence

$$\frac{I_t}{I_i} = \frac{\lambda}{2\pi n_2 l I_i}(k - k_0)l. \tag{13.22}$$

According to the above equation, I_t / I_i is a linear function of $(k - k_0)l$. There
are lots of constants on the right-hand-side of eqn (13.22) which are irrelevant,
but note the presence of I_i. Each straight line in Fig. 13.22 corresponds to a
different value of I_i. For each value of I_i the permissible values of I_t are given
by the intersections of the straight line with the curve. For OJ, a low value of
I_i (i.e. high value of the slope), there is only one solution. As I_i increases there
are two solutions for OK, three solutions for OL, two solutions for OM and,
again, only one solution for ON. Using this graphical method, the I_t versus I_i
curve can be constructed. In the present case it looks roughly like that shown
in Fig. 13.23(a). Physically what happens is that as I_i reaches the value of I_{il},
the value of I_t will jump from I_{t1} to I_{t2} (see Fig. 13.23(b)) and then will follow
the upper curve. In the reverse direction, when I_i decreases, I_t will suddenly
jump from I_{t3} to the lower curve at I_{t4} and will then follow the lower curve.

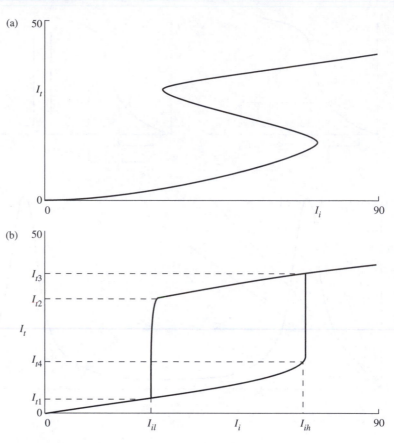

Fig. 13.23

(a) The I_t versus I_i relationship as determined by the construction in Fig. 13.22. (b) The I_t versus I_i characteristics as it would be measured.

What is bistability good for? Quite obviously, just as in the case of the similarly looking ferrite hysteresis loop, we can make memory elements out of them. By adding for example a switching beam to a holding beam the device can switch from a low output state to a high output state.

Summarizing, our non-linear cavity has yielded components both for logical arithmetic and storage. The hope is that one day they will be parts of all-optical computers. Their main advantage in the applications discussed in the present section is speed. The physical mechanism causing the non-linearity is fast. Switching speeds of the order of 1 ps have been measured.

13.10 Optical switching

MEMS were mentioned in Section 9.24. They represent a new way of doing things. Parts of the structures produced that way can actually move, so it is possible, for example, to produce movable mirrors which can redirect a beam of light. But that is exactly the thing we need for optical switching. We need it badly. The present practice is rather cumbersome. It may be likened to the plight of the traveller who wants to travel from Oxford to Cambridge in the comfort of a railway carriage. He can certainly take a train from Oxford to Paddington Station, London but there he is forced to disembark. He must then travel by tube to Liverpool Station from where he is allowed the luxury of boarding another train. The journey by tube is a nuisance. Signals travelling in optical fibres face the same problems. They can rarely reach their destination without change of a rather brutal nature. Within picoseconds of their arrival they are unceremoniously converted into electronic form, interrogated as for their final destination, reconverted into optical form and finally bundled into the appropriate optical output fibre. Obviously, this conversion–reconversion business is a nuisance. Actually, it is more than a nuisance. Electrons generate heat and as the density of elements increases the point might have already been reached when there is no easy way to keep the temperature rise to an acceptable limit. It is as if the carriages in the tube that take you from Paddington to Liverpool Station would be not only uncomfortable but unbearably hot.

So let us see how such an interchange will be done in the future. The movable mirror whose construction was briefly discussed in Section 9.24 can of course be constructed in two-dimensional arrays. A fine example of a 6 × 6 array, in which the angular position of each element can be controlled, is shown in Fig. 13.24(a). Two such arrays may then be used in an optical cross-connect switch, as shown in Figure 13.24(b). The inputs and outputs are two square bundles of optical fibres equipped with collimators to produce small diameter free-space beams. Each input beam is arranged to strike one of the MEMS mirrors in the first array. This mirror may turn to point at any mirror in the second array, which may then turn to route the beam to its corresponding output fibre. Switches of this type may be extremely large, with up to 1000 inputs and 1000 outputs.

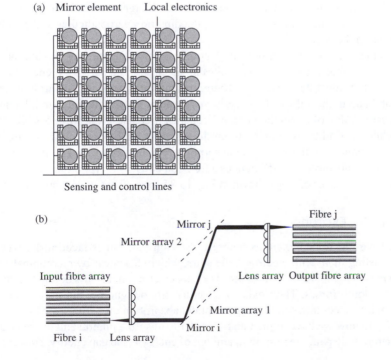

Fig. 13.24
(a) A 6 × 6 mirror array of dual axis MEMS torsion mirror. (b) Optical cross-connect constructed from two mirror arrays. After R.R.A. Syms, *J. Lightwave Technol.* **20**, 1084, 2002.

What about speed? Is mechanical movement not bound to be slow? Yes, it is, but it does not matter. It is a massively parallel operation. We can switch simultaneously hundreds of beams. It is optics. The beams can cross each other without any cross-talk. And, besides, a single mirror can switch an enormous amount of information from fibre A to fibre B, may be a 100 Gigabits or may be even more in the future. Who would worry then about a switching time that might be a few milliseconds?

13.11   Electro-absorption in quantum well structures

This is a fairly new phenomenon with potential for device applications. We shall include it not only because it might become a winner (it is rather hazardous to predict which device will prove to be commercially competitive) but because it is such a good illustration of a number of physical principles discussed in this course. We shall touch upon such topics as the confinement of electrons and holes to a certain region by the erection of potential barriers, the modification of semiconductor absorption characteristics when excitons (bound electron–hole pairs) are taken into account, what happens to excitons in a potential well, how an electric field influences the energy levels and, in particular, how it affects confined excitons (known as the Quantum Confined Stark Effect) and, finally, how these varied phenomena can be exploited in devices.

I have already talked a lot about quantum wells. One of the examples we looked at was made of GaAs and AlGaAs. For our present purpose, it is import- ant that the wells are wide enough for tunnelling to be negligible but narrow enough so that the electrons and holes know that they are not in an infinitely thick material.

The energy levels in such two-dimensional wells were discussed in Section 12.7. We know that the available energies may be represented by a set of sub- bands. The momentum is quantized in the direction perpendicular to the walls but not in the other two.

Let us consider now optical absorption in such a quantum well material. It is still true that when a photon excites an electron–hole pair, both energy and momentum must be conserved. Therefore, the transitions will occur between a sub-band in the valence band and a sub-band in the conduction band, having the same value of n. As the photon frequency increases, it is still possible, for a while, to find transitions between the same two sub-bands, and as long as the sub-bands are the same, the absorption remains constant. The theoretical plot of the absorption coefficient as a function of $hf - E_g$ (photon energy–gap energy) is a series of steps shown in Fig. 13.25 for a 30 nm wide quantum well.

13.11.1   Excitons

We have mentioned excitons before in Section 9.4, rather facetiously, when contrasting them with electron–hole pairs, which disappear by recombination. Let me say again what excitons are. They are electron–hole pairs bound together by Coulomb forces. Their existence can usually be ignored, but they must be taken into account when looking at optical absorption on a fine scale.

Up to now we have argued that, in order to absorb a photon and to create an electron–hole pair, the minimum amount of energy is the gap energy. However,

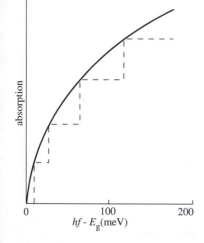

Fig. 13.25

Absorption as a function of excess photon energy for bulk material (continuous lines) and 30 nm thick quantum well material (dotted lines)

if we take excitons into account, then we may realize that photons incident with an energy somewhat less than the gap energy may also be absorbed because creating a bound electron–hole pair (i.e. an exciton) needs less energy than creating a free electron–hole pair. The difference between the two is the binding energy. How can we determine this binding energy? We may simply argue that the binding energy between an electron and a hole is the same type as that between an electron and a proton, and we may then use the hydrogen atom formulae into which we need to substitute the actual values of dielectric constant and effective mass. In bulk GaAs the binding energy may be calculated to be 4.2 meV, and the orbit of the exciton in its lowest state (called the 1s state in analogy with the hydrogen atom) is about 30 nm.

Does this mean that we see a sharply defined energy level below the gap? No, because we have to take into account that the lifetime of an exciton is not more than a few hundred femtoseconds and, therefore, the line will be considerably broadened. The outcome of all this is that, in bulk materials at room temperature, the excitonic resonance can hardly be noticed.

13.11.2 Excitons in quantum wells

Let us now consider the effect of excitons upon the absorption spectrum in a quantum well material. If the width of the well is larger than the exciton orbit then, obviously, the excitons will hardly be affected. But if the width of the well is less than the diameter of the lowest orbit, then the exciton has no other option but to get squashed in the direction perpendicular to the layers. What will happen in the other directions? A good indication can be obtained by solving the problem of the two-dimensional hydrogen atom. It turns out that the diameter will reduce by a factor of 4. So we may come to the conclusion that the size of the exciton will be reduced in all three dimensions, and consequently its binding energy will be increased. Calculations yield a binding energy of 10 meV, indicating that the exciton absorption effect is much stronger. We may expect to see exciton peaks associated with each absorption step.

Experiments show that our expectations are correct. For a GaAs quantum well of 10 nm width, the measured absorption spectrum is shown in Fig. 13.26. Both the peaks and the steps may be clearly distinguished. Do not worry about the double peak. It is simply due to the fact that there are both heavy and light holes present, which have not so far been mentioned because they do not affect the basic argument.

13.11.3 Electro-absorption

This means optical absorption in the presence of an electric field. Will the excitons know that an electric field is present? They certainly will. An electric field will try to move the two particles in opposite directions. Since there is a tendency anyway for the two particles to part company, they can more easily do so in the presence of an electric field. Consequently, the exciton is more likely to be ionized (this is called field-ionization), its lifetime will be shorter, and the absorption line will be broader. All this is true for the bulk material. But will the argument change in any way for a quantum well material? If the electric field is applied in the direction of the layers, then the same argument will still roughly

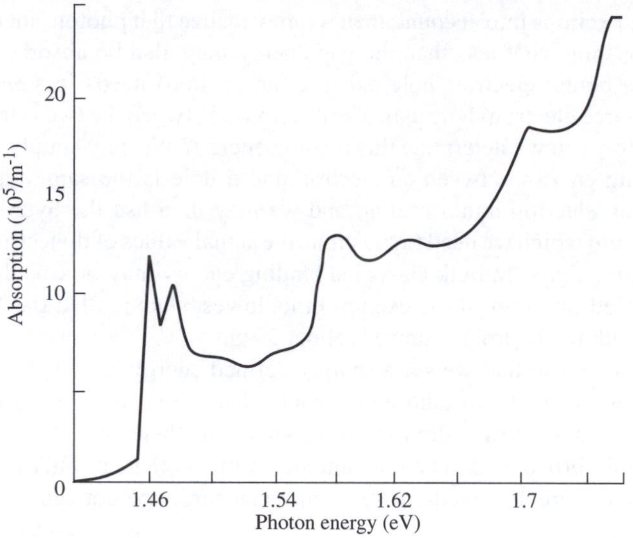

Fig. 13.26

Measured absorption spectrum of a GaAs quantum well of 10 nm thickness.

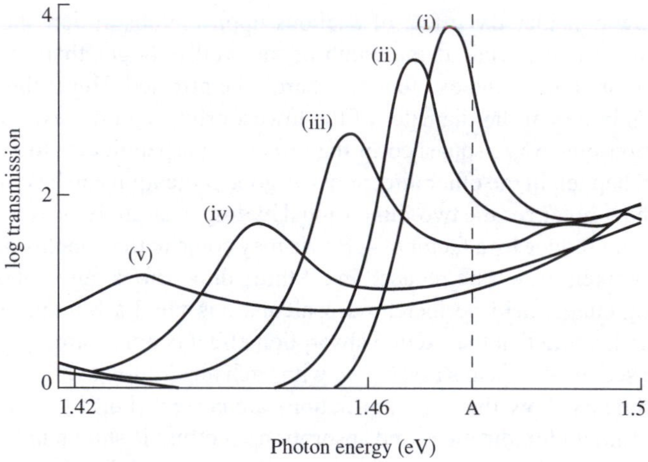

Fig. 13.27

Measured absorption spectrum of a quantum well for electric field applied perpendicularly to the layers, (i) $1.6 \times 10^4 \, \mathrm{V\,m^{-1}}$, (ii) $10^5 \, \mathrm{V\,m^{-1}}$, (iii) $1.4 \times 10^5 \, \mathrm{V\,m^{-1}}$, (iv) $1.8 \times 10^5 \, \mathrm{V\,m^{-1}}$, (v) $2.2 \times 10^5 \, \mathrm{V\,m^{-1}}$.

hold: the chances of being able to ionize the exciton with the aid of the electric field must be high. However, when the electric field is applied perpendicular to the layers, then the good intentions of the electric field in trying to separate the particles are frustrated by the presence of the walls. Provided the well is narrow enough, the exciton is not field-ionized. Thus, the exciton resonance is still there albeit with reduced amplitude due to the increased separation of the electron–hole pair. There is also a shift in the position of the resonance due to the electrostatic energy (this is nothing else but the energy of a dipole in an electric field, something we have discussed before).

The expectations based on the above qualitative argument are borne out by the experimental observations, as shown in Fig. 13.27, where the absorption spectrum measured for a GaAs–AlGaAs quantum well structure is plotted against photon energy. This phenomenon is known as the Quantum Confined Stark Effect. The excitons are obviously quantum confined, and the Stark effect is the description used for the shift and splitting of energy levels in an electric field.

Talking of electro-absorption, I should also mention the Franz–Keldysh effect, which is usually observed at electric fields considerably higher than that needed to see the effect upon exciton resonance. It bears some similarity to the Schottky effect, in which electrons excited thermally to high enough energy levels could tunnel across a barrier made thin by the presence of a high electric field. For the Franz–Keldysh effect, the energy is provided by an input photon with an energy less than the bandgap energy, and then tunnelling can do the rest to provide an electron–hole pair.

13.11.4   Applications

If I had to classify the Quantum Confined Stark Effect in literary terms, I would not quite know where to place it. Perhaps melodrama would be the right category, considering the touching affection between electrons and holes. If we consider, however, how they stave off brutal intervention by the electric field, with their backs against the potential wall, and how quickly all these things happen, then melodrama might give way to a thriller. And that is certainly the category to which our ultimate question belongs: 'can these effects be used for something?'

Well, if the attenuation of the device depends on the electric field, then the amount of light across it may be modulated by changing the applied voltage. We need relatively high fields which, we know, may be achieved by placing the multiple quantum well structure inside a junction as shown in Fig. 13.28. It is a reverse bias p–i–n junction, in which the p$^+$ and n$^+$ materials are made of AlGaAs, and the quantum well provides the intrinsic part of the junction.

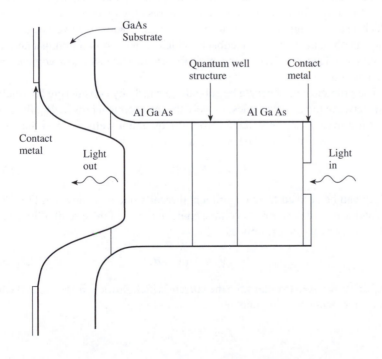

Fig. 13.28

A quantum well structure inside a reverse biased p–i–n junction used as a light modulator.

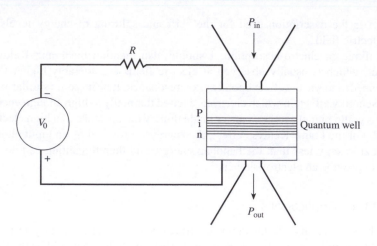

Fig. 13.29

The device of Fig. 13.24 in a circuit in
which it can exhibit bistability.

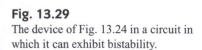

It is grown on a GaAs substrate, which needs to be etched away, since it is not
transparent at the wavelength where the mechanism is suitable for modulation.

What else can we do with this device? It can be used for light detection in
the same manner as an ordinary p–i–n junction. The added advantage of the
multiple quantum well structure is that the resulting photocurrent is strongly
dependent on the applied voltage and on the wavelength. It can therefore be
used, admittedly in a very narrow range, for measuring wavelength.

The most interesting application is, however, for a bistable device shown
schematically in Fig. 13.29. The wavelength of the incident light is chosen, so
that decreasing reverse-bias voltage gives increasing optical absorption (line
A in Fig. 13.27). In the absence of optical input, no current flows, and hence
all the applied voltage appears across the quantum wells. As light is incident,
a photocurrent flows, hence the voltage across the quantum wells decreases,
which leads to higher absorption, which in turn leads to lower voltage, etc.
There is obviously positive feedback, which under certain circumstances may
be shown to lead to switching to a high absorption state with low voltage across
the junction.

The problem can of course be solved rigorously by considering the relationship between the four variables, namely the input optical power, P_{in}, the output
optical power, P_{out}, the voltage across the quantum wells, V, and the current
flowing in the circuit, I. First, we need

$$P_{out} = P_{out}(P_{in}, V, \lambda), \qquad (13.23)$$

which can be derived from experimental results like that shown in Fig. 13.27.
Second, we need the current–voltage characteristics of the circuit of Fig. 13.25,
which may be written simply as

$$V = V_0 - IR. \qquad (13.24)$$

And third, we need to determine the current which flows in response to the light
power incident upon the junction.

$$I = I(V, P_{in}, \lambda). \qquad (13.25)$$

From eqns (13.23–25) we may then derive the P_{out} versus P_{in} curve which, in many cases, will exhibit bistability similar to that shown in Fig. 13.23.

Let us now summarize the advantages of multiple quantum well structures for the applications discussed. The main advantage is compatibility, that is the voltages are compatible with the electronics, and the wavelengths are compatible with laser diodes. In addition, the materials are compatible with those used both in electronics and for laser diodes, so the devices are potential candidates for components in integrated opto-electronic systems.

I have only mentioned GaAs–AlGaAs structures, but there are, of course, others as well. The rules are clearly the same, which apply to the production of heterojunction lasers. The lattice constants must be close, and the bandgaps must be in the right range. Interestingly, some of the combinations offer quite new physics, for example in an InAs–GaSb quantum well, the electrons are confined in one layer and the holes in the other one.

As you may have gathered, I find the topic of quantum wells quite fascinating, so perhaps I spent a little more time on it than its present status would warrant. I hope you will forgive me.

Exercises

13.1. Light of frequency ν and intensity I_0 is incident upon a photoconductor (Fig. 13.30) which has an attenuation coefficient α. Assuming that only electrons are generated show that the excess current due to the input light is

$$\Delta I = e\frac{b}{c}\frac{\eta I_0}{h\nu}\tau_e \mu_e\, V\frac{1 - e^{-\alpha d}}{\alpha},$$

where τ_e is the electron lifetime and η is the quantum efficiency (average number of electrons generated per incident photon).

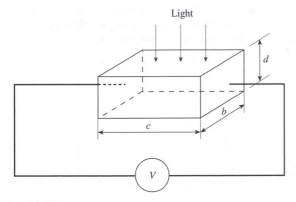

Fig. 13.30

13.2. The photoconductive gain is defined as

$$G = \frac{\text{Number of photocarriers crossing the electrodes per unit time}}{\text{Number of photocarriers generated per unit time}}.$$

Find an expression for it for the photoconductor discussed in Exercise 13.1.

13.3. In a p–i–n diode the so-called intrinsic region is usually a lightly doped n-type region. Determine the electric field and potential distribution for a reverse bias of U_r when the impurity densities of the three regions are N_A, N_{D1} and N_{D2} (see Fig. 13.31).

When this device is used as a photodetector with light incident from the left, the p$^+$ region must be made extremely thin. Why?

(Hint: Assume that the depletion region is all in the lightly doped n region. Neglect the built-in voltage.)

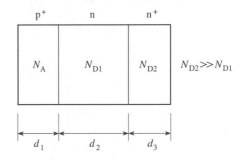

Fig. 13.31

13.4. A volume hologram is recorded in a photosensitive material with a refractive index of 1.52 at a wavelength of

514.5 nm by two beams incident from air at angles $\pm 5°$ in the geometry of Fig. 13.3.

(i) What is the grating spacing in the material?
(ii) What should be the incident angle if the hologram is to be replayed at 633 nm?
(iii) If the hologram is to be replayed at the second Bragg angle what will that angle be at the two above mentioned wavelengths?

13.5. A volume hologram is recorded by two beams incident perpendicularly upon a photosensitive medium from opposite sides (the so called reflection geometry).

(i) What will be the grating spacing at wavelengths 514.5 nm and 633 nm? Take $n = 1.52$.
(ii) Which configuration do you think will be more sensitive to replay wavelength, the transmission type or the reflection type?

13.6. Some crystals have the property that they are isotropic in the absence of an electric field but become aniostropic when an electric field is applied. The dielectric constant tensor in such a material in the XYZ coordinate system (Fig. 13.32) is as follows

$$\epsilon = \epsilon_0 \begin{bmatrix} \epsilon_{XX} & 0 & 0 \\ 0 & \epsilon_{YY} & 0 \\ 0 & 0 & \epsilon_{ZZ} \end{bmatrix}$$

where

$$\epsilon_r = \epsilon_{XX} = \epsilon_{YY} = \epsilon_{ZZ}.$$

The electro-optic coefficient may now be assumed to have a component which relates, in the same coordinate system, the change in the ϵ_{XY} component of the dielectric tensor to the Z component of the electric field as

$$-\Delta\epsilon_{XY} = \epsilon_r^2 r_{XYZ} \mathcal{E}_Z.$$

(i) Find the dielectric tensor in the presence of an electric field applied in the Z-direction.
(ii) Transform this dielectric tensor into the xyz coordinate system. Note that x, y is 45° clockwise from XY.
(iii) Will the input wave, shown in Fig. 13.28, be affected by the applied electric field?

(Note: This is the actual tensor of a $Bi_{12}SiO_{20}$ crystal cut in a certain way, apart from optical activity which we have disregarded here.)

13.7. The differential equations for the amplitudes of waves in two coupled waveguides are given in eqns (13.16) and (13.17).

(i) Find the solution with boundary conditions $A_1 = A_{10}$ and $A_2 = 0$ at $z = 0$.
(ii) Show that the solution reduces to those of eqns (13.18) and (13.19) when $k_1 = k_2 = k$.
(iii) If the length of the interaction region is 1 cm what should be the value of the coupling constant in order to achieve complete power transfer from waveguide 1 into waveguide 2?
(iv) By how much should v_2 the phase velocity in waveguide 2 be different from that in waveguide 1 in order to reduce the power coupled into waveguide 2 by a factor 2? Take $L = 1\,\text{cm}$, $K = \pi/2\,\text{cm}^{-1}$, $\lambda = 633\,\text{nm}$ and, initially, $v_1 = v_2 = v = 10^8\,\text{m s}^{-1}$.

13.8. An electromagnetic wave with an electric field $\mathcal{E}_i$ is incident perpendicularly on the Fabry–Perot resonator shown in Fig. 13.33. The amplitude reflectivities and transmittivities of the mirrors are r_1, t_1 and r_2, t_2 respectively, the wave number is k and the gain coefficient of the medium filling the resonator is γ. Derive an expression (by summing an infinite geometrical series) for the transmitted electric field, $\mathcal{E}_t$.

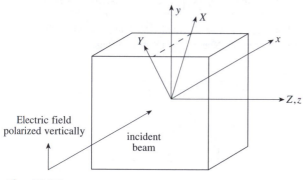

Fig. 13.32

Fabry–Perot resonator

Fig. 13.33

Superconductivity

I go on for ever.
Tennyson *The Brook*

14.1 Introduction

Superconductivity was a scientific curiosity for a long time, but by now there are some actual applications like producing high magnetic fields for Magnetic Resonance Imaging, and there are lots of potential applications. The big prize would of course be superconductive lines for power transmission which would eliminate losses. But quite apart from applications, I feel that some acquaintance with superconductivity should be part of modern engineering education. Superconductivity is, after all, such an extraordinary phenomenon, so much in contrast with everything we are used to. It is literally out of this world. Our world is classical, but superconductivity is a quantum phenomenon—a quantum phenomenon on macroscopic scale. The wavefunctions, for example, that lead an artificial existence in quantum mechanics proper appear in superconductivity as measurable quantities.

The discovery of superconductivity was not very dramatic. When Kamerlingh Onnes succeeded in liquefying helium in 1908, he looked round for something worth measuring at that temperature range. His choice fell upon the resistivity of metals. He tried platinum first and found that its resistivity continued to decline at lower temperatures, tending to some small but finite value as the temperature approached the absolute zero. He could have tried a large number of other metals with similar prosaic results. But he was in luck. His second metal, mercury, showed quite unorthodox behaviour. Its resistivity (as shown in Fig. 14.1) suddenly decreased to such a small value that he was unable to measure it—and no one has succeeded in measuring it ever since. The usual technique is to induce a current in a ring made of superconducting material and measure the magnetic field due to this current. In a normal metal the current would decay in about 10^{-12} s. In a superconductor the current can go round for a considerably longer time—measured not in picoseconds but in years. One of the longest experiments was made somewhere in the United States; the current was going round and round for three whole years without any detectable decay. Unfortunately, the experiment came to an abrupt end when a research student forgot to fill up the Dewar flask with liquid nitrogen—so the story goes anyway.

Thus, for all practical purposes we are faced with a real lossless phenomenon. It is so much out of the ordinary that no one quite knew how to approach the problem. Several phenomenological theories were born, but its real cause remained unknown for half a century. Up to 1957 it defied all attempts; so much so, that it gave birth to a new theorem, namely that 'all theories of superconductivity are refutable'. In 1957, Bardeen, Schrieffer, and Cooper produced a theory (called

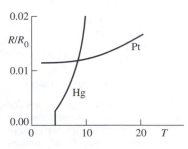

Fig. 14.1

The resistance of samples of platinum and mercury as a function of temperature (R_0 is the resistance at 0°C).

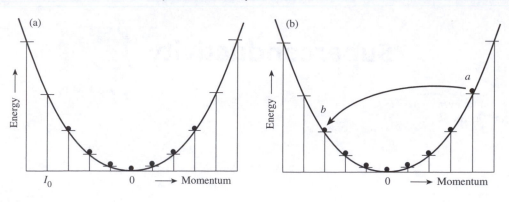

Fig. 14.2

A one-dimensional representation of the energy–momentum curve for seven electrons in a conductor. (a) All electrons in their lowest energy states, the net momentum is zero. (b) There is a net momentum to the right as a consequence of an applied electric field.

the BCS theory) that managed to explain all the major properties of super-conductivity for all the superconductors known at the time. Unfortunately, the arrival of a host of new superconducting materials has cast fresh doubts on our ability ever to produce a complete theory. For the time being BCS is the best theory we have. The essence of the theory is that superconductivity is caused by electron–lattice (or, using more sophisticated language, by electron–phonon) interaction and that the superconducting electrons consist of ordinary electrons paired up.

There is not much point in going into the details of this theory; it is far too complicated, but a rough idea can be provided by the following qualitative explanation due to Little.

Figure 14.2(a) shows the energy–momentum curve of an ordinary conductor with seven electrons sitting discreetly in their discrete energy levels. In the absence of an electric field the current from electrons moving to the right is exactly balanced by that from electrons moving to the left. Thus, the net current is zero.

When an electric field is applied, all the electrons acquire some extra momentum, and this is equivalent to shifting the whole distribution in the direction of the electric field, as shown in Fig. 14.2(b). Now what happens when the electric field is removed? Owing to collisions with the vibrating lattice, with impurity atoms, or with any other irregularity, the faster electrons will be scattered into lower energy states until the original distribution is re-established. For our simple model, it means that the electron is scattered from the energy level, a, into energy level, b.

In the case of a superconductor, it becomes energetically more favourable for the electrons to seek some companionship. Those of opposite momenta (the spins incidentally must also be opposite, pair up to form a new particle called a *superconducting electron* or, after its discoverer, a *Cooper pair*.*This link between two electrons is shown in Fig. 14.3(a) by an imaginary mechanical spring.

We may ask now a few questions about our newly born composite particle. First of all, what is its velocity? The two constituents of the particle move with

* In actual fact, the first man to suggest the pairing of electrons was R.A. Ogg. According to Gamow's limerick:

> In Ogg's theory it was his intent
> That the current keep flowing, once sent;
> So to save himself trouble,
> He put them in double,
> And instead of stopping, it went.

Ogg preceded Cooper by about a decade, but his ideas were put forward in the language of an experimental chemist, which is unforgivable. No one believed him, and his suggestion faded into oblivion. This may seem rather unfair to you, but that is how contemporary science works. In every discipline there is a select band of men whose ideas are taken up and propagated, so if you want to invent something great, try to associate yourself with the right kind of people.

Do not try to make any contributions to theoretical physics unless you are a trained theoretical physicist, and do not meddle in theology unless you are a bishop.

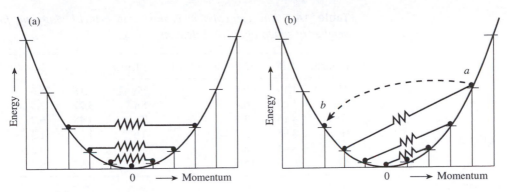

Fig. 14.3

The energy–momentum curve for seven electrons in a superconductor. Those of opposite momenta pair up—that is represented here by a mechanical spring. (a) All pairs in their lowest energy states, the net momentum is zero. (b) There is a net momentum to the right as a consequence of an applied electric field.

v and $-v$, respectively; thus the velocity of the centre of mass is zero. Remembering the de Broglie relationship ($\lambda = h/p$) this means that the wavelength associated with the new particle is infinitely long. And this is valid for *all* superconducting electrons.

It does not quite follow from the above simple argument (but it comes out from the theory) that all superconducting electrons behave in the same way. This is, for our electrons, a complete break with the past. Up to now, owing to the rigour of the Pauli principle, all electrons had to be different. In superconductivity they acquire the right to be the same—so we have a large number of identical particles all with infinite wavelength; that is, we have a quantum phenomenon on a macroscopic scale.

An applied electric field will displace all the particles again, as shown in Fig. 14.3(b), but when the electric field disappears, there is no change. Scattering from energy level a to energy level b is no longer possible because then the electrons both at b and c would become pairless, which is energetically unfavourable. One may imagine a large number of simultaneous scatterings that would just re-establish the symmetrical distribution of Fig. 14.3(a), but that is extremely unlikely. So the asymmetrical distribution will remain; there will be more electrons going to the right than to the left, and this current will persist forever—or, at least, for three years.

14.2   The effect of a magnetic field

14.2.1   The critical magnetic field

One of the applications of superconductivity coming immediately to mind is the production of a powerful electromagnet. How nice it would be to have high magnetic fields without any power dissipation. The hopes of the first experimenters were soon dashed. They found out that above a certain magnetic field the superconductor became normal. Thus, in order to have zero resistance, not only the temperature but also the magnetic field must be kept below a certain

Table 14.1 *The critical temperature and critical magnetic field of a number of superconducting elements*

Element	$T_c(K)$	$H_0 \times 10^{-4}$ (A m)$^{-1}$	Element	$T_c(K)$	$H_0 \times 10^{-4}$ (A m^{-1})
Al	1.19	0.8	Pb	7.18	6.5
Ga	1.09	0.4	Sn	3.72	2.5
Hg α	4.15	3.3	Ta	4.48	6.7
Hg β	3.95	2.7	Th	1.37	1.3
In	3.41	2.3	V	5.30	10.5
Nb	9.46	15.6	Zn	0.92	0.4

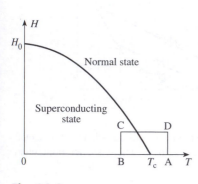

Fig. 14.4
The critical magnetic field as a function of temperature.

threshold value. Experiments with various superconductors have shown that the dependence of the critical magnetic field on temperature is well described by the formula,

$$H_c = H_0 \left\{ 1 - \left(\frac{T}{T_c} \right)^2 \right\}. \tag{14.1}$$

This relationship is plotted in Fig. 14.4. It can be seen that the material is normal above the curve and superconducting below the curve. H_0 is defined as the magnetic field that destroys superconductivity at absolute zero temperature. The values of H_0 and T_c for a number of superconducting elements are given in Table 14.1. Alloys could have both much higher critical temperatures and much higher critical magnetic fields. They will come later.

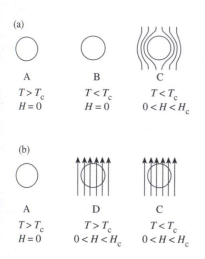

Fig. 14.5
The magnetic states of a superconductor while tracing the (a) ABC and (b) ADC paths in Fig. 14.4.

14.2.2 The Meissner effect

We have seen that below a certain temperature and magnetic field a number of materials lose their electrical resistivity completely. How would we expect these materials to behave if taken from point A to C in Fig. 14.4 by the paths ABC and ADC respectively? At point A there is no applied magnetic field and the temperature is higher than the critical one [Fig. 14.5(a)]. From A to B the temperature is reduced below the critical temperature; so the material loses its resistivity, but nothing else happens. Going from B to C means switching on the magnetic field. The changing flux creates an electric field that sets up a current opposing the applied magnetic field. This is just Lenz's law, and in the past we have referred to such currents as eddy currents. The essential difference now is the absence of resistivity. The eddy currents do not decay; they produce a magnetic field that completely cancels the applied magnetic field inside the material. Thus, we may regard our superconductor as a perfect diamagnet.

Starting again at A with no magnetic field [Fig. 14.5(b)] and proceeding to D puts the material into a magnetic field at a constant temperature. Assuming that our material is non-magnetic (superconductors are in fact slightly paramagnetic above their critical temperature, as follows from their metallic nature), the magnetic field will penetrate. Going from D to C means reducing the temperature at constant magnetic field. The material becomes superconducting at some point, but there is no reason why this should imply any change in the magnetic field distribution. At C the magnetic field should penetrate just as well as at D.

Thus, the distribution of the magnetic field at C depends on the path we have chosen. If we go via B, the magnetic field is expelled; if we go via D, the magnetic field is the same inside as outside. The conclusion is that for a perfect conductor (meaning a material with no resistance) the final state depends on the path chosen. This is quite an acceptable conclusion; there are many physical phenomena exhibiting this property. What is interesting is that superconductors do *not* behave in this expected manner. A superconductor cooled in a constant magnetic field will set up its own current and expel the magnetic field when the critical temperature is reached.

The discovery of this effect by Meissner in 1933 showed superconductivity in a new light. It became clear that superconductivity is a new kind of phenomenon that does not obey the rules of classical electrodynamics.

14.3 Microscopic theory

The microscopic theory is well beyond the scope of an engineering undergraduate course and, indeed, beyond the grasp of practically anyone. It is part of quantum field theory and has something to do with Green's functions and has more than its fair share of various operators. We shall not say much about this theory, but we should just like to indicate what is involved.

The fundamental tenet of the theory is that superconductivity is caused by a second-order interaction between electrons and the vibrating lattice. This is rather strange. After all, we do know that thermal vibrations are responsible for the presence of resistance and not for its absence. This is true in general; the higher the temperature the larger the electrical resistance. Below a certain temperature, however, and for a select group of materials, the lattice interaction plays a different role. It is a sort of intermediary between two appropriately placed electrons. It results in an apparent attractive force between the two electrons, an attractive force larger than the repulsive force, owing to the Coulomb interaction. Hence, the electron changes its character. It stops obeying Fermi–Dirac statistics, and any number of electrons (or more correctly any number of electron pairs) can be in the same state. Besides the atom laser (Section 12.14) this is another example of a Bose–Einstein condensation.

Do we have any direct experimental evidence that superconductivity is caused by electron–lattice interaction? Yes, the so-called isotope effect. The critical temperature of a superconductor depends on the total mass of the nucleus. If we add a neutron (that is, use an isotope of the material) the critical temperature decreases.

A simple explanation of the interaction between two electrons and the lattice is shown in Fig. 14.6(a). The first electron moving to the right causes the positive lattice ions to move inwards, which then attract the second electron. Hence, there is an indirect attraction between the two electrons.

Are there any other kind of interactions resulting in superconductivity? Nobody knows for certain, but it may be worthwhile to describe briefly one of the mechanisms proposed for explaining the behaviour of the recently discovered oxide superconductors. It is electron attraction mediated by spin waves. As may be seen in Fig. 14.6(b), an electron with a certain spin disrupts the spin of an ion, which causes the spin of its neighbouring ion to flip, which then attracts a second electron of opposite spin. It has been suggested recently

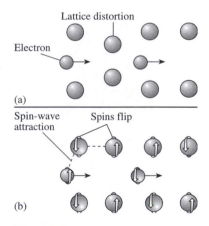

Fig. 14.6
Interactions leading to Bose–Einstein condensation (a) between the lattice and electrons, (b) between spin waves and electrons.

that induced magnetic fluctuations may also be responsible for the pairing mechanism.

14.4 Thermodynamical treatment

Let us look again at Fig. 14.4. Above the curve our substance behaves in the normally accepted way. It has the same sort of properties it had at room temperature. Its magnetic properties are the same, and its electric properties are the same; true, the electrical resistivity is smaller than at room temperature, but there is nothing unexpected in that. However, as soon as we cross the curve, the properties of the substance become qualitatively different. Above the curve the substance is non-magnetic, below the curve it becomes diamagnetic; above the curve it has a finite electrical resistance, below the curve the electric resistance is zero.

If you think about it a little you will see that the situation is very similar to that you have studied under the name of 'phase change' or 'phase transition' in thermodynamics. Recall, for example, the diagram showing the vaporization of water (Fig. 14.7). The properties of the substance differ appreciably above and below the curve, and we do not need elaborate laboratory equipment to tell the difference. Our senses are quite capable of distinguishing steam from water. It is quite natural to call them by different names and refer to the state above the curve as the liquid phase, and to the state below the curve as the vapour phase. Analogously, we may talk about normal and superconducting phases when interpreting Fig. 14.4.

Thus, the road is open to investigate the properties of superconductors by the well-established techniques of thermodynamics. Well, is the road open? We must be careful; thermodynamics can be applied only if the change is reversible. Is the normal to superconducting phase change reversible? Fortunately, it is. Had we a perfect conductor instead of a superconductor the phase change would *not* be reversible, and we should not be justified in using thermodynamics. Thanks to the Meissner effect, thermodynamics *is* applicable.

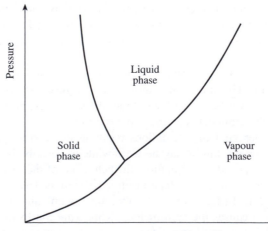

Fig. 14.7
The pressure against temperature diagram for water.

Let us now review the thermodynamical equations describing the phase transitions. There is the first law of thermodynamics:

$$\mathrm{d}E = \mathrm{d}Q - \mathrm{d}W$$
$$= T\,\mathrm{d}S - P\,\mathrm{d}V. \tag{14.2}$$

Then there is the Gibbs function (to which we shall also refer as the Gibbs free energy) defined by

$$G = E + PV - TS. \tag{14.3}$$

An infinitesimal change in the Gibbs function gives

$$\mathrm{d}G = \mathrm{d}E + P\,\mathrm{d}V + V\,\mathrm{d}P - T\,\mathrm{d}S - S\,\mathrm{d}T, \tag{14.4}$$

which, using eqn (14.2) reduces to

$$\mathrm{d}G = V\,\mathrm{d}P - S\,\mathrm{d}T. \tag{14.5}$$

Thus, for an isothermal, isobaric process

$$\mathrm{d}G = 0,$$

that is the Gibbs function does not change while the phase transition takes place.

In the case of the normal-to-superconducting phase-transition the variations of pressure and volume are small and play negligible roles, and so we can just as well forget them but, of course, we shall have to include the work due to magnetization.

In order to derive a relationship between work and magnetization let us investigate the simple physical arrangement shown in Fig. 14.8. You know from studying electricity that work done on a system in a time $\mathrm{d}t$ is

$$\mathrm{d}W = -UI\,\mathrm{d}t, \tag{14.6}$$

Further, using Faraday's law, we have

$$U = NA\frac{\mathrm{d}B}{\mathrm{d}t}. \tag{14.7}$$

From Ampère's law

$$HL = NI. \tag{14.8}$$

We then get

$$\mathrm{d}W = -NA\frac{\mathrm{d}B}{\mathrm{d}t}I\,\mathrm{d}t = -NIA\,\mathrm{d}B = -HLA\,\mathrm{d}B = -VH\,\mathrm{d}B. \tag{14.9}$$

According to eqn (11.3)

$$B = \mu_0(H + M).$$

Therefore,

$$\mathrm{d}W = -VH\mu_0(\mathrm{d}H + \mathrm{d}M)$$
$$= -\mu_0 VH\,\mathrm{d}H - \mu_0 VH\,\mathrm{d}M. \tag{14.10}$$

The first term on the right-hand side of eqn (14.10) gives the increase of energy in the vacuum, and the second term is due to the presence of the magnetic

E is the internal energy, W the work, S the entropy, P the pressure, V the volume, and Q the heat.

U is the voltage and I the current, and the negative sign comes from the accepted convention of thermodynamics that the work done *on* a system is negative.

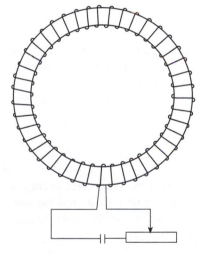

Fig. 14.8

The magnetization of magnetic material in a toroid (for working out the magnetic energy).

A is the cross-section of the toroid, N the number of turns, and L the mean circumference of the toroid.

material. Thus, the work done *on* the material is

$$dW = -\mu_0 VH\,dM. \tag{14.11}$$

Hence, for a paramagnetic material the work is negative, but for a diamagnetic material (where M is opposing H) the work is positive, which means that the system needs to do some work in order to reduce the magnetic field inside the material.

Now to describe the phase transition in a superconductor, we have to define a 'magnetic Gibbs function'. Remembering that $P\,dV$ gives positive work (an expanding gas does work) and $H\,dM$ gives negative work, we have to replace PV in eqn (14.3) by $-\mu_0 VHM$. Our new Gibbs function takes the form,

$$G = E - \mu_0 VHM - TS, \tag{14.12}$$

and

$$dG = dE - \mu_0 VH\,dM - \mu_0 VM\,dH - T\,dS - S\,dT. \tag{14.13}$$

Taking account of the first law for magnetic materials (again replacing pressure and volume by the appropriate magnetic quantities)

$$dE = T\,dS + \mu_0 VH\,dM. \tag{14.14}$$

Equation (14.13) reduces to

$$dG = -S\,dT + \mu_0 VM\,dH. \tag{14.15}$$

This is exactly what we wanted. It follows immediately from the above equation that for a constant temperature and constant magnetic field process

G remains constant while the superconducting phase transition takes place.

$$dG = 0. \tag{14.16}$$

For a perfect diamagnet

$$M = -H, \tag{14.17}$$

which substituted into eqn (14.15) gives

$$dG = -S\,dT + \mu_0 VH\,dH. \tag{14.18}$$

Integrating at constant temperature, we get

$G_s(0)$ is the Gibbs free energy at zero magnetic field, and the subscript s refers to the superconducting phase.

$$G_s(H) = G_s(0) + \tfrac{1}{2}\mu_0 H^2 V. \tag{14.19}$$

Since superconductors are practically non-magnetic above their critical temperatures, we can write for the normal phase

$$G_n(H) = G_n(0) = G_n. \tag{14.20}$$

In view of eqn (14.16) the Gibbs free energy of the two phases must be equal at the critical magnetic field, H_c, that is

$$G_s(H_c) = G_n(H_c). \tag{14.21}$$

Substituting eqns (14.19) and (14.20) into eqn (14.21) we get

$$G_n = G_s(0) + \tfrac{1}{2}\mu_0 H_c^2 V. \tag{14.22}$$

Comparison of eqns (14.19) and (14.22) clearly shows that at a given temperature (below the critical one) the conditions are more favourable for the superconducting phase than for the normal phase, provided that the magnetic field is below the critical field. There are three cases:

(i)   If $H < H_c$ then $G_n > G_s(H)$.   (14.23)

(ii)   If $H > H_c$ then $G_n < G_s(H)$.   (14.24)

(iii)   If $H = H_c$ then $G_n = G_s(H_c)$.   (14.25)

Now our substance will prefer the phase for which the Gibbs free energy is smaller; that is in case (i) it will be in the superconducting phase, in case (ii) in the normal phase, and in case (iii) just in the process of transition.

If the transition takes place at temperature, $T + dT$, and magnetic field, $H_c + dH_c$, then it must still be valid that

$$G_s + dG_s = G_n + dG_n, \tag{14.26}$$

whence

$$dG_s = dG_n. \tag{14.27}$$

This, using eqn (14.15), leads to

$$-S_s\, dT - \mu_0 V M_s\, dH_c = -S_n\, dT - \mu_0 V M_n\, dH_c. \tag{14.28}$$

But, as suggested before,

$$M_s = -H_c \quad \text{and} \quad M_n = 0, \tag{14.29}$$

and this reduces eqn (14.28) to

$$S_n - S_s = -\mu_0 V H_c \frac{dH_c}{dT}. \tag{14.30}$$

The latent heat of transition may be written in the form,

$$L = T(S_n - S_s), \tag{14.31}$$

and with the aid of eqn (14.30) this may be expressed as

$$L = -\mu_0 T V H_c \frac{dH_c}{dT}. \tag{14.32}$$

After much labour we have, at last, arrived at a useful relationship. We now have a theory that connects the independently measurable quantities L, V, H_c, and T. After measuring them, determining dH_c/dT from the $H_c - T$ plot, and substituting their values into eqn (14.32), the equation should be satisfied; and it is satisfied to a good approximation thus giving us experimental proof that we are on the right track.

It is interesting to note that L vanishes at two extremes of temperature, namely, at $T = 0$ and at $T = T_c$ where the critical magnetic field is zero. A transition which takes place with no latent heat is called a second-order phase transition. In this transition entropy remains constant, and the specific heat is discontinuous.

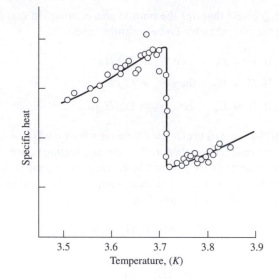

Fig. 14.9

Temperature dependence of the specific heat of tin near the critical temperature (after Keesom and Kok, 1932).

Neglecting the difference between the specific heats at constant volume and constant pressure, we can write, in general for the specific heat,

$$c = T \frac{\mathrm{d}S}{\mathrm{d}T}. \tag{14.33}$$

Substituting from eqn (14.30)

$$c_n - c_s = T \left(\frac{\mathrm{d}S_n}{\mathrm{d}T} - \frac{\mathrm{d}S_s}{\mathrm{d}T} \right)$$

$$= -VT\mu_0 \left\{ -H_c \frac{\mathrm{d}^2 H_c}{\mathrm{d}T^2} + \left(\frac{\mathrm{d}H_c}{\mathrm{d}T} \right)^2 \right\}. \tag{14.34}$$

At $T = T_c$ where $H_c = 0$,

$$c_n - c_s = -\left\{ VT\mu_0 \left(\frac{\mathrm{d}H_c}{\mathrm{d}T} \right)^2 \right\}_{T=T_c}. \tag{14.35}$$

This is negative because the experimentally established $H_c - T$ curves have finite slopes at $T = T_c$. It follows that in the absence of a magnetic field the specific heat has a discontinuity. This is borne out by experiments as well, as shown in Fig. 14.9, where the specific heat of tin is plotted against temperature. The discontinuity occurs at the critical temperature $T_c = 3.72$ K.

14.5 Surface energy

The preceding thermodynamical analysis was based on perfect diamagnetism, that is, we assumed that our superconductor completely expelled the magnetic field. In practice this is not so, and it can not be so. The currents that are set up to exclude the magnetic field must occupy a finite volume, however small that might be. Thus, the magnetic field can also penetrate the superconductor to a small extent. But now we encounter a difficulty. If the magnetic field can penetrate to a finite distance, the Gibbs free energy of that particular layer will decrease, because it no longer has to perform work to exclude the magnetic field. The magnetic field is admitted, and we get a lower Gibbs free energy. Carrying this argument to its logical conclusion, it follows that the optimum arrangement for minimum Gibbs free energy (of the whole solid at a given temperature) should look like that shown in Fig. 14.10, where normal and superconducting layers alternate. The width of the superconducting layers, s, is small enough to permit the penetration of magnetic field, and the width of the normal region is even smaller, $n \ll s$. In this way the Gibbs free energy of the superconducting domains is lower because the magnetic field can penetrate, while the contribution of the normal domains to the total Gibbs free energy remains negligible because the volume of the normal domains is small in comparison with the volume of the superconducting domains.

Thus, a consistent application of our theory leads to a superconductor in which normal and superconducting layers alternate. Is this conclusion correct? Do we find these alternating domains experimentally? For some superconducting materials we do; for some other superconducting materials we do not. Incidentally, when the first doubts arose about the validity of the simple thermodynamical treatment, all the experimental evidence available at the time suggested that no break-up could occur. We shall restrict the argument to this historically authentic case for the moment. Theory suggests that superconductors should break up into normal and superconducting domains; experiments show that they do not break up. Consequently, the theory is wrong. The theory cannot be completely wrong, however, for it predicted the correct relationship for specific heat. So instead of dismissing the theory altogether, we modify it by introducing the concept of *surface energy*. This would suggest that the material does not break up because maintaining boundaries between the normal and superconducting domains is a costly business. It costs energy.*Hence, the simple explanation for the absence of domains is that the reduction in energy resulting from the configuration shown in Fig. 14.10 is smaller than the energy needed to maintain the surfaces.

* This is really the same argument that we used for domains in ferromagnetic materials. On the one hand, the more domains we have the smaller is the external magnetic energy. On the other hand, the more domains we have, the larger is the energy needed to maintain the domain walls. So, the second consideration will limit the number of domains.

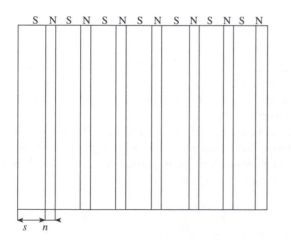

Fig. 14.10
Alternating superconducting and normal layers.

The introduction of surface energy is certainly a way out of the dilemma, but it is of limited value unless we can give some quantitative relationships for the maintenance of a wall. The answer was given at about the same time by Pippard and (independently) by Landau and Ginzburg. We shall discuss the latter theory because it is a little easier to follow.

14.6   The Landau–Ginzburg theory

With remarkable intuition Landau and Ginzburg suggested (in 1950) a formulation that was later (1958) confirmed by the microscopic theory. We shall give here the essence of their arguments, though in a somewhat modified form to fit into the previous discussion.

1. In the absence of a magnetic field, below the critical temperature, the Gibbs free energy* is $G_s(0)$.

*From now on, for simplicity, all our quantities will be given per unit volume.

2. If a magnetic field H_a is applied and is expelled from the interior of the superconductor, the energy is increased by $\frac{1}{2}\mu_0 H_a^2$ per unit volume. This may be rewritten with the aid of flux density as $\frac{1}{2}(1/\mu_0)B_a^2$. If we now abandon the idea of a perfect diamagnet, the magnetic field can penetrate the superconductor, and the flux density at a certain point is B instead of zero. Hence the flux density expelled is not B_a but only $B_a - B$, and the corresponding increase in the Gibbs free energy is

$$\frac{1}{2}\frac{1}{\mu_0}(B_a - B)^2. \tag{14.36}$$

3. All superconducting electrons are apparently doing the same thing. We, therefore, describe them by the same wave function, ψ, where

$$|\psi(x, y, z)|^2 = N_s, \tag{14.37}$$

the density of superconducting electrons. In the absence of an applied magnetic field the density of superconducting electrons is everywhere the same.

4. In the presence of a magnetic field the density of superconducting electrons may vary in space, that is $\nabla\psi \neq 0$. But, you may remember, $-i\hbar\nabla\psi$ gives the momentum of the particle. Hence, the kinetic energy of our superconducting electrons,

$$KE = \frac{1}{2m}|-i\hbar\nabla\psi|^2, \tag{14.38}$$

will add to the total energy. It follows then that the appearance of alternating layers of normal and superconducting domains is energetically unfavourable because it leads to a rapid variation of ψ, giving a large kinetic energy contribution to the total energy.

[†] For a discussion, see *The Feynman lectures on Physics*, vol. 3, pp. 21–5.

Equation (14.38) is not quite correct. It follows from classical electrodynamics[†] that in the presence of a magnetic field the momentum is given by $\mathbf{p} - e\mathbf{A}$, where $\mathbf{A}$ is the magnetic vector potential. Hence, the correct formula for the kinetic energy is

$2e$ is the charge on a superconducting electron.

$$KE = \frac{1}{2m}|-i\hbar\nabla\psi - 2e\mathbf{A}\psi|^2. \tag{14.39}$$

We may now write the Gibbs free energy in the form

$$G_s(B) = G_s(0) + \frac{1}{2\mu_0}(B_a - B)^2 + \frac{1}{2m}|-i\hbar\nabla\psi - 2e\mathbf{A}\psi|^2. \quad (14.40)$$

5. The value of the Gibbs function at zero magnetic field should depend on the density of superconducting electrons, among other things. The simplest choice is a polynomial of the form,

$$G_s(0) = G_n(0) + a_1|\psi|^2 + a_2|\psi|^4, \quad (14.41)$$

where the coefficients may be determined from empirical considerations. At a given temperature the density of superconducting electrons will be such as to minimize $G_s(0)$, that is

$$\frac{\partial G_s(0)}{\partial|\psi|^2} = 0, \quad (14.42)$$

leading to

$$|\psi|^2 \equiv |\psi_0|^2 = -\frac{a_1}{2a_2}. \quad (14.43)$$

Substituting this value of $|\psi|^2$ back into eqn (14.41) we get

$$G_s(0) = G_n - \frac{a_1^2}{4a_2}. \quad (14.44)$$

Let us go back now to eqn (14.22) (rewritten for unit volume)

$$G_s(0) = G_n - \tfrac{1}{2}\mu_0 H_c^2. \quad (14.45)$$

Comparing the last two equations, we get

$$H_c = -a_1/(2a_2\mu_0)^{1/2}. \quad (14.46)$$

It is assumed that $a_2 > 0$ and $a_1 < 0$.

According to experiment, H_c varies linearly with temperature in the neighbourhood of the critical temperature. Thus, for this temperature range we may make eqn (14.46) agree with the experimental results by choosing

$$a_1 = c_1(T - T_c) \quad \text{and} \quad a_2 = c_2. \quad (14.47)$$

c_1 and c_2 are independent of temperature.

If you now believe that eqn (14.41) was a reasonable choice for the $G_s(0)$, we may substitute it into eqn (14.40) to get our final form for Gibbs free energy

$$G_s(B) = G_n(0) + a_1|\psi|^2 + a_2|\psi|^4$$
$$+ \frac{1}{2\mu_0}(\nabla \times \mathbf{A} - B_a)^2 + \frac{1}{2m}|i\hbar\nabla\psi - 2e\mathbf{A}\psi|^2, \quad (14.48)$$

where the relationship

$$\mathbf{B} = \nabla \times \mathbf{A} \quad (14.49)$$

has been used.

The arguments used above are rather difficult. They come from various sources (thermodynamics, quantum mechanics, electrodynamics, and actual

measured results on superconductors) and must be carefully combined to give
an expression for the Gibbs free energy.

6. The Gibbs free energy for the entire superconductor may be obtained by
integrating eqn (14.48) over the volume

$$\int_V G_s(B) \, dV.$$

The integrand contains two undetermined functions $\psi(x, y, z)$ and $A(x, y, z)$
which, according to Ginzburg and Landau, may be obtained from the condition
that the integral should be a minimum.

The problem belongs to the realm of variational calculus. Be careful; it is
not the minimum of a *function* we wish to find. We want to know how A and ψ
vary as functions of the coordinates x, y, and z in order to minimize the above
definite integral.

We shall not solve the general problem here but shall restrict the solution
to the case of a half-infinite superconductor that fills the space to the right of
the $x = 0$ plane. We shall also assume that the applied magnetic field is in
the z-direction and is independent of the y- and z-coordinates, reducing the
problem to a one-dimensional one, where x is the only independent variable.

In view of the above assumptions,

A_y is the only component of **A**, and
will be simply denoted by A.

$$B_z = \frac{dA_y}{dx}. \tag{14.50}$$

Since $\nabla\psi$ is a vector in the x-direction, it is perpendicular to **A**, so that

$$\mathbf{A} \cdot \nabla\psi = 0. \tag{14.51}$$

* Taking ψ real reduces the mathematical
labour and, fortunately, does not restrict
the generality of the solution.

Under these simplifications the integrand takes the form*

$$G_s(B) = G_n + a_1\psi^2 + a_2\psi^4 + \frac{1}{\mu_0}\left(B_a - \frac{dA}{dx}\right)^2$$

$$+ \frac{1}{2m}\left\{\hbar^2\left(\frac{\partial\psi}{\partial x}\right)^2 + 4e^2A^2\psi^2\right\}. \tag{14.52}$$

The solution of the variational problem is now considerably easier. As shown
in Appendix IV, $\psi(x)$ and $A(x)$ will minimize the integral if they satisfy the
following differential equations:

$$\frac{\partial G_s(B)}{\partial\psi} - \frac{d}{dx}\frac{\partial G_s(B)}{\partial(\partial\psi/\partial x)} = 0 \tag{14.53}$$

and

$$\frac{\partial G_s(B)}{\partial A} - \frac{d}{dx}\frac{\partial G_s(B)}{\partial(\partial A/\partial x)} = 0. \tag{14.54}$$

Substituting eqn (14.52) into eqn (14.53) and performing the differentiations,
we get

$$2a_1\psi + 4a_2\psi^3 + \frac{1}{2m}8e^2A^2\psi - \frac{d}{dx}\frac{1}{2m}\hbar^2 2\frac{\partial\psi}{\partial x} = 0, \tag{14.55}$$

which after rearrangement yields

$$\frac{d^2\psi}{dx^2} = \frac{m}{\hbar^2} 2a_1 \left(1 + \frac{2e^2}{a_1 m} A^2\right) \psi + \frac{4m}{\hbar^2} a_2 \psi^3. \tag{14.56}$$

Similarly, substituting eqn (14.52) into eqn (14.54) we get

$$\frac{d^2 A}{dx^2} = \frac{4e^2 \psi^2 \mu_0}{m} A, \tag{14.57}$$

which must be solved subject to the boundary conditions

$$B = B_a = \mu_0 H_a, \quad d\psi/dx = 0 \quad \text{at } x = 0 \tag{14.58}$$

$$B = 0, \quad \psi^2 = \psi_0^2, \quad d\psi/dx = 0 \quad \text{at } x = \infty. \tag{14.59}$$

The boundary conditions for the flux density simply mean that at the boundary with the vacuum the flux density is the same as the applied flux density, and it declines to zero far away inside the superconductor. The condition for $d\psi/dx$ comes from the more stringent general requirement that the normal component of the momentum should vanish at the boundary. But since in the one-dimensional case $\mathbf{A}$ is parallel to the surface, $\mathbf{A} \cdot \mathbf{i}_x$ is identically zero, and the boundary condition reduces to the simpler $d\psi/dx = 0$. Since A is determined except for a constant factor, we can prescribe its value at any point. We shall choose $A(\infty) = 0$.

Introducing the new parameters

$$\lambda^2 = \frac{m}{4e^2 \psi_0^2 \mu_0} \tag{14.60}$$

and

$$k = \lambda^2 \frac{2^{3/2} e H_c \mu_0}{\hbar}, \tag{14.61}$$

and making use of eqns (14.43) and (14.46), we can rewrite eqns (14.56) and (14.57) in the forms

$$\frac{d^2\psi}{dx^2} = \frac{k^2}{\lambda^2} \left\{-\left(1 - \frac{A^2}{2H_c^2 \lambda^2 \mu_0^2}\right) \psi + \frac{\psi^3}{\psi_0^3}\right\} \tag{14.62}$$

and

$$\frac{d^2 A}{dx^2} - \frac{1}{\lambda^2} \frac{\psi^2}{\psi_0^2} A = 0. \tag{14.63}$$

In the absence of a magnetic field, $A \equiv 0$; eqn (14.62) gives $\psi = \psi_0$, as it should. In the presence of a magnetic field the simplest approximation we can make is to take $\kappa = 0$, which still gives $\psi = \psi_0$. From eqn (14.63)

$$A = A(0)e^{-x/\lambda}, \tag{14.64}$$

leading to

$$B = -\frac{1}{\lambda} A(0)e^{-x/\lambda} = B_a e^{-x/\lambda}. \tag{14.65}$$

Thus, we can see that the magnetic flux density inside the superconductor decays exponentially, and λ appears as the penetration depth.

Better approximations can be obtained by substituting

$$\psi = \psi_0 + \varphi \qquad (14.66)$$

into eqns (14.62) and (14.63) and solving them under the assumption that φ is small in comparison with ψ_0. Then ψ also varies with distance, and B has a somewhat different decay; but these are only minor modifications and need not concern us.

The main merit of the Landau–Ginzburg theory is that by including the kinetic energy of the superconducting electrons in the expression for the Gibbs free energy, it can show that the condition of minimum Gibbs free energy leads to the expulsion of the magnetic field. The expulsion is not complete, as we assumed before in the simple thermodynamic treatment; the magnetic field can penetrate to a distance, λ, which is typically of the order of 10 nm.

Thus, after all, there can be no such thing as the break-up of the superconductor into alternating normal and superconducting regions—or can there? We have solved eqn (14.62) only for the case when κ is very small. There are perhaps some other regions of interest. It turns out that another solution exists for the case when

$$\psi \ll \psi_0 \quad \text{and} \quad B = B_a. \qquad (14.67)$$

So we claim now that there is a solution when the magnetic field can penetrate the whole superconducting material, and this happens when the density of superconducting electrons is small. Then (choosing for this case the vector potential zero at $x = 0$)

$$A(x) = B_a x, \qquad (14.68)$$

and neglecting the last term in eqn (14.62) we get

$$\frac{d^2\psi}{dx^2} = -\frac{\kappa^2}{\lambda^2}\left(1 - \frac{B_a^2 x^2}{2H_c^2\lambda^2\mu_0^2}\right)\psi. \qquad (14.69)$$

Now this happens to be a differential equation that has been thoroughly investigated by mathematicians. They maintain that a solution exists only when

$$B_a = \mu_0 H_c \kappa \sqrt{2}/(2n+1). \qquad (14.70)$$

The maximum value of B_a occurs at $n = 0$, giving

$$B_a = \mu_0 H_c \kappa \sqrt{2}. \qquad (14.71)$$

When $\kappa > 1/\sqrt{2}$ the magnetic field inside the superconductor may exceed the critical field. You may say this is impossible. Have we not defined the critical field as the field that destroys superconductivity? We have, but that was done on the basis of diamagnetic properties. We defined the critical field only for the case when the magnetic field is expelled. Now we say that in certain materials for which $\kappa > 1/\sqrt{2}$, superconductivity may exist up to a magnetic field, H_{c2}. The new critical magnetic field is related to the old one by the relationship,

$$H_{c2} = \kappa\sqrt{2}H_c. \qquad (14.72)$$

Up to H_{c1} the superconductor is diamagnetic, as shown in Fig. 14.11, where $-M$ is plotted against the applied magnetic field. Above H_{c1} the magnetic field begins to penetrate (beyond the 'diamagnetic' penetration depth)

n is an integer; otherwise ψ diverges as $x \to \infty$.

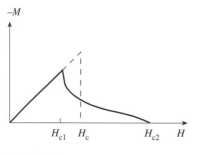

Fig. 14.11

Magnetization curves for type I and type II superconductors. The area under both magnetization curves is the same.

and there is complete penetration at H_{c2}, where the material becomes normal. Materials displaying such a magnetization curve are referred to as type II superconductors, while those expelling the magnetic field until they become normal (dotted lines in Fig. 14.11) are called type I superconductors.

A two-dimensional analysis of a type II superconductor shows that the intensity of the magnetic field varies in a periodic manner with well-defined maxima as shown in Fig. 14.12(a). Since the current and the magnetic field are uniquely related by Maxwell's equations, the current is also determined. It is quite clear physically that the role of the current is either 'not to let in' or 'not to let out' the magnetic field.

(a)

(b)

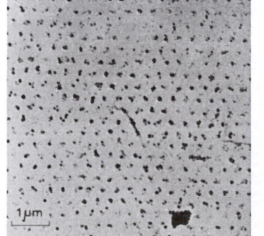

Fig. 14.12
(a) The lines of current flow for a two-dimensional type II superconductor. The magnetic field is maximum at the centres of the smaller vortices and minimum at the centres of the larger ones. (After Abrikosov 1957.) (b) Triangular vortex structure measured on the surface of a lead–indium rod at 1.1 K (after Essmann and Träuble 1967).

The density of superconducting electrons is zero at the maxima of the magnetic field. Thus, in a somewhat simplified manner, we may say that there is a normal region surrounded by a supercurrent vortex. There are lots of vortices; their distance from each other is about 1 μm. The vortex structure determined experimentally by Essmann and Träuble is shown in Fig. 14.12(b). It has a triangular structure, somewhat different from that calculated by Abrikosov.

The preceding treatment of the theories of superconductivity is not a well-balanced one, neither historically nor as far as their importance is concerned. A comprehensive review would be far too lengthy, so we have just tried to follow one line of thought.

14.7   The energy gap

As you know from electromagnetic theory, such optical properties as reflectivity and refractive index are related to the bulk parameters, resistivity, and dielectric constant. Thus, zero resistivity implies quite radical optical properties, which are not found experimentally. Nothing untoward happens below the critical temperature. Hence, we are forced to the conclusion that, somewhere between zero and light frequencies, the conductivity is restored to its normal value. What is the mechanism? Having learned band theory, we could describe a mechanism that *might* be responsible; this is the existence of an energy gap. When the frequency is large enough, there is an absorption process, owing to electrons being excited across the gap. Pairing of electrons is no longer advantageous; all traces of superconductivity disappear. This explanation happens to be correct and is in agreement with the predictions of the BCS theory.

The width of the gap can be deduced from measurements on specific heat, electromagnetic absorption, or tunnelling. Typical values are somewhat below one milli-electronvolt. The gap does not appear abruptly; it is zero at the critical temperature and rises to the value of $3.5\,kT_c$ at absolute zero temperature. The temperature variation is very well predicted by the BCS theory, as shown in Fig. 14.13 for these superconductors.

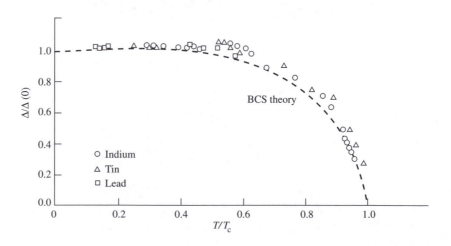

Fig. 14.13

The temperature variation of the energy gap (related to the energy gap at $T = 0$) as a function of T/T_c.

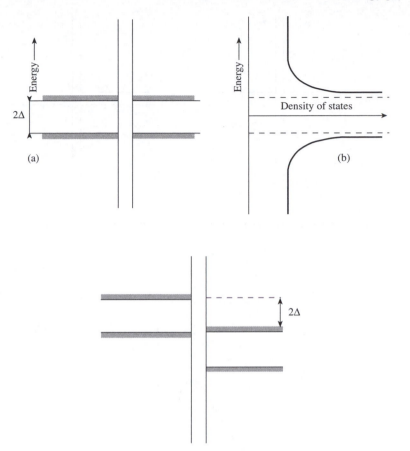

(a)   (b)

Fig. 14.14
(a) Energy diagram for two identical superconductors separated by a thin insulator. (b) The density of states as a function of energy.

Fig. 14.15
The energy diagram of Fig. 14.14(a) when a voltage, $2\Delta/e$ is applied.

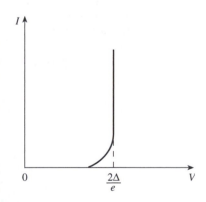

Fig. 14.16
The current as a function of voltage for a junction between two identical superconductors separated by a thin insulator.

When we put a thin insulator between two identical superconductors, the energy diagram [Fig. 14.14(a)] looks very similar to those we encountered when studying semiconductors. The essential difference is that in the present case the density of states is high just above and below the gap, as shown in Fig. 14.14(b). An applied voltage produces practically no current until the voltage difference is as large as the gap itself; the situation is shown in Fig. 14.15. If we increase the voltage further, electrons from the left-hand side may tunnel into empty states on the right-hand side, and the current rises abruptly as shown in Fig. 14.16.

An even more interesting case arises when the two superconductors have different gaps. Since the Fermi level is in the middle of the gap (as for intrinsic semiconductors) the energy diagram at thermal equilibrium is as shown in Fig. 14.17(a). There are some electrons above the gap (and holes below the gap) in superconductor A but hardly any (because of the larger gap) in superconductor B. When a voltage is applied, a current will flow and will increase with voltage (Fig. 14.18) because more and more of the thermally excited electrons in superconductor A can tunnel across the insulator into the available states of superconductor B. When the applied voltage reaches $(\Delta_2 - \Delta_1)/e$ [Fig. 14.17(b)], it has become energetically possible for all thermally excited electrons to tunnel across. If the voltage is increased further, the current

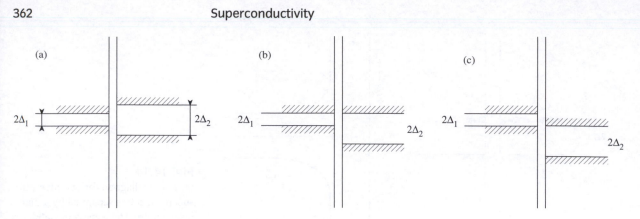

Fig. 14.17

Energy diagrams for two different superconductors separated by a thin insulator. (a) $U = 0$, (b) $U = (\Delta_2 - \Delta_1)/e$, (c) $U = (\Delta_1 + \Delta_2)/e$.

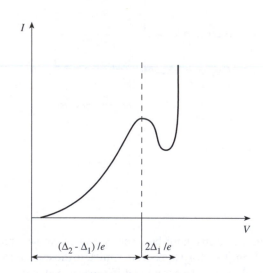

Fig. 14.18

The current as a function of voltage for a junction between two different superconductors separated by a thin insulator. There is a negative resistance region of $(\Delta_2 - \Delta_1)/e < U < (\Delta_1 + \Delta_2)/e$.

decreases because the number of electrons capable of tunnelling is unchanged, but they now face a lower density of states. When the voltage becomes greater than $(\Delta_2 + \Delta_1)e$ the current increases rapidly because electrons below the gap can begin to flow.

Thus, a tunnel junction comprised of two superconductors of different energy gaps may exhibit negative resistance, similarly to the semiconductor tunnel diode. Unfortunately, the superconducting tunnel junction is not as useful because it works only at low temperatures.

The tunnelling we have just described follows the same principles we encountered when discussing semiconductors. There is, however, a tunnelling phenomenon characteristic of superconductors, and of superconductors alone; it is the so-called superconducting or Josephson tunnelling (discovered theoretically by Josephson, a Cambridge graduate student at the time) which takes place when the insulator is very thin (less than 1.5–2 nm).

It displays a number of interesting phenomena, of which we shall briefly describe four.

1. For low enough currents there can be a current across the insulator without any accompanying voltage; the insulator turns into a superconductor. The reason is that Cooper pairs (*not* single electrons) tunnel across.

2. For larger currents there can be finite voltages across the insulator. The Cooper pairs descending from the higher potential to the lower one may radiate their energy according to the relationship,

$$\hbar\omega = (2e)U_{AB}. \tag{14.73}$$

U_{AB} is the d.c. voltage between the two superconductors, and ω is the angular frequency of the electromagnetic radiation.

Thus, we have a very simple form of a d.c. tuneable oscillator that could work up to infrared frequencies. It is not, however, very practical because, besides the low temperatures needed, it can produce only a minute amount of power.

3. A direct transition may be caused between the Josephson characteristics and the 'normal' tunnelling characteristics by the application of a small magnetic field (Fig. 14.19).

4. When two Josephson junctions are connected in parallel [Fig. (14.20)] the maximum supercurrent that can flow across them is a periodic function of the magnetic flux,

$$I_{max} = 2I_J \left| \cos \frac{\pi\,\Phi}{\Phi_0} \right|. \tag{14.74}$$

I_J is a constant depending on the junction parameters, Φ is the enclosed magnetic flux, and Φ_0 is the so-called flux quantum equal to $h/2e = 2 \times 10^{-15}$ Wb.

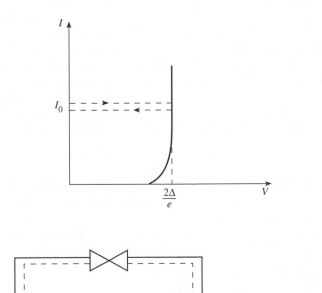

Fig. 14.19
The current as a function of voltage for a junction which may display both 'normal' and Josephson tunnelling. I_0 is the current flowing without any accompanying voltage. The application of a small magnetic field causes a transition between the Josephson and 'normal' tunnelling characteristics. Once this extra magnetic field is removed, the voltage returns to zero.

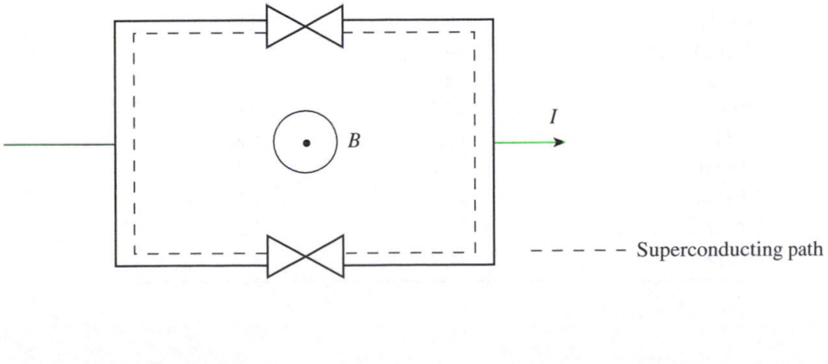

- - - - - Superconducting path

Fig. 14.20
Two Josephson junctions in parallel connected by a superconducting path.

14.8 Some applications

14.8.1 High-field magnets

For the moment the most important practical application of superconductivity is in producing a high magnetic field. There is no doubt that for this purpose a superconducting solenoid is superior to conventional magnets. A magnetic flux density of 20 T can be produced by a solenoid not larger than about 12 cm × 20 cm. A conventional magnet capable of producing one-third of that flux density would look like a monster in comparison and would need a few megawatts of electric power and at least a few hundred gallons of cooling water per minute.

What sort of materials do we need for obtaining high magnetic fields? Obviously, type II superconductors—they remain superconducting up to quite high magnetic fields. However, high magnetic fields are allowed only at certain points in the superconductor that are surrounded by current vortices. When a d.c. current flows (so as to produce the high magnetic field in the solenoid) the vortices experience a $\mathbf{J} \times \mathbf{B}$ force that removes the vortices from the material. To exclude the high magnetic field costs energy, and the superconductor consequently becomes normal, which is highly undesirable. The problem is to keep the high magnetic fields inside. This is really a problem similar to the one we encountered in producing 'hard' magnetic materials, where the aim was to prevent the motion of domain walls. The remedy is similar; we must have lots of structural defects; that is we must make our superconductor as 'dirty' and as 'non-ideal' as possible. The resulting materials are, by analogy, called hard superconductors. Some of their properties are shown in Table 14.2.

There is, however, a further difficulty with vortices. Even if they do not move out of the material, *any* motion represents ohmic loss, causing heating, and making the material become normal at certain places. To avoid this, a good thermal conductor and poor electrical conductor, copper—yes, copper—is used for insulation, so that the heat generated can be quickly led away.

It must be noted that it is not particularly easy to produce any of these compounds, and different techniques may easily lead to somewhat different values of T_c and H_c. The two superconductors used in practical devices are the

Table 14.2 *The critical temperature and critical magnetic field (at $T = 4.2$ K) of the more important hard superconductors*

Material	T_c (K)	$H_c \times 10^{-7}$ (A m^{-1})
Nb–Ti	9	0.9
$Pb_{0.9}Mo_{5.1}S_6$	14.4	4.8
V_3Ga	14.8	1.9
NbN	15.7	0.8
V_3Si	16.9	1.8
Nb_3Sn	18.0	2.1
Nb_3Ga	20.2	2.6
$Nb_3(Al_{0.7}Ge_{0.3})$	20.7	3.3
Nb_3Ge	22.5	2.9

ductile Nb–Ti alloy and the brittle intermetallic compound Nb_3Sn, the latter one being used at the highest magnetic fields.

14.8.2   Switches and memory elements

The use of superconductors as switches follows from their property of becoming normal in the presence of a magnetic field. We can make a superconducting wire resistive by using the magnetic field produced by a current flowing in another superconducting wire. Memory elements based on such switches have indeed been built, but they were never a commercial success.

Superconducting memory elements based on the properties of Josephson junctions have a much better chance. As we have mentioned before, and may be seen in Fig. 14.19, the junction has two stable states, one with zero voltage and the other one with a finite voltage. It may be switched from one state into the other one by increasing or decreasing the magnetic field threading the junction. The advantage of this Josephson junction memory is that there is no normal to superconducting phase transition necessary, only the *type* of tunnelling is changed, which is a much faster process. Switching times as short as 10 ps have been measured.

Will it ever be worthwhile to go to the trouble and expense of cooling memory stores to liquid helium temperatures? So far computer manufacturers have been rather reluctant (understandably, it is a high risk business) to introduce superconducting memories. It is difficult to predict, but the latest members of the family, Rapid Single Flux Quantum (RSFQ) devices, might have a chance to be introduced in practice some time in the future when high speed will be the principal requirement. The basic architecture is a ring containing a Josephson junction. A large number of such rings coupled magnetically make up the device that can serve both as a memory element (it stores a single flux quantum) and a logic device. The latter property is due to the fact that voltage pulses can travel from element to element extremely rapidly along such line. The highest speed observed so far at which these devices can operate is 770 GHz. Apart from speed a further advantage is the quantized nature of the storage mechanism providing protection both against noise and cross talk.

14.8.3   Magnetometers

A further important application of Josephson junctions is in a magnetometer called SQUID (Superconducting Quantum Interference Device). Its operation is based on the previously mentioned property that the maximum supercurrent through the two junctions in parallel is dependent on the magnetic flux enclosed by the loop. It follows from eqn (14.74) that there is a complete period in I_{max}, while Φ varies from 0 to Φ_0. Thus, if we can tell to an accuracy of 1% the magnitude of the supercurrent, and we take a loop area of $1\,cm^2$, the smallest magnetic field that can be measured is 10^{-12} T. Commercially available devices (working on roughly the same principle) can offer comparable sensitivities.

Although the Josephson junction does many things superlatively well, like other topics in superconductivity, its applications (so far) are few. However, it is worth mentioning two sensitive magnetometer applications which would be quite impossible with classical devices.

A major preoccupation of the military is to keep watch on nuclear submarines. The difficulty is that water is such a good absorber of microwaves, light, and sound, which are traditionally used to locate targets. However, underwater caches of superconducting magnetometers can detect small perturbations of the Earth's magnetic field as a submarine arrives in the locality. They have to be connected to a surface buoy containing a transmitter which informs boffins in bunkers what is going past.

A more definite and much safer application is one which Oxford's Laboratory of Archaeology works on and publishes freely. The silicaceous and clay-like materials in pottery are mildly paramagnetic. When they are fired in kilns, the high temperature destroys the magnetism, and as they cool the permanent dipoles re-set themselves in the local magnetic field of the Earth. When an archaeologist uncovers an old kiln, he can measure this magnetism in the bricks and thus find the direction of the Earth's field when the kiln was last fired. The variation of the Earth's field and angle of dip has been determined for several thousand years at some places. Thus, it is possible to date kilns by accurate measurements. As large ceramic articles have to be kilned standing on their bases, accurate measurements of the dip angle can also date cups and statues, if their place of origin is known. With very sensitive magnetometers, this measurement can be done on a small, unobtrusive piece of ceramic removed from the base of the statue. It is a method considerably used by major museums.

14.8.4   Metrology

We mentioned earlier that one can determine one of our fundamental constants (velocity of light) with the aid of lasers. It turns out that Josephson junctions may be used for determining another fundamental constant. The relevant formula is eqn (14.73). By measuring the voltage across the junction and the frequency of radiation, h may be determined. As a result, the accepted value of Planck's constant changed recently from 6.62559 to 6.626196×10^{-34} J s.

14.8.5   Suspension systems and motors

Frictionless suspension systems may be realized by the interaction between a magnetic flux produced externally and the currents flowing in a superconductor. If the superconductor is pressed downwards, it tries to exclude the magnetic field, hence the magnetic flux it rests on is compressed, and the repelling force is amplified. Noting further that it is possible to impart high speed rotation to a suspended superconducting body, and that all the conductors in the motor are free of resistance, it is quite obvious that the ideal of a hundred-percent-efficient motor can be closely approximated.

14.8.6   Radiation detectors

The operation of these devices is based on the heat provided by the incident radiation. The superconductor is kept just above its critical temperature, where the resistance is a rapidly varying function of temperature. The change in resistance is then calibrated as a function of the incident radiation.

14.8.7 Heat valves

The thermal conductivity of some superconductors may increase by as much as two orders of magnitude, when made normal by a magnetic field.

This phenomenon may be used in heat valves in laboratory refrigerations systems designed to obtain temperatures below 0.3 K.

14.9 High-T_c superconductors

There were always hopes that superconductors will, one day, break out of their low temperature habitat and will have a significant impact upon the design and operation of a wide range of devices. It was felt intuitively that Nature can not possibly be so mean as to tuck away such a tremendously important phenomenon into a dark corner of physics. Well, the break-out towards higher temperatures did take place in the month of January, 1986. Müller and Bednorz of the IBM Zurich Laboratories found a ceramic, barium–lanthanum–copper oxide, with a critical temperature of 35 K. 'How did you come to the idea', I asked Professor Müller, 'that oxide superconductors will have high critical temperatures?' 'Simple,' he said and produced the diagram shown in Fig. 14.21, 'the line of maximum critical temperature against time for traditional superconductors (dotted line) intersected the extrapolated line for oxide superconductors (continuous line) in 1986. We were bound to succeed.'

Progress was not particularly fast, mainly because 35 K sounded too good to be true. Many experts regarded the claim with some scepticism. It took just about a year until the next step. In February 1987, nearly simultaneously, Chu in Houston and Zha Zhong-xian in Beijing produced a new superconducting ceramic, yttrium–barium–copper oxide (YBCO) with critical temperatures between 90 and 100 K, well above 77 K, the boiling point of nitrogen. Those reports really did open the floodgates. Scientists streamed into the field, and scientific reports streamed out. So where are we now, concerning maximum

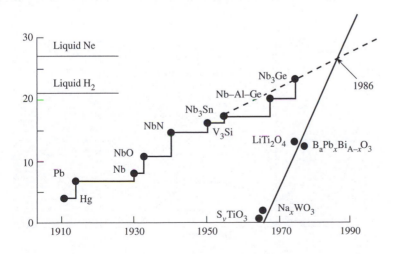

Fig. 14.21

The maximum critical temperature against time for traditional and oxide superconductors.

critical temperature? The record, reigning for a number of years was 125 K, achieved by an oxide with the chemical formula $Tl_2Ba_2Ca_2Cu_3O_{10}$, known as TBCCO. The latest figure is about 10 K higher. The compound is mercury barium calcium copper oxide (HBCCO).

What is the basic structure of these superconductors? The first one discovered, $La_{2-x}Ba_xCuO_4$, contains single CuO_2 planes separated by layers which provide a charge reservoir, and the same is true for the two materials mentioned so far, YBCO and TBCCO.

How do they work? Copper-oxide is an insulator, so that is not much good for the purpose. It needs dopants for creating carriers which will then flow along the CuO_2 planes. The carriers may, for example, be provided by Ba for holes and by Ce for electrons. Note also that many of the properties of these compounds are highly anisotropic, which may be measured on single crystal specimen. The electrical resistivity perpendicular to the CuO_2 layers may be 10^5 times as large as along the in-plane layers. The temperature-dependence of resistivity is also different: in the perpendicular plane resistivity increases with temperature as in a metal, but in-plane resistivity decreases with temperature as in a semiconductor. There are also different phases of these materials which depend on the doping level.

A generic phase diagram of cuprate superconductors is shown in Fig. 14.22. As many as five different phases may be seen, starting with an antiferromagnetic insulator. In a certain range of doping (roughly between 0.1 and 0.2 holes per copper oxide) and below a certain temperature they are superconductors, above that temperature they are metals having rather odd properties. In fact, theoreticians believe that it would be easy to work out the physics of the transition to superconductivity once the properties of the metallic phase are understood. And that is not the case as yet.

There has been no proper theory developed either for cuprates or for the other main type of oxide superconductors based on $BaBiO_3$ compounds.* However, a consensus exists concerning some aspects of the theory. There is no doubt that pairing is involved, and the effective charge is $2e$. It is also known that the pairs are made up of electrons with opposite momenta, just as was shown in Fig. 14.3(a). Interaction of the electrons with the lattice might play a role, but it is certainly not the full story. Another possible mechanism is pairing

* We should perhaps add here a new class of superconductors, whose discovery has made it even more difficult to devise a theory. They are based on the fullerene C_{60} mentioned in Chapter 5. Some of their representatives are K_3C_{60}, Rb_3C_{60}, and $Rb_{2.7}Tl_{2.3}C_{60}$, with critical temperatures of 19 K, 33 K, and 42 K, respectively.

Fig. 14.22

Generic phase diagram of the cuprate superconductors. The doping level is measured relative to the insulating parent compound.

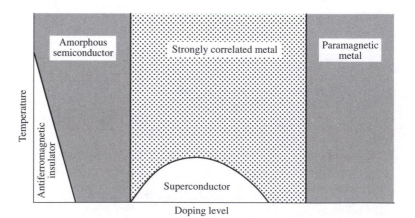

by spin waves, as already referred to earlier in this chapter. An important experiment is to measure the magnetic flux in superconducting rings containing Josephson junctions. With conventional superconductors, the enclosed flux is always an integer multiple of the flux quantum. With cuprate superconductors, the enclosed flux turns out to be an odd multiple of the half flux quantum. This is no proof for spin-wave pairing, but if pairing is by spin waves, then this is one of the conditions that must be satisfied.

In which other directions could one look for a theory? One might possibly rely on the analogy between the quantum Hall effect and superconductivity in cuprates. Two-dimensional effects and sudden loss of resistance are characteristic to both. An attempt on such lines has been made at explaining the quantum Hall effect by a theory which treats electrons as some kind of composite bosons.

Most theoreticians believe that an energy gap always exists, and for cuprate superconductors the relationship between gap energy and critical temperature is $2\Delta(0) \approx 6\,kT_c$ in contrast to $3.5\,kT_c$, which we have come across for low temperature superconductors. No one entertains great hopes that a theory which would be able to predict the critical temperatures of various compounds will be forthcoming in the near future. The theoretical interest will be sustained, however, very likely for decades. Pairing mechanisms have become popular. Neutron stars are supposed to have pairing condensations, and it is also believed that quark condensations began just one second after the Big Bang, although experimental evidence is lacking for the moment.

How are these superconductors produced? Being ceramics, they were first produced by mixing, grinding, and baking of powdered reagents. Single crystal samples, as described above, greatly helped in advancing the understanding of their properties, but they are not suitable for mass applications. For devices one needs them as thin films. For high-field magnets they have to be in the form of wires. Thin films are mostly made by sputtering and pulsed laser deposition (a pulsed excimer laser evaporates the material which is already available in a stoichiometric mixture of its constituents) and they are polycrystalline. But if superconductive properties are anisotropic, how will they survive in a polycrystalline material? The answer is that any departure from the single crystal form is deleterious but not necessarily disastrous. Josephson tunnelling comes to our aid in the sense that superconducting electrons may tunnel across disoriented grain boundaries, provided the angle of disorientation is small.

What about applications? What has become of the dazzling prospects of levitated trains, electromagnetically propelled ships, and magnetic energy storage devices? Not in the near future, is the answer. Some applications, however, are bound to come quite soon, since there are obvious economic benefits to working at 77 K (using liquid nitrogen) in contrast to 4.2 K (using liquid helium). Liquid nitrogen costs only as much as a cheap beer, whereas liquid helium is in the class of a reasonably good brandy, so maintaining the samples at the right temperature will be much cheaper. The application that is closest is probably in electronic devices, and the property used is the lack of electrical resistance. So total heat dissipated is reduced, which is good and particularly good in preventing thermally activated damage like corrosion and electromigration of atoms. In heavy current engineering, the most likely candidates for applications are underground cables. The present cables are made of copper and are cooled by oil. The future ones replacing them will most likely be made of high-T_c materials cooled by liquid nitrogen. Highly rated transformers and coils for rotors in motors and generators are also close contenders.

At microwave frequencies, superconductors can no longer offer zero resistivities. However, their lower resistance is still a major advantage in microwave resonators. There were already some applications using conventional superconductors, but chances have very much improved with the advent of high T_c superconductors. We would just like to mention one successful device, the disk resonator shown in Fig. 14.23(a). The resonance occurs in the same manner as in the Fabry–Perot resonator discussed in Section 12.5. The main difference is that, in the present case, it is possible to excite a mode which leads to very low losses, since the current disappears at the edges. The calculated current distribution is shown in Fig. 14.23(b). At a frequency of 4.7 GHz and a temperature of 60 K, the measured Q (quality) factor was close to 20 000 in contrast to 600, the Q factor achievable by copper. The superconductor was one of the TBCCO family deposited in thin film form by DC sputtering.

Will high T_c superconductors make a big difference in the performance of high-field magnets? They probably will in due course, but there are lots of problems at present. It is difficult to reach high critical current densities because of the granular nature of these materials already mentioned. If we have to rely on Josephson tunnelling across grain boundaries, that means that the current cannot exceed the critical current which makes the tunnelling normal (cf. Section 14.7).

The greatest success so far has been achieved with BSCCO, which may have the composition of $Bi_2Sr_2CaCu_2O_8$ (known as Bi-2212, $T_c \approx 85$ K) or $Bi_2Sr_2Ca_2Cu_3O_{10}$ (known as Bi-2223, $T_c \approx 110$ K). It does not have particularly good properties at 77 K, as may be seen in Fig. 14.24. The critical current declines very rapidly with magnetic field (dotted lines). However, at 4.2 K (i.e. well below its critical temperature) BSCCO has properties superior to traditional superconductors. It still has a critical current density of 10^5 A cm^{-2}

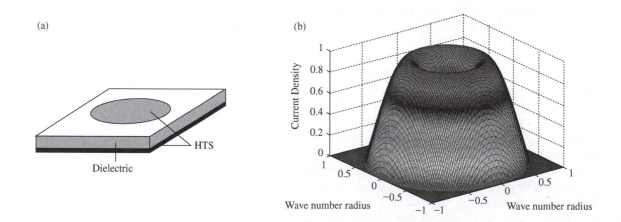

Fig. 14.23
(a) A microwave disc resonator (b) current distribution on the surface of the disc.

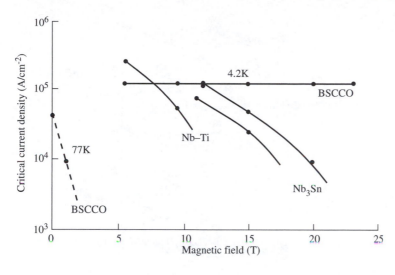

Fig. 14.24
Critical current densities as a function of magnetic field at 77 K (- - -) and at 4.2 K(- - -) for BSCCO, Nb–Ti and Nb_3Sn.

over 20 T. These results were obtained with tapes with a high degree of crystallographic alignment. Wires with this performance are not available as yet, but a practical device capable of producing 20 to 25 tesla is clearly feasible.

A further useful property of BSCCO is that their critical currents are fairly independent of temperature in the 4–20 K range; hence instead of being dipped into liquid helium, they could be kept in the right temperature range by refrigerators.

So what is the conclusion? The initial euphoria has evaporated, but it still seems very likely that many useful devices will appear in the next decade. I have to repeat what I said in Section 9.1. Revolutions are few and far between.

14.10 New superconductors

The phenomenon of superconductivity never ceases to surprise us. Since the last edition a number of new superconductors have been discovered which are most reluctant to fit into the general framework: The situation reminds me of Pope's well known epitaph intended for Newton

Nature, and Nature's laws lay hid in the night
God said, Let Newton be! and all was light

and of Squire's addition to it a couple of centuries later

It did not last: the Devil howling "Ho!
Let Einstein be! restored the status quo.

Well, this is what happened to superconductivity. After the formulation of the BCS theory in 1957 all was light for a long time. But then, in 1986 our confidence in understanding the physics was shattered by the arrival of high T_c oxide superconductors. So, we could say at the time, there are conventional superconductors and oxide superconductors and one day we shall understand how those in the latter family work. But nowadays nothing can be taken for granted. The old type of intermetallic compounds reappear with much higher critical temperature, organic materials join the club and it turns out that an

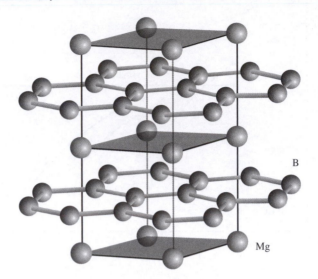

B

Mg

Fig. 14.25
Crystal structure of MgB_2

applied magnetic field is not necessarily a bad thing. We do not really know any more what the limits are, what is achievable and what it not. The status quo of ignorance has been restored.

Let us start with magnesium boride, a simple intermetallic compound with a crystalline structure shown in Fig. 14.25. The boron atoms arrange themselves in two-dimensional hexagonal sheets, like graphite, within a cubic structure of magnesium. What is extraordinary about it is its critical temperature well above that of other intermetallic compounds. It does obey though BCS theory in one respect: it has an isotope effect. The critical temperature is 40.2 K for atomic weight 10 and 39.2 K for atomic weight 11. It differs, however, from other metallic superconductors by not having a high charge carrier density. There is an energy gap but it is of a different kind. Two superimposed energy gaps have to be assumed to explain its properties.

Another recently discovered superconductor is $PuCoGa_5$ which has a high critical temperature of 18 K and in which induced magnetic fluctuations of the electrons are supposed to be responsible for the superconducting transition. Clearly, this is not an oxide superconductor but could the superconducting mechanism be close to that of oxides? Will there be similar compounds found with higher critical temperatures? The answers are not known at the moment. A further interesting feature of the $PuCoGa_5$ superconductor is its extremely high upper critical field estimated at 35 T. The tentative explanation is the radioactivity of plutonium 239 which is responsible for pinning the flux lines by creating line defects.

Let us now come to the effect of a magnetic field. We have been happy to accept so far that the critical temperature is reduced by applying a magnetic field and a high enough magnetic field will completely destroy superconductivity. This is not surprising at all. Cooper pairs are made up by electrons of opposite momenta and opposite spins. Therefore a magnetic field, whether applied or internal due to the ferromagnetic line up of dipoles, may be expected to be harmful because it effects differently the spin up and the spin down state. So the clear conclusion is that superconductivity might coexist with antiferromagnetism but never with ferromagnetism! Well, the discovery of superconductivity in UGe$_2$ proved it otherwise. If the material is kept all the time above the Curie temperature so that its magnetic state is paramagnetic then, however low is the temperature, no superconducting state exists. On the other hand, below the Curie temperature, in the ferromagnetic state, there is a range of pressures for which superconductivity is present below a critical temperature. This is so much against the grain that a new theory is needed. The tentative answer is that some other type of Cooper pairs must exist in which electrons of opposite momenta but identical spins pair up, and then an applied magnetic field might actually be helpful. The likely reason why these materials (there is a number of them) have only recently been discovered is their anisotropic nature. If anisotropic, then the state will crucially depend on the electron momenta in various directions which can be seriously altered by impurity scattering. Hence, superconductivity exists only when the material is made pure enough-and up to now the technology was just not available.

The last family I wish to mention here is that of organic superconductors. It is a fashionable subject nowadays. What is known is that the molecule has to be long, that the molecules must be close to each other so that electrons and holes can hop from one to the next one, and they should be stacked in two dimensions. They have some unusual properties the most outrageous of them is that the superconducting state can actually be brought on by applying a magnetic field. We know (see Fig. 11.29) that on the application of a magnetic field the electronic bands split into a spin up and spin down band which have somewhat different momenta. When two electrons of different spins pair up the resulting momentum will be non-zero. Could that cause the various anomalies observed? It remains to be shown. One thing is sure, neither experimenters nor theoreticians will have a quiet time in the next few decades.

Exercises

14.1. It follows from eqn (1.15) that in the absence of an electric field the current density declines as

$$J = J_0 \exp(-t/\tau)$$

where τ is the relaxation time related to the conductivity by eqn (1.10).

In an experiment the current flowing in a superconducting ring shows no decay after a year. If the accuracy of the measurement is 0.01%, calculate a lower limit for the relaxation time and conductivity (assume 10^{28} electrons m^{-3}). How many times larger is this conductivity than that of copper?

14.2. What is the maximum supercurrent that can be passed through a 2 mm diameter lead wire at 5 K (use data from Table 14.1).

14.3. In the first phenomenological equations of superconductivity, proposed by F. and H. London in 1935, the current density was assumed to be proportional to the vector potential and div $\mathbf{A} = 0$ was chosen. Show that these assumptions lead to a differential equation in $\mathbf{A}$ of the form of eqn (14.63).

14.4. The parameter λ defined in eqn (14.60) may be regarded as the penetration depth for $\kappa \cong 0$. A typical value for the measured penetration depth is 60 nm. To what value of ψ_0^2 does it correspond?

14.5. The energy diagram for a tunnel junction between two identical superconductors is shown in Figs 14.13 and 14.14. The superconducting density of states [sketched in Fig. 14.13(b)] is given as

$$C \frac{E}{\sqrt{E^2 - \Delta^2}}$$

where C is a constant and E is the energy measured from the Fermi level (middle of the gap). Show that at $T = 0$ the tunnelling current is zero when $U < 2\Delta/e$, and the tunnelling current is proportional to

$$\int_{\Delta}^{eU-\Delta} \frac{eU - E}{[(eU - E)^2 - \Delta]^{1/2}} \frac{E}{[E^2 - \Delta^2]^{1/2}} dE$$

for $U > 2\Delta/e$.

14.6. A lead-insulator-tin superconducting tunnel junction has a current–voltage characteristic at 1 K similar to that shown in Fig. 14.17 with the current maximum at $U = 0.52$ mV and the point of sudden upsurge at $U = 1.65$ mV.

(i) Find the energy gaps in lead and tin at zero temperature.
(ii) At what temperature will the current maximum disappear?

14.7. If a microwave cavity made of tin is cooled to 1 K can you expect the losses to be substantially less than at 4 K?

At what frequency would you expect superconductive effects completely to disappear in tin held at 1 K?

14.8. What is the frequency of the electromagnetic waves radiated by a Josephson junction having a voltage of 650 μV across its terminals?

Epilogue

Eigentlich weiss man nur, wenn man wenig weiss,
mit dem Wissen wächst der Zweifel.
                    Johann Wolfgang von Goethe

The Republic of Science shows us an association of independent initiatives,
combined towards an indeterminate achievement . . . its continued existence
depends on its constant self-renewal through the originality of its followers.
                    Michael Polanyi article in *Minerva* 1962

I hope these lectures have given you some idea how the *electrical properties* of materials come about and how they can be modified and exploited for useful ends. You must be better equipped now to understand the complexities of the physical world and appreciate the advances of the last few decades. You are, I hope, also better equipped to question premises, to examine hypotheses, and to pass judgment on things old and new. If you have some feeling of incompleteness, if you find your knowledge inadequate, your understanding hazy, don't be distressed; your lecturers share the same feelings.

The world has changed a lot since the first edition of this book. The quiet optimism reigning thirty-three years ago is no longer the order of the day. Nowadays people tend to be either wildly optimistic, envisaging all the wealth our automatic factories will produce, or downright pessimistic, forecasting the end of civilization, as we run out of energy and raw materials. The optimists take it for granted that the engineers will design the automatic factories for them, and even the pessimists have some lingering hopes that the engineers will somehow come to the rescue. It is difficult indeed to see any alternative group of people who could effect the desired changes. I greatly admire physicists. Their discoveries lie at the basis of all our engineering feats, but I don't think they can do much in the present situation. We don't need to pry any more into the secrets of Nature, we need to make them work for us. The current research of geneticists, microbiologists and biologists may well produce a new species of supermen, but it is unlikely that we can wait for them. We cannot put much trust into politicians either. They will always (they have to) promise a better future, but the power to carry out their promises is sadly missing. There is no escape. The responsibility is upon your shoulders. Some of you will, no doubt, opt for management, but I hope many of you will employ your ingenuity in trying to find solutions to the burning engineering problems of the day. You are more likely to succeed if you aim high. And it is more fun too.

I would like to end by quoting a passage written about the joys of inventive engineering by the late Professor Kompfner, the inventor of two important microwave tubes, and designer responsible for the electronics of the first communications satellites, with whom we were privileged to work for a few years in this laboratory.

'The feeling one experiences when he obtains a new and important insight, when a crucial experiment works, when an idea begins to grow and bear fruit, these mental states are indescribably beautiful and exciting. No material reward can produce effects even distantly approaching them. Yet another benefit is that an inventor can never be bored. There is no time when I cannot think of a variety of problems, all waiting to be speculated about, perhaps tackled, perhaps solved. All one has to do is to ask questions, why? how? and not be content with the easy, the superficial answer.'

Added in 1993 for the 5th edition. The Cold War is over, we can now afford to beat many of our weapons into ploughshares and devote more and more of our resources to the solution of our never-diminishing scientific and technological problems. Oddly enough, what we witness now is not renewed enthusiasm towards the value of science, rather the contrary. In a recently published book, by someone who is a representative of this trend, the following lines appeared. 'The spectacular successes of modern scientific method have made it the final arbiter of truth in our age. Scientists claim that they and only they hold the key to the meaning, purpose, and justification of human life. But what of human values? Are we in danger of losing our souls?'

This is, of course, utter nonsense that any of you could refute. What do these new prophets hope to achieve? They hope to reverse the forward march of science; they want to get rid of a discipline they do not understand. Some of them, I presume, want to go back to the good old pre-scientific times, forgetting that the land was then ravaged by cholera, typhoid, and the bubonic plague, and life expectancy was half of that we have today. Some others blame science for the decline of the True Faith and wish to resurrect the ideals of Torquemada for dealing with the problem. It looks as though there is still a battle to be fought for reason.

Appendix I: Organic semiconductors

A1.1 Introduction

The range of materials suitable for producing various electronic devices has been steadily increasing. It is no great surprise that organic materials have also put in a claim to be represented. They were never completely disregarded, but having reached prominence in the last few years (writing in 2003), they can no longer be ignored. It may indeed be argued that attempts to produce organic light sources are bound to succeed. After all, it has been known a long time that fireflies glow fairly brightly by a process called bio-luminescence. Enzymes in the fly cause reactions that excite organic molecules. When they revert to their ground state, light is emitted, with a very high quantum efficiency, but of course, not much power.

As far as this book is concerned the main problem is that organic chemists and engineers use different technologies and talk a different language. Electrical conduction (or semiconduction) in an organic material is explained by chemists in terms quite different from that of an engineer. They prefer to look at the individual rather than at the crowd. They regard the properties of the individual molecules as paramount and will maintain that the properties of the solid follow from there. It would have been therefore difficult to treat the properties of organics alongside inorganic materials. The best solution seems to be to devote a separate section to them in which the technology and applications follow immediately after the principles.

A1.2 Fundamentals

Some properties of organic materials have though been touched upon in Section 5.3. We have come across σ bonds and π bonds and we have even gone that far as to give the chemical formula of acetylene. Since it was not more than a very gentle introduction, it seems best to start here at the beginning, review some properties of carbon bonding and say a few words about hydrogen atoms which are well known for their predilection to join carbon atoms.

Carbon has six electrons—$1s^2 2s^2 2p^2$. The two inner electrons do not participate in the bonding but the four outer ones do. In simple compounds like methane, CH_4 or ethane, C_2H_6 (Fig. A1.1), will the 2s electrons behave differently from the 2p electrons? The experimental evidence is that they do not have separate properties. Some kind of compromise is apparently taking place. One of the 2s electrons is 'promoted' to a 2p state by taking up some energy, then the three 2p orbitals and the remaining 2s orbital create

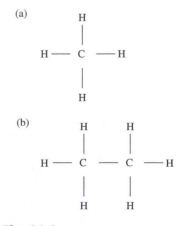

Fig. A1.1
(a) Methane and (b) ethane. Saturated hydrocarbons with all σ bonds.

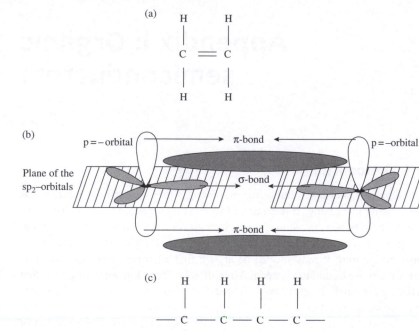

Fig. A1.2
(a) Ethylene, an unsaturated
hydrocarbon, (b) A more picturesque
presentation of the scheme: σ bonds
at 120° in a plane from each C which
have their remaining electron orbital
perpendicular forming a π bond,
(c) Polyethylene, a saturated polymer.
All single bonds.

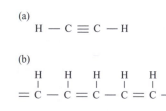

Fig. A1.3
(a) Acetylene, a linear molecule with
triple bond, (b) polyacetylene,
a conjugated chain polymer.

four tetrahedrally disposed equal orbits called sp^3 hybrids. The extra energy
required for this hybridisation is compensated by the energy gained by making tetrahedral bonds to ligands [e.g. 4H in methane, Fig. A1.1(a)]. Each
hybridized orbital contains a single unpaired electron, which can pair with
a hydrogen 1s electron to form a bond. This is the σ bond. Everything is stable,
all the electrons of the carbon atom have been used up. No electron has been
left over.

It is also possible for one s and two p orbitals to form three sp^2 orbitals (sp^2
hybridisation) which are planar trigonal orbitals (120° separation in a plane).
In ethylene, C_2H_4 [Fig. A1.2(a)] each C forms three σ bonds with sp^2 hybrids
to the other C and 2H. Now, each carbon atom has one more electron left over.
Having no role to play in the horizontal plane, each one orbits in the vertical
plane. They are, however, not independent of each other. They form a looser
bond known as a π bond [Fig. A1.2(b)]. This is what happens in ethylene. The
double line between the two carbon atoms signifies a double bond: one is a σ
bond, the other one is a π bond.

Now let us reduce the H/C ratio to one. The lowest member of this tribe
is C_2H_2, acetylene, well known as the welding gas which burns at a high
temperature. Shown in Fig. A1.3(a), it is a linear molecule which has two sp
hybrids on each C forming σ bonds each linking to the other C and one H. There
are two electrons left over on each carbon atom. What will they do? They will
again form two π bonds. Thus, in acetylene the bond between the two carbon
atoms is a triple bond, one σ bond and two π bonds.

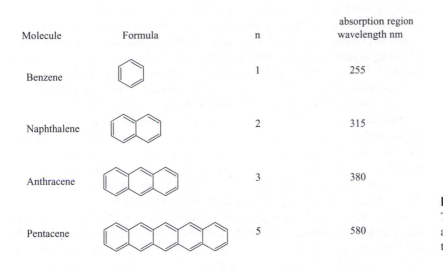

Molecule	Formula	n	absorption region wavelength nm
Benzene		1	255
Naphthalene		2	315
Anthracene		3	380
Pentacene		5	580

Fig. A1.4
The benzene series, showing optical absorption progressing from the uv to the visible.

Let us next look at benzene, a planar six atom ring molecule, with a chemical formula, C_6H_6. (Note that we are now following the notation of Organic chemists and omit the H atoms, see Fig. A1.4.) Each carbon atom makes σ bonds to two adjacent carbons and to one hydrogen. Its fourth electron orbits at right angles to the plane and forms a π bond with one of the adjacent carbons. These six electrons in π bonds form an electron cloud on both sides of the plane ring, which contribute to the stability of the benzene molecule. The standard formula of this family is $C_{4n+2}H_{2n+4}$. With $n = 2$, 3 and 5 (shown also in Fig. A1.4) they are known as naphtalene, anthracene and pentacene. All these molecules are flat and there are $4n + 2\pi$ bonds per molecule, one for each carbon atom, which influence the electronic properties. The σ network is quite stable but the electrons in the π bonds are less tightly bound.

We should now briefly return to the single π bond of ethylene. The two electrons are coupled, hence the energy levels split. The lower one is called a bonding state the higher one an antibonding state. Under normal conditions, the lower one is occupied, the higher one is empty. When there are more double bonds in a molecule, the individual energy levels split further. The 3, 5 and 7 double bonds in benzene, naphtalene and anthracene will cause three-fold, five-fold and seven-fold splits. As the number of double bonds increases, one can talk about a band of energies, that is a band structure.

There is no need to use rings if a long structure is required. One can do the same thing with linear molecules, for example, ethylene. It can mate up with other C_2H_4 molecules (which can be called monomers) to form a long chain polymer with the carbon chain potentially many thousands long. To do this, a chemical process involves breaking the π bond between the two carbon atoms and inserting a CH_2 trio, as shown in Fig. A1.2(c). This is reasonably called polymerisation and the resulting polymer in this case is called polyethylene, more usually polythene. It is a well known inert plastic, chemically very stable. The π bonds have gone, polythene has no interesting electronic properties.

A similar process can be done by chemical processing acetylene to break the triple bond and putting in a CH pair as shown in Fig. A1.3(b). The 'new' carbon is σ bonded to the two carbons and the 'new' hydrogen, but has a spare orbital to complete a π bond with only one of two adjacent carbons. Hence, we get a polymer (known as polyacetylene) in which single and double bonds alternate. Polymers with such structures are known as conjugated polymers. Having many π bonds, there is now, as mentioned above, a band structure consisting of the 'bonding' lower band and the 'antibonding' higher band.

Once we know that there is a band gap, that is more or less under our control, the possibility of constructing a light emitting device immediately arises. There must be some mechanism of exciting the electrons into a higher state and then they can give up their energy by emitting a photon. Organic chemists describe the process as a transition between lowest unoccupied molecular orbital (LUMO) and highest occupied molecular orbital (HOMO).

So how can we make an organic light emitting device (OLED)? We need to construct a p–n junction, apply an electric field across it and when the electrons and holes combine they will emit light. Let us assume for the moment that there are p and n-type materials (doping will be described a little later), what kind of electrodes would we need? Indium–tin oxide (ITO) is often used as the anode partly because it is transparent (suitable for bringing out the light) and partly because it has a high work function. It serves to inject holes into the HOMO levels. The cathode is made of magnesium or calcium, low work function materials, suitable to inject electrons into the LUMO levels.

What happens when the electrons and holes meet at the junction? It is not just a straight descent from the upper band into the lower band. There are exciton intermediaries of two kinds. When the electron–hole pair has opposite spins it is a singlet if they have parallel spins it is known as a triplet. On average one singlet and three triplets are formed for every four electron–hole pairs. Singlets decay fast (order of ns) and emit a photon, triplets decay slowly (order of ms) and generate heat. Thus normally an OLED cannot have higher than 25% efficiency. There are though hopes that, by including heavy metal elements into the compound, the triplets can also be persuaded to help the radiative process.

A1.3 Technology

Having described the basic principles let us now come to hard realities. The conjugated chain semiconducting polymer that we have described is a hard, inflexible, insoluble plastic. However, organic chemistry can change this. For inorganic semiconductors (e.g. Si), we know that doping with impurities of up to 1 part per million can produce enormous ranges of both p and n-type of conductivities. Conjugate polymers like polyacetylene can also be 'doped' by replacing some of the chain of H shown in Fig. A1.3(b) by n or p-type groups or more obviously by elements, for example Na, Ca as electron sources or sulphur for holes. Also the whole nature of the polymer can be changed—even making it soluble. This type of doping involves relatively massive impurities— from a solid fraction of a per cent, up to 40 per cent. Bucket chemistry, in a very controlled environment. The impurities are usually added to the molten material, sometimes during the polymerization process. The facility of making a semiconductor like polyacetylene or polyanaline soluble is an important step in device (transistor or photodiode) manufacture. A thin film is needed (otherwise the voltages to be applied would have to be impracticable) and this can be made easily by spinning a drop of the solution on a suitable substrate. After drying, films of less than $1\,\mu$m thick are easily and reproducibly obtained. Another trick is that the solution can be mixed with a photochemical which becomes an insulator on exposure to UV light. This has been exploited by a group at Philips Research Labs. who used a photochemical mixed with polyaniline which was exposed to UV through a mask to define conducting channels separated by insulating barriers. This is a final stage in making a transistor or integrated circuit which the Philips group has done, using double layers of semiconducting polymers with three-dimensional interconnections. So far the electron mobility in polymers is about $10^{-7}\,\mathrm{m^2V^{-1}s^{-1}}$, a million times less than Si. A lot of work is being done to increase that figure.

Research on polymer electro-luminescence was first reported by the National Physical Lab in 1983, when feeble blue light from poly(N-vinyl carbazole), (PVK) was obtained. Interest was not great until Professor R. Friend's group at Cambridge started publishing results on poly p-phenylene vinylene (PPV), which eventually by 1998 achieved an efficiency of $20\,\mathrm{lm\,W^{-1}}$. This device consists of: (i) a PPV layer doped to conduct electrons; (ii) a relatively undoped PPV layer that luminesces; and (iii) a PPV layer doped to conduct holes. Further advance has been made by J. Kido of Yamagata Unviersity who with a single polymer film doped with several laser dyes (see Section 12.6.3) managed to obtain an external quantum efficiency of 1% and a luminance of $4000\,\mathrm{cd\,m^{-2}}$.

The other strand of development, away from polymers, began when it was realised that the quite small and stable organics like Alq [Fig. A1.5(a)] could be easily evaporated in vacuum to form a thin film of about $0.2\,\mu$m. The high fields required for luminescence could now be obtained with a few volts. A simple diode, originally in the Kodak Labs and refined elsewhere is shown in Fig. A1.5(b). Alq is an electron conductor whilst the other active layer, a diamine with a formula mercifully abbreviated to NBP (it contains naphyl and phenyl), conducts with holes. The diode is fabricated by coating a glass substrate with a thin layer of ITO. The hole conducting layer NPB is next evaporated onto this electrode in a vacuum (about 10^{-6} Torr) followed by a layer of Alq, having a combined thickness of about $0.2\,\mu$m. Finally, the cathode, an alloy of about 10:1 Mg–Ag is evaporated from separate tantalum boats. With about

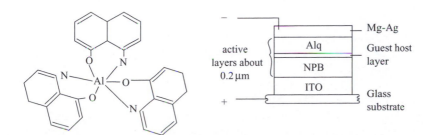

Fig. A1.5
(a) Aluminium quinolate (Alq).
(b) Schematic of thin film small molecule OLED.

6 V applied to the diode, a current of typically $7\,mA/mm^2$ flows and gives a luminance*of green light of $4300\,cd\,m^{-2}$

Recent advances in OLEDs was stimulated by the 'guest–host' doped emitter system. This consists of adding a third (very thin about 50 nm) layer between the two active layers of a flourescent dye†.This central layer becomes the main recombination region. Whilst the two active layers are optimised to conduct electrons and holes, the third region is optimised for luminous efficiency. It has been found possible to enhance the green emission natural to an Alq system, and also by careful choice of other dyes to shift the photo emission into the red. So coloured display panels can be made, by evaporating the layers through masks to form pixels. Luminous efficiency is mainly limited by non radiative decay of excited states. the use of phosphorescent dye has put the efficiency up to $40\,lm\,W^{-1}$. OLEDs have achieved operational stability of up to 10 000 hours. This performance is adequate for a number of passive display applications. A large number of laboratories are working on detailed interface problems, and the small molecule, evaporated OLED could have even more widespread uses in the medium term future.

The main asset of organics is their flexibility. It is unlikely that in the near future they would be able to compete with inorganic LEDs in efficiency, but they can on price and in applications where flexibility is an important requirement. There will surely be demand for light sources which are large, cool, cheap and can be fixed to curved surfaces. The ultimate would be a (say) 1 m by 2 m television screen which can hang on a wall in the sitting room and when needed in another room it can be rolled up and swiftly moved.

Finally, FETs. They have also been produced from organic materials. To begin with, it was only the active material (the channel between source and drain) that was made of a polymer, but later it turned out to be possible to use heavily doped polymers also for the electrodes. The advantage of an all-polymer transistor is its low cost. These transistors are inherently slow on account of the low mobilities of the organic materials but it may not matter. The main application envisaged for these transistors is as small plastic memory chips attached to various consumer products (mounted possibly on anti-theft stickers) which is used for storing all kinds of information about the product, a lot more than is contained in present bar codes. A further major advantage would be that these memory chips could be remotely interrogated by a radio frequency indentification system.

Appendix II: Artificial materials

A2.1   Introduction

All the materials discussed in the main text of this book were produced by nature. Well, not by nature alone, we did certainly help nature here and there. We combined the elements in a manner which led to a variety of new properties. We managed to persuade some crystals to grow under circumstances when they were most reluctant to do so. We produced structures with the thickness of a single atom, but we were always restricted by the ways atoms were willing to arrange themselves. An artificial material, on the other hand, comes about by taking an entirely innocuous dielectric, immersing into it some microscopic structure, and lo and behold, its electrical behaviour radically changes.

The idea of producing artificial materials is not new. It has been known for about three scores of years that inclusion of bits of metal in a dielectric will lead to some modest change in the dielectric constant. There were, however, some new developments in the last decade which we should no longer ignore. They could be roughly divided into two classes. In one case the distance between the inclusions is comparable with the wavelength, in the other case it is small relative to the wavelength. The first one is known as the subject of photonic band gaps and the second one as metamaterials. The first one is based on the Bragg effect which came repeatedly into our treatment (X-ray and electron diffraction, the Ziman model of electronic band gaps, volume holography, phase conjugation, external Bragg reflector lasers, etc.). It is more difficult to define the principles behind the recent work on metamaterials. What is crucial there is the properties of the individual elements, their resonant nature but the interaction between them is also important.

A2.2   Photonic band gaps

As we know, electrons in a semiconductor have allowed and forbidden energies. We have seen and discussed that umpteen times. Why do electrons behave that way? We have discussed that too. It is essentially due to the wave-like nature of the electron. When they see a periodic potential in a periodic medium, they respond. But why only electrons? Could not photons do the same thing if they find themselves in a periodic medium? Yes, of course, we discussed that too in relation to the Bragg effect. So the idea is obvious. Put photons into a periodic medium and they will have allowed and forbidden energies which, in this context, means that the propagation of the electromagnetic waves in that medium is allowed or forbidden. The modern term for it is photonic bandgaps. A simple structure which can produce a (not very good) bandgap is shown in Fig. A2.1. It consists of a set of dielectric rods.

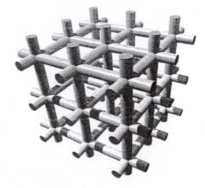

Fig. A2.1
An example of a photonic band gap material made from a set of dielectric rods.

The discipline started in the 1990s. Why so late? If physicists of long time ago managed to figure out the mysteries of X-ray diffraction, why did they not think about building materials exhibiting photonic band gaps? They must have thought about the possibility, but how to do the experiments? The evidence for electronic band structures could be provided by relatively simple measurements on semiconductors. The X-ray measurements on various crystal structures did show that there was perfect reflection of the incident wave at some incident angles, but not for all angles. One could easily conclude that nature does not like* photonic band gaps. It was relatively easy to build them by optical means in volume holography but those methods gave reflections only in one direction. For a photonic band gap, perfect reflection must occur within a range of wavelengths from whichever direction the electromagnetic wave comes. So there was no clear guidance on how such a material could be built and at the same time there was some legitimate doubt whether photonic band gap materials exist at all. If in doubt try numerical simulations. After all it is only Maxwell's equations which need to be solved. That was indeed the way forward. Serious investigations could only start when technology was advanced enough to produce the samples at optical wavelengths and computers were powerful enough to solve the problem numerically†. And that leads us to the early 1990s. The pioneers were Eli Yablonovitch and Sajeev John. The favoured technological solution was to drill holes in a dielectric rather than to put together a structure of rods. Holes of submicrometre dimensions had to be drilled. Half a million holes later there was still no success. But success eventually came in the form of the diamond structure that was shown in Fig. 5.3. The holes had to be drilled so as to follow the directions of the chemical bonds.

What are the applications of photonic bandgap materials? They can be used whenever there is a need for electromagnetic waves propagating in any direction to be reflected. They are singularly suitable for constructing resonant cavities. Replace a few elements of a photonic bandgap material by one capable of lasing, pump the laser at a wavelength for which the bandgap material is transparent and the whole laser device is ready. This is actually the way to produce very small lasers where very small means that its dimensions are submicrometre. Another application is for guiding light. If we have a cylindrical photonic bandgap material and we clear the area around the axis, then an optical wave can propagate there without being able to spread outwards in the radial direction. This is because a wave propagating in any but the axial direction will be reflected. These photonic bandgap waveguides promise lower losses for guiding waves to a long distance. They might very well turn out to be practical in less than a decade.

A2.3   Metamaterials

The subject may be reckoned to have started by Pendry recalling a 1968 paper by Veselago which claimed that it is possible to realize in a limited frequency band a material which has negative relative permittivities and permeabilities, and as a consequence has a negative index of refraction. How to realize negative permittivity? Well, it has been known for a long time that plasmas have negative effective permittivity below the plasma frequency [it does follow from the analysis in Section 1.5, see eqn (1.53)]. Pendry showed that a set of metallic rods behaves as a plasma and the plasma frequency can be very low (microwaves or below) provided the rods have sufficiently small diameters. Pendry also showed that the permeability can be radically influenced by the inclusion of a set of metallic split ring resonators (Fig. A2.2). How can we determine the resonant properties of this element? If we want a good approximation, we could represent it with a distributed circuit model similar to that of transmission lines. However, if all we want to have is some idea of the factors influencing the value of the resonant frequency then we can simply state that there are (i) three capacitances, one gap capacitance in each ring and the inter-ring capacitance, and (ii) three inductances, the self inductance of the inner and of the outer ring and the mutual inductance between them.

How will this structure affect the permeability? When an electromagnetic wave is incident it will induce currents in the rings which will radiate and then adding this re-radiation to the incident wave we find that the effective velocity of the wave changed from which an equivalent permeability can be defined. Performing all the relevant calculations we find that the effective permeability varies as a function of frequency as shown in Fig. A2.3. This is actually for the lossy case. In the absence of losses, the effective permeability changes from plus infinity to minus infinity at the equivalent resonant frequency. In the presence of losses the maximum achievable permeability is limited, but there is still a sudden change from positive values to negative values. The imaginary part of the effective permeability* is maximum at resonance and sharply declines away from resonance.

As mentioned above, negative effective permittivity can be achieved by an array of thin metallic rods. What happens when a structure is put together in which the rod and the split ring are combined (Fig. A2.4) to represent a single element? Then both the permittivity and the permeability may be negative within a certain frequency band providing Veselago's negative refractive index material. They are supposed to exhibit a reverse Snell's law, reverse Doppler shift and reverse Cerenkov radiation. Experiments to prove these have already been done but the subject is still controversial.

Fig. A2.2
A doubly split metallic ring as a metamaterial element. Dimension in the third dimension is usually small. The metal is evaporated in a substrate.

* We have not so far mentioned the imaginary part of the permeability in this course but it is the same kind of thing as the imaginary part of the permittivity which we discussed in Section 10.5 and we have also seen the resonant behaviour of permittivity in Section 10.6 due to effects at the molecular level.

Fig. A2.4
An array in which each element consists of a metal rod in front of a split ring.

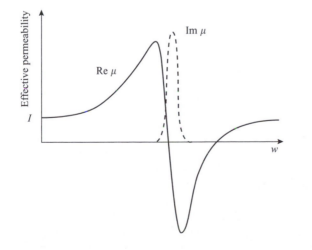

Fig. A2.3
Typical variation of the effective permeability of a metamaterial with frequency.

What will be these materials good for? A potential application is in MRI where the need is to collect the radio frequency magnetic field radiated from the sample (mostly the human body) without disturbing the d.c. magnetic field. Hence the need for a resonant guide that is only magnetic in the vicinity of the frequency of the nuclear magnetic resonance.

Appendix III: Physical constants

List of frequently used symbols and units

Quantity	Usual symbol	Unit	Abbreviation
Current	I	ampere	A
Voltage	U, V	volt	V
Charge	q	coulomb	C
Capacitance	C	farad	F
Inductance	L	henry	H
Energy	E	joule	J
(1 eV is the electron energy change caused by 1 volt acceleration $(= 1.60 \times 10^{-19}\,\text{J}))$		electron volt	eV
Resistance	R	ohm	Ω
Power	P	watt	W
Electric field	$\mathcal{E}$	volt/metre	V m^{-1}
Magnetic field	H	ampere/metre	A m^{-1}
Electric flux density	D	coulomb/(metre)^2	C m^{-2}
Magnetic flux density	B	tesla	T
Frequency	f	hertz	Hz
Wavelength	λ	metre	m
Temperature	T	Kelvin	K

List of physical constants

Symbol	Quantity	Value
m	mass of electron	$9.11 \times 10^{-31}\,\text{kg}$
e	charge of electron	$1.60 \times 10^{-19}\,\text{C}$
e/m	charge/mass ratio of electron	$1.76 \times 10^{11}\,\text{C kg}^{-1}$
m_p	mass of proton	$1.67 \times 10^{-27}\,\text{kg}$
h	Planck's constant	$6.62 \times 10^{-34}\,\text{J s}$
$\hbar$	$h/2\pi$	$1.05 \times 10^{-34}\,\text{J s}$
N	Avogadro's number	$6.02 \times 10^{26}\,\text{molecules/kg mole}$
k	Boltzmann's constant	$1.38 \times 10^{-23}\,\text{J K}^{-1}$
		$8.62 \times 10^{-5}\,\text{eV K}^{-1}$
kT	(at 293 K)	$0.025\,\text{eV}$
		$4.05 \times 10^{-21}\,\text{J}$

Continued

Symbol	Quantity	Value
c	velocity of light	$3.00 \times 10^8 \, \mathrm{m\,s^{-1}}$
ϵ_0	permittivity of vacuum	$8.85 \times 10^{-12} \, \mathrm{F\,m^{-1}}$
μ_0	permeability of vacuum	$4\pi \times 10^{-7} \, \mathrm{H\,m^{-1}}$
hc/e	wavelength of photon having 1 eV energy	1240 nm
μ_{mB}	Bohr magneton	$9.27 \times 10^{-24} \, \mathrm{A\,m^2}$

Visible light (part of electromagnetic spectrum)

Colour	Wavelength	Frequency	Quantum energy
violet	400 nm	750 THz	3.1 eV
red	700 nm	430 THz	1.8 eV

Prefixes

a atto = 10^{-18}	μ micro = 10^{-6}	M Mega = 10^6
f femto = 10^{-15}	m milli = 10^{-3}	G Giga = 10^9
p pico = 10^{-12}	c centi = 10^{-2}	T Tera = 10^{12}
n nano = 10^{-9}	k kilo = 10^3	P Peta 10^{15}
		E exa 10^{18}

Definitions of light intensity

Candela: Luminous intensity in the normal direction from a black body $1 \, \mathrm{m^2}$ in area at a temperature of the solidification of platinum (2042 K).

Lumen: Luminous flux emitted in 1 steradian solid angle by a point source having an intensity of 1 cd.

These definitions shake one's faith in S.I. units, however it was even worse in the days of candle power. The following figures for light sources at the end of the last millennium given one a reference or target for new light sources.

Device	Luminous efficiency ($\mathrm{lm\,W^{-1}}$)	Brightness ($\mathrm{cd\,m^{-2}}$)
Tungsten electric light bulb	17	
Quartz halogen bulb	25	
Fluorescent light	70	8000
C.R.T. spot		300

Appendix IV: Variational calculus. Derivation of Euler's equation

The problem of variational calculus is to find the form of the function $y(x)$ which makes the integral

$$\int_{x_1}^{x_2} F(x, y(x), y'(x)) \, dx \tag{A.1}$$

an extremum (maximum or minimum), where F is a given function of x, y, and y'.

Assume that $y(x)$ does give the required extremum, and construct a curve $y(x) + \eta(x)$ which is very near to $y(x)$ and satisfies the condition $\eta(x_1) = 0 = \eta(x_2)$. For the new curve our integral modifies to

$$\int_{x_1}^{x_2} F(x, y(x) + \eta(x), y'(x) + \eta'(x)) \, dx. \tag{A.2}$$

Now making use of the fact that η and η' are small, we may expand the integral to the first order, leading to

$$\int_{x_1}^{x_2} \left\{ F(x, y(x), y'(x)) + \frac{\partial F}{\partial y}(x, y(x), y'(x))\eta(x) \right.$$
$$\left. + \frac{\partial F}{\partial y'}(x, y(x), y'(x))\eta'(x) \right\} dx. \tag{A.3}$$

Integrating the third term by parts (and dropping the parentheses showing the dependence on the various variables) we get

$$\int_{x_1}^{x_2} \frac{\partial F}{\partial y'} \eta' \, dx = \left[\eta \frac{\partial F}{\partial y'} \right]_{x_1}^{x_2} - \int_{x_1}^{x_2} \eta \frac{\partial}{\partial x} \frac{\partial F}{\partial y} \, dx = -\int_{x_1}^{x_2} \eta \frac{\partial}{\partial x} \frac{\partial F}{\partial y'} \, dx, \tag{A.4}$$

which substituted back into eqn (A.3) gives

$$\int_{x_1}^{x_2} \left[F + \left(\frac{\partial F}{\partial y} - \frac{\partial}{\partial x} \frac{\partial F}{\partial y'} \right) \eta \right] dx. \tag{A.5}$$

We may argue now that $y(x)$ will make the integral an extremum, if the integral remains unchanged for small variations of η. This will occur when the coefficient of η vanishes, that is, when

$$\frac{\partial F}{\partial y} - \frac{\partial}{\partial x}\frac{\partial F}{\partial y'} = 0. \tag{A.6}$$

The above equation is known as the Euler differential equation, the solution of which gives the required function $y(x)$.

When F depends on another function $z(x)$ as well, an entirely analogous derivation gives one more differential equation in the form

$$\frac{\partial F}{\partial z} - \frac{\partial}{\partial x}\frac{\partial F}{\partial z'} = 0. \tag{A.7}$$

These two differential equations, (A.6) and (A.7), are used in Section 14.6 for obtaining the vector potential and the wave function which minimize the Gibbs free energy.

Appendix V: Suggestions for further reading

(a) *General texts*

R.P. Feynman *et al.*, *The Feynman lectures on physics* (Addison Wesley, 1989).
C. Kittel, *Introduction to solid state physics*, 7th edn (Wiley, 1995).
H.M. Rosenberg, *The solid state*, 3rd edn (O.U.P., 1988).
(b) *Topics arising in specific chapters*

Chapter 4
W. Heitler, *Elementary wave mechanics* (O.U.P., 1956).

Chapter 5
A.H. Cottrell, *Introduction to the modern theory of metals* (Institue of Metals, 1988).

Chapter 7
J.M. Ziman, *Electrons in metals* (Taylor and Francis, 1962).

Chapter 8
J.P. McKelvey, *Solid state and semiconductor physics* (Harper and Row, 1966).
Bernard Gil (Ed) Low Dimensional Nitride Semiconductors O.U.P (2002)

Chapter 9
A.S. Grove, *Physics and technology of semiconductor devices* (Wiley, 1967).
J.E. Carroll, *Physical models for semiconductor devices* (Arnold, 1974).
S.M. Sze, *Semiconductor devices, physics and technology* (Wiley, 1985).
B.G. Streetman, *Solid state electronic devices*, 3rd edn (Prentice Hall, 1990).

Chapter 10
P.J. Harrop, *Dielectrics* (Butterworth, 1972).

Chapter 11
J.P. Jakubovics, *Magnetism and magnetic materials*, 2nd edn (Institute of Metals, 1994).
J.D. Livingston, *Driving force* (Harvard University Press, 1996).

Chapter 12
G.H.B. Thompson, *Physics of semiconductor laser devices* (Wiley, 1980).
Amnon Yariv, *Optical electronics in modern communications*, 5th edn (O.U.P., 1997).

Chapter 13
J. Wilson and J.E.B. Hawkes, *Optoelectronics: an introduction*, 3rd edn (Prentice Hall, 1997).
A. Yariv and P. Yeh, *Optical waves in crystals, propagation and control of laser radiation* (Wiley, 1984).
L. Solymar and D.J. Cooke, *Volume holography and volume gratings* (Academic Press, 1981).

Chapter 14

L. Solymar, *Superconductive tunnelling and applications* (Chapman and Hall, 1972).

E. Rhoderick and A.C. Rose-Innes, *Introduction to superconductivity*, 2nd edn (Heinemann, 1978).

Appendix 1

M. Pope and C.E. Swenberg, *Electronic process in organic crystals a polymers* (O.U.P., 1999).

Answers to exercises

Chapter 1

1.1. 2.66×10^{-13} s.

1.2. $m^* = 0.015 m_0, \mu = 46.7\,\text{m}^2\,\text{V}^{-1}\,\text{s}^{-1}, \omega_c \tau = 15 \gg 1$. The resonance may be regarded as sharp.

1.3. $\mu_e = 0.14\,\text{m}^2\,\text{V}^{-1}\,\text{s}^{-1}, \mu_h = 0.014\,\text{m}^2\,\text{V}^{-1}\,\text{s}^{-1}$.

1.4. (i) $N_e = 2.5 \times 10^{28}\,\text{m}^{-3}$, (ii) $\mu_e = 5.32 \times 10^{-3}\,\text{m}^2\text{V}^{-1}\text{s}^{-1}$, (iii) $m^* = 0.99 m_0$, (iv) $\tau = 2.98 \times 10^{-14}$ s, (v) 0.98.

1.5. 1.09×10^{-4} m, 4.67×10^{-8} m, 2.45×10^{-5} m.

1.6. (i) There may be a misalignment of contacts, (ii) 1.04×10^{21} m^{-3}, (iii) $\sigma = 40.3\,\text{S}\,\text{m}^{-1}$, (iv) $\mu = 0.24\,\text{m}^2\,\text{V}^{-1}\,\text{s}^{-1}$.

1.7. The carriers will recombine.

1.8. $R = -(1/e)(N_e \mu_e^2 - N_h \mu_h^2)/(N_e \mu_e + N_h \mu_h)$, not necessarily.

1.9. Reflection $= \left| \dfrac{1 - (1 - \omega_p^2/\omega^2)^{1/2}}{1 + (1 - \omega_p^2/\omega^2)^{1/2}} \right|^2$.

For $\omega < \omega_p$ transmission $= 0$

For $\omega > \omega_p$ transmission $= 4(1 - \omega_p^2/\omega^2)^{1/2}/[1 + (1 - \omega_p^2/\omega^2)^{1/2}]^2$.

1.10. Transmission

$$= \left[\cosh^2(k_{2i}d) + \frac{1}{4}\left(\frac{k_{2i}}{k_1} - \frac{k_1}{k_{2i}} \right)^2 \sinh^2(k_{2i}d) \right]^{-1},$$

$8.62 \times 10^{-5}, 8.67 \times 10^{-47}$.

1.11. $\bar{\bar{\epsilon}}_{\text{eqv}} = \epsilon \begin{bmatrix} a_{11} & a_{12} & 0 \\ a_{12} & a_{22} & 0 \\ 0 & 0 & a_{33} \end{bmatrix}$

$a_{11} = 1 - \dfrac{\omega_{pe}^2}{\omega_c^2 - \omega^2} - \dfrac{\omega_{pi}^2}{\omega_c^2 - \omega^2}, a_{12} = \text{i}\dfrac{\omega_c(\omega_{pe}^2 - \omega_{pi}^2)}{\omega(\omega_c^2 - \omega^2)}$

$a_{21} = -a_{12}, a_{22} = a_{11}, a_{33} = 1 + \dfrac{\omega_{pe}^2}{\omega^2} + \dfrac{\omega_{pi}^2}{\omega^2}$.

Chapter 2

2.1. (i) $6.22 \times 10^{-9}/a^{1/2}$ m, (ii) 7.27×10^{-11} m, (iii) 4.53×10^{-15} m.

2.3. Resolution $\approx \lambda = 5.48 \times 10^{-12}$ m, 27.2 eV, lens aberrations, voltage stability.

2.4. Maxima at $\theta = 0, \sin^{-1} 0.42, \sin^{-1} 0.83$.

2.6. $\hbar \omega_p = 15.8$ eV.

Chapter 3

3.2. $0, \dfrac{a^2}{3}\left[1 - (-1)^n \dfrac{6}{n^2\pi^2} \right], 0, \dfrac{\hbar^2 n^2 \pi^2}{4a^2} = \hbar^2 k_n^2 = 2m E_n$.

3.3. $0, a^2/3, 0, p^2 = 2mE$.

3.5. For $E < V_2$ transmitted current is zero. For $E > V_2$ $J_2/J_{\text{inc}} = (k_{2r}/k_1)|2k_1/(k_1 + k_{2r})|^2$. The equations are formally the same as in exercise 1.6.

3.6. $J_3/J_1 = [\cosh^2(k_{2i}d) + \frac{1}{4}(k_{2i}/k_1 - k_1/k_{2i})^2 \sinh^2(k_{2i}d)]^{-1}$, $0.136, 6.58 \times 10^{-13}$.

3.7. With the approximation $J_3/J_1 = \exp(-2k_{2i}d)$ the values are 0.055 and 2.57×10^{-13}.

3.8. All the power is reflected.

3.9. $E = (\hbar^2 \pi^2/2m)(n^2/L_x^2 + m^2/L_y^2); 13E_0/9, 25E_0/9, 40E_0/9, 5E_0, 52E_0/9$ where $E_0 = \hbar^2 \pi^2/2m L_x^2$.

3.10. $E \approx 1.9 \times 10^{-19}$ J.

3.11. $\begin{vmatrix} 0 & \cos k_2 a & \sin k_2 a \\ \cos k_1 b & -\cos k_2 b & -\sin k_2 b \\ k_1 \sin k_1 b & -k_2 \sin k_2 b & k_2 \cos k_2 b \end{vmatrix} = 0.$

3.12. $P_d = (-\text{i}/4\omega\mu)(\mathcal{E}_x^*(\partial\mathcal{E}_x/\partial z) - \mathcal{E}_x(\partial\mathcal{E}_x/\partial z)^*)$.

3.13. $E_n = (n + \frac{1}{2})\hbar\omega_0$; zero point energy at $n = 0$.

Chapter 4

4.1. 1.22×10^{-7} m.

4.2. 4.13×10^6 m s^{-1}.

4.3. $\Delta\lambda = 6.75 \times 10^{-5}$ nm.

4.5. $\langle r \rangle = 3/2c_0$.

4.6. (i) quantum numbers are $n = 2, l = 0, m_l = 0$, (ii) $c_1 = -c_0/2$, (iii) $E = -me^4/32\epsilon_0^2 h^2$, (iv) $A = (c_0^3/8\pi)^{1/2}$, (v) $r = (3 + \sqrt{5})/c_0$.

4.7. The energy is the same for all the wavefunctions with $n = 2$.

4.8. -54.4 eV; mutual repulsion of the electrons.

4.9. $-\dfrac{\hbar^2}{2m}(\nabla_1^2 + \nabla_2^2 + \nabla_3^2)\psi + \dfrac{1}{4\pi\epsilon_0}$

$\times \left[-\dfrac{3e^2}{r_1} - \dfrac{3e^2}{r_2} - \dfrac{3e^2}{r_3} + \dfrac{e^2}{r_{12}} + \dfrac{e^2}{r_{13}} + \dfrac{e^2}{r_{23}} \right]\psi$

$= E\psi$.

Chapter 5

5.2. $F = -3q^2 d^2/2\pi\epsilon_0 r^4$.

5.3. The atoms can be pulled apart when the applied force is larger than the maximum interatomic force.

5.4. 3.12×10^{-10} m.

5.6. $(-q^2/2\pi\epsilon_0)\log 2$.

Chapter 6

6.1. 0.270, 1262 K.

6.2. 3.16 eV.

6.3. All of them.

6.4. $0.928\,E_F$.

6.5. $Nh^2/4\pi m$.

6.7. (i) $A_0 = 1.2 \times 10^6\,\mathrm{A\,m^2\,K^{-2}}$, (ii) $7.1 \times 10^{-4}\,\mathrm{K^{-1}}$, (iii) by about 90%.

6.8. $1.02 \times 10^{-7}\,\mathrm{A}$.

6.9. $7.27\,\mathrm{eV}$; $394\,\mathrm{J\,kg^{-1}\,K^{-1}}$, 0.5%.

6.10. $I \sim (E_{F2} - E_0)^{1/2} E_F^{1/2}\,\mathrm{eV}$.

6.11. 0, 0, 1, 1, 10 is a possibility.

Chapter 7

7.2. $1.11 m_0$.

7.3. $m^* = \hbar^2/2Aa^2 \cos ka$.

7.6. $E = (a_{xx}k_x^2 + a_{yy}k_y^2 + a_{zz}k_z^2 + a_{yz}k_yk_z)/\hbar^2$.

7.7. $2V_0/\pi$.

7.8. The width of the allowed band is $h^2(2n - 1)/8ma^2 - 2V_0 w/a$.

7.9. The electron.

Chapter 8

8.1. $0.043\,\mathrm{eV}$.

8.2. $\langle E \rangle = 3kT/2$.

8.3. (i) $0.74\,\mathrm{eV}$, (ii) $m_0/2$.

8.4. $0.66\,\mathrm{eV}$; $1.88 \times 10^{-6}\,\mathrm{m}$.

8.6. $N_h/N_e = 2$.

8.8. (i) $\rho = 20.3\,\mathrm{k\Omega\,m}$ (ii) $0.45\,\mathrm{k\Omega\,m}$ (iii) $\alpha = -7.94 \times 10^{-2}\,\mathrm{K^{-1}}$.

8.9. $N_A = 2.45 \times 10^{19}\,\mathrm{m^{-3}}, N_D = 1.47 \times 10^{20}\,\mathrm{m^{-3}}$.

8.10. $N_A = 1.25 \times 10^{19}\,\mathrm{m^{-3}}$.

8.11. 2.0×10^{-5}.

8.12. (i) $N_A = 2.52 \times 10^{22}\,\mathrm{m^{-3}}$, (ii) 1/2, (iii) $N_A = 2.52 \times 10^{22}\,\mathrm{m^{-3}}$.

8.13. (i) $7.31 \times 10^{22}\,\mathrm{m^{-3}}$, (ii) $1.89 \times 10^{-6}\,\mathrm{m^{-3}}$, (iii) $7.31 \times 10^{22}\,\mathrm{m^{-3}}$, (iv) $1.38 \times 10^{17}\,\mathrm{m^{-3}}$, (v) $171.3\,\mathrm{K}$, (vi) $0.295\,\mathrm{eV}$.

8.16. 2.51.

8.18. $\dfrac{\partial N_h}{\partial t} = \dfrac{N_{hn} - N_h}{\tau_p} - \dfrac{1}{e}\nabla; J_h$.

8.19. (i) 4, (iii) $m^*/m_0 = 0.16, 0.21, 0.34, 0.51$, (iv) No, (v) $4.6 \times 10^{-11}\,\mathrm{s}$ and $5.8 \times 10^{-11}\,\mathrm{s}$, (vi) $N_{h1}/N_{h2} = 0.41$, (vii) The deep level is so sparsely populated that the resonance is not observed.

Chapter 9

9.3. $1.4\,\mathrm{k\Omega\,m}$.

9.4. $x_n = (\epsilon_s/\epsilon_i)d_i + [(\epsilon_s/\epsilon_i)^2 d_i^2 + 2\epsilon_s\epsilon_0 U_0/eN_D]^{1/2}$.

9.5. (i) $d = [6U_0\epsilon d_0/eN_D]^{1/3}$ (ii) $d = [4U_0\epsilon/eN_D + d_0^2/3]^{1/2}$.

9.6. $0.35\,\mathrm{V}$ for Ge, $0.77\,\mathrm{V}$ for Si.

9.7. $1.81 \times 10^{19}\,\mathrm{m^{-3}}, 8.55 \times 10^{16}\,\mathrm{m^{-3}}$.

9.8. (i) $kT \log[(1 - \alpha)/\alpha]$, (ii) $kT \log[\beta/(1 - \beta)]$, (iii) $T = 197\,\mathrm{K}$.

9.9. $U_0 = 19.5\,\mathrm{V}$.

9.10. $N_h - N_{hn} = N_{hn}[\exp(eU_1/kT) - 1]\exp[-x/(D_h\tau_p)^{1/2}]$.

9.11. $0.94\,\mathrm{mm}$.

9.12. $J_h = e(D_h/\tau_p)^{1/2}N_{hn}[\exp(eU_1/kT) - 1]\exp[-x/(D_h\tau_p)^{1/2}]$.

9.13. $I_0 = e[(D_e/\tau_h)^{1/2}N_{ep} + (D_n/\tau_p)^{1/2}N_{hn}]A$.

9.14. (i) $\sigma_0 = 2e(\mu_e + \mu_h)(m_e^*m_h^*)^{3/4}(2\pi kT/h^2)^{3/2}$, (ii) 49.2%, (iv) I_0 increases by 123%.

9.16. (i) $I_0 = 8.7 \times 10^{-19}\,\mathrm{mA}, T = 495\,\mathrm{K}$, (ii) $N_A = 1.1 \times 10^{23}\,\mathrm{m^{-3}}$, (iii) $U_0 = 2.0\,\mathrm{V}, A = 1.49 \times 10^{-7}\,\mathrm{m^2}$.

Chapter 10

10.2. $1.43 \times 10^{-40}\,\mathrm{F\,m^2}$.

10.3. $1.23\,\mathrm{V\,m^{-1}}$.

10.5. $\tau_0 = 3.77 \times 10^{-14}\,\mathrm{s}, H = 1.01\,\mathrm{eV}$.

10.6. $a = \tau, b = \epsilon_s, c = \tau\epsilon_\infty$.

10.8. The capacitance is reduced by 22%. The breakdown voltage is reduced from $1000\,\mathrm{V}$ to $4.5\,\mathrm{V}$.

10.9. $\mathrm{Im}\delta = e\omega^2(\mu\mathcal{E}_0 + \upsilon_s)/2N_{e0}\mu\upsilon_s^2 c$.

10.10. $\epsilon_r' = 1 + \omega_p^2(\omega_1^2 - \omega^2)/[(\omega_1^2 - \omega^2)^2 + (\omega\gamma/m)^2], \epsilon'' = \omega_p^2(\omega\gamma/m)/[(\omega_1^2 - \omega^2)^2 + (\omega\gamma/m)^2]\omega_1^2 = k/m - \omega_p^2/3$.

10.11. $2.68 \times 10^{12}\,\mathrm{Hz}$.

Chapter 11

11.1. $\chi_m \approx 10^{-5}$.

11.2. $(\mu_m)_{ind} = 1.96 \times 10^{-29}\,\mathrm{A\,m^2}$.

11.3. (i) $E = BIab \cos\theta$ (ii) $\bar{\mu}_m = $ area of the loop $\times I\hat{n}$ where $\hat{n}$ is a unit vector normal to the plane of the loop.

11.6. $\theta = 633\,\mathrm{K}, C = 4.98 \times 10^{-2}, 0.46$.

11.7. $\chi_m = 2.08$.

11.8. $T = 3.88\,\mathrm{K}$.

11.9. (i) $2.8 \times 10^9\,\mathrm{Hz}$, (ii) $4.3 \times 10^6\,\mathrm{Hz}$.

Chapter 12

12.1. (a) (i) 5.0×10^{-4}, (ii) 4.9×10^{-27}; (b) (i) $= 1.16 \times 10^{30}$, (ii) 0.27; $\nu = 4.23 \times 10^{12}\,\mathrm{Hz}$ at $T = 293\,\mathrm{K}$.

12.3. $N/N_0 = 2.08 \times 10^{-15}, 339\,\mathrm{\mu W}, \lambda = 122\,\mathrm{nm}$, not in the visible range.

12.4. $N_3 - N_2 = 3.93 \times 10^{21}\,\mathrm{m^{-3}}$.

12.5. 0.953.

12.6. 2×10^{-4}.

12.7. $\upsilon/2l$.

12.8. (ii) $\Delta \nu = 4.66 \times 10^9$ Hz, (iii) 46.

12.10. $J = 8.2 \times 10^6$ A m^{-2}.

Chapter 13

13.2. $\sigma = \tau_e \mu_e V / c^2$.

13.3. $\mathcal{E}(0) = e N_{D1} d_2 / 2\epsilon - V / d_2$.

13.4. (i) 1.94 µm; (ii) 6.16°; (iii) 10.03, 12.40.

13.5. (i) 169.2 nm, 208.2 nm; (ii) the reflection type.

13.6. (ii) $\epsilon_{xx} = \epsilon_r + \epsilon_r^2 r_{XYZ} \mathcal{E}_z, \epsilon_{xy} = 0, \epsilon_{xz} = 0, \epsilon_{yx} = 0,$
$\epsilon_{yy} = \epsilon_r - \epsilon_r^2 r_{XYZ} \mathcal{E}_z, \epsilon_{yz} = 0, \epsilon_{zx} = 0, \epsilon_{zy} = 0, \epsilon_{zz} = \epsilon_r.$

13.7. (i) $A_1 = A_{10} \exp[i(k_1 + k_2)z/2][\cos \phi z + i((k_1 - k_2)/2\phi) \sin \phi z]$

$A_2 = (i\kappa A_{10}/\phi) \exp(i(k_1 + k_2)z/2] \sin \phi z$
$\phi = [(k_1 - k_2)^2/4 + \kappa^2]^{1/2}.$
(iii) $\kappa = \pi/2$ cm^{-1}, (iv) ± 842.2 m s^{-1}.

13.8. $t_1 t_2 \exp(\gamma/2 + ik)L[1 - r_1 r_2 \exp(\gamma/2 + ik)L]^{-1} \mathcal{E}_i.$

Chapter 14

14.1. 3.15×10^{11} s, 1.5×10^{24}.

14.2. 0.053 A.

14.4. 1.97×10^{27} m^{-3}.

14.6. (i) 1.13 meV, 2.17 meV; (ii) 3.72 K.

14.7. 5.24×10^{11} Hz.

14.8. 3.14×10^{11} Hz.

Index

A

Abrikosov 359–60
acceptor levels 125–6, 313
acceptors 124–7, 134, 136, 156–7, 160, 163, 190, 192
acetylene 72, 377–8, 380
acoustics 72, 226, 302
Acoustic wave amplifier 226–7, 235
acousto-optic interaction 323–5, 329
AlGaAs 151, 191–2, 207, 285–8, 326, 336, 338–9, 341
alkali metals 11, 19, 67
allowed energy band 100–1, 105, 121
Alnico 247, 250–2
amorphous magnetic materials 247–8
   semiconductors 138–9, 232, 312
Ampère's law 34, 246, 249, 349
analogy 30–1, 41, 68, 71, 78–9, 107, 128, 136–7, 172, 188, 195, 229, 232, 266, 275, 364, 369
antiferromagnetic resonance 261
antiferromagnetism 259
antimony 124
anti-Sod's law 94
archaeology 70, 366
argon laser 281, 309
arsenic 124, 133
artificial materials 383–6
atom laser 82, 308–9
avalanche
   breakdown 182, 195, 223
   diode 181–2
   photodiode 311
average
   value of operators 45
   velocity 2–3, 9, 36, 130
axial modes 295–6

B

backward diode 181
band bowing 152, 291
bandgap engineering 290–1
band-pass filter 228
band theory 97–118, 124, 126, 213, 360
Bardeen 166, 344

barium borate (BBO) 298
barium
   lanthanum copper oxide 367
   titanate 225, 317
base 166–71, 189, 202, 207
BCS theory 344, 360, 371–2
Bednorz 367
benzene 218, 379
beryllium 58, 113, 115
bias
   forward 161–6, 169, 172–3, 176, 178, 180–1, 186, 189, 193, 195, 207, 284, 286–7
   reverse 161–2, 165–6, 168, 176–9, 182–6, 188, 192, 194, 195, 207, 311, 339–40
Big Bang 369
bio-luminescence 377
bistability 331–2, 334, 340–1
Black art 88, 315
black body radiation 272
bloomed lens 217
Bohr
   magneton 254, 257, 388
   radius 53, 60
Boltzmann
   distribution 2, 83
   statistics 160, 215, 256, 272
bonds 58, 63–79
   carbon, see carbon bonds
   covalent 67–70, 72, 97, 120, 125, 134–5, 139, 208
   ionic 66–7, 71, 135, 215, 218
   metallic 67, 71, 258
   Van der Waals 66, 70–1, 208
bosons 82, 308, 347, 369
bowing factor 291
Bragg cell 323
Bragg reflection 102, 231, 287, 292, 307, 318, 324, 329, 384
Brattain 166
breakdown 151, 181–2, 184, 194–6, 223–4, 235
   avalanche 182, 195, 223
   dielectric 223–4

   thermal 223–4
   Zener 182, 184
Brewster angle 280
Brobeck 264
BSCCO 370–1
built-in voltage 159, 165, 189, 311–12
bulk elastic modulus 65, 70
buried layer 202

C

cadmium sulphide 135, 136, 141, 151, 226, 311, 314
carbon 1, 58, 61, 67–9, 70–2, 232, 250–1, 255, 291, 378–9
   bonds 68, 72, 377–9, 380
carbon dating 70
carbon dioxide 70
   laser 281, 303
carrier lifetime 136, 139, 146–7
cavity dumping 296
charge-coupled devices (CCD) 166, 192–5
chemical bond 58, 60, 63–4
chemical laser 283
chromium 61, 62, 87, 248, 255, 257
Clausius–Mosotti equation 223
CMOS 187–8
cobalt 59, 61, 87, 249–52, 255, 259
coercivity 245, 248, 250, 251
collector 166–71, 184, 202
collisions 2–4, 58, 129–30, 137, 142, 196, 223, 271, 276, 280–1, 344
collision time 3, 10, 15, 129–30, 136, 138, 143
communications 13, 31, 135, 184, 230–1, 283, 291, 299, 302, 305, 311, 325, 327, 376
compact disc 287, 302–3
conduction band 116–17, 120–8, 131–3, 136–8, 143–6, 152, 157–8, 160, 166, 171–2, 174–6, 178, 180–2, 189–91, 196, 200, 223, 284, 288, 293, 313, 319, 336

conductivity, electrical 2–4, 7, 9, 18–19, 36, 97, 128, 210
conjugated polymer 380
contact
    ohmic 175, 184, 195, 285
    potential 95, 160
continuity equation 48, 155, 212, 235
Cooper 344
    pair 344, 363, 372–3
copper 4, 18, 59, 119, 147, 149, 235, 263, 274, 364
    oxide 367, 368
Coulomb
    blockade 209
    force 47, 64, 163, 336
    island 209
    staircase 209
coupled modes 72–7
covalent bond 67–70, 72, 97, 120, 125, 134–5, 139, 208, 259, 261
critical magnetic field 345–7, 350–1, 358, 364
critical temperature 346–7, 352, 354–5, 360, 364, 366–73
crystal growth 147–8
crystallography 1, 87, 243–4, 372
cubic lattice 1, 32, 219
cuprate superconductors 368–9
curie
    constant 241, 270
    temperature 241–2, 270, 373
current gain parameter 170
cyclotron
    frequency 14–15, 262
    resonance 13–16, 19, 108, 142–3, 151, 155, 262

D

Davisson and Germer 22, 24–5
de Broglie 23–4, 29, 32, 38, 139, 345
    wavelength 24, 32, 92, 128, 345
Debye 221
Debye equations 218, 221, 234
deep depletion mode 177, 192
degenerate semiconductors 178, 284
density
    gradient 137, 211, 319
    of states 81–4, 96, 110, 121–2, 196, 208, 213, 270, 288–90, 361–2, 373
depletion
    region 158, 160, 163, 165, 172, 176–7, 184–5, 190, 211, 311, 341
    mode 187
Dewar flask 142, 343
diamond 67–71, 113, 116, 120, 134, 136–7, 218–19

dielectric
    constant 7, 21, 125, 173, 211, 213, 216– 17, 219, 221, 223, 225, 229, 235–6, 298, 318–20, 323, 326, 331, 337, 342, 360, 383–4
    materials 213–36
    mirror 216–17, 281, 330–1
diffusion 137, 140, 158, 202–3, 205, 235, 312, 316, 319
    coefficient 137
    current 168, 175–7, 211–12
    equation 137
    reactance 168
diode
    avalanche 181–2
    backward 181
    Gunn 196–8
    photo 311–13
    tunnel 178–81, 191, 208, 362
    varactor 183–4, 229, 297
    Zener 181–2
direct-gap 144, 146, 154, 284, 314
directional coupler 327–9
dislocations 2, 152–3, 314–16
dispersion equation 8, 13–14, 17, 235
divalent metals 115–16
domain wall 244, 353
donor level 124–5, 127–8, 157, 212, 313
donors 124, 126–8, 131, 134, 136, 144, 154, 160, 165, 211, 313, 319–20
Doppler broadening 276, 309
double heterojunction laser 286–7
drain 184–6, 192, 204, 207–9, 229, 267, 382
drift velocity 3–4, 6, 139, 140, 198
Drude 18
Drude model 2, 18
dye laser 281–2, 297–8

E

eddy currents 149, 246–7, 346
erbium doped fibre amplifiers (EDFA) 299
effective mass 16, 20, 32, 108–10, 114, 118, 120–1, 123, 125, 127, 129–30, 138–9, 141–3, 153–5, 166, 190, 196, 288, 337
    negative 109, 114
    table 141
    tensor 110, 118, 121
effective number of electrons 111, 113–14, 116, 120
eigen values 55
Einstein 82, 211, 273, 308, 309, 347, 371
    Coefficients 373–4
    relationship 211

electrical conductivity 2–4, 7, 9, 18–19, 36, 97, 128, 210
electrical motors 269, 366, 370
electrical noise 3, 184, 228, 299–300, 312
electro-absorption 336–41
electron
    affinity 172, 314
    microscope 25–6, 32
    spin-resonance 261, 270
    volt 47, 53–4, 387
electronegative atoms 314
electro-optic material 316–17
electrostriction 228–9, 246
emitter 87, 166–71, 184, 189, 202
energy band 97, 105–6, 108, 111–12, 116, 118
    allowed 100, 105–6, 121
    forbidden 100, 105–6, 118, 121
energy gap 104, 115–17, 120, 123–4, 127–8, 133, 135, 136, 139, 143–6, 153, 178, 188–90, 198, 207, 213, 223, 284, 286, 288, 290–2, 312–13, 319
    superconducting 360–3, 369, 372
energy, surface 352–4
enhancement mode 187
entropy 349, 351
epitaxial growth 149–50, 192, 198, 201–3, 264
erbium 299
etalon see Fabry–Perot
etching 153, 202, 205, 315
Euler's equation 389
exchange interaction 259–60
excimer laser 203, 283, 369
exciton 163, 302, 336–9, 380
exclusion principle 56, 58, 60
extrinsic semiconductor 120, 124–8, 130, 144, 155, 166

F

Fabry–Perot etalon 278, 295, 331, 333, 342
Fermi
    function 84, 95, 116, 120–2, 124, 125, 178, 213, 271
    level 83–5, 91, 94–5, 116, 118, 121, 123, 126–8, 130–1, 144, 153–4, 157, 161, 166, 171–2, 174–8, 180, 191, 209, 211–13, 270, 361
        extrinsic semiconductor 126
        intrinsic semiconductor 126, 128
        superconductor 361
Fermi–Dirac distribution 81–4, 347
ferrimagnetic resonance 261
ferrimagnetism 260

ferrites 248–50, 264, 267–9, 334
ferroelectrics 229–30
ferroelectric Random Access Memory
    229–39
ferromagnetic resonance 261
ferromagnetism 237, 259–60, 373
FET *see* field-effect transistor
Feynman 72–7, 354
    model 105–8, 112, 115, 118, 129
field-effect transistor 184–8, 190, 229,
    267, 382
field emission 91
    microscope 91–2
filters 228, 231, 329
Flatland 204
floating zone purification 149
forbidden energy band 100, 105–6, 118,
    121
Fourier transform 31, 221, 276
four wave mixing 321
FRAM *see* Ferroelectric Random Access
    Memory
Franz-Keldysh effect 339
free carrier absorption 285, 313
free electron theory 80–95
fullerenes 71, 208, 368

G

Gabor xiii, 306
gallium 124, 135, 152, 264
    arsenide 135, 138, 141, 154, 219, 286,
        291, 336
    nitride 135, 141, 151, 152–3, 291,
        314–16
    phosphide 135, 141, 313–14
garnets 69, 260, 264–6, 279
gas
    dynamic laser 282–3
    lasers 217, 304
    sensors 200
gate 184–8, 192, 204, 207–10, 229, 267,
    332–3
gauge factor 199
germanium 20, 63, 68, 97, 120, 124, 131,
    134, 135, 137, 142, 149, 211–12,
    309
    covalent bond 68, 97
    crystal structure 137
    cyclotron resonance 143, 155
    energy gap 116, 119, 128
    impurities in 119, 128
    indirect gap 137, 145
    melting point 135
    mobility 130–1, 152
    resistivity 154
    table 59, 61, 68, 124, 130, 134–5, 141,
        211, 313–14

g-factor 253, 261
Giaever 361
giant magneto-resistance 266–7
Gibbs
    free energy 349–50, 351, 353–5, 356,
        358, 390
    function 348–9, 349–50, 355
Gilbert 237, 241, 247
GINA alloys 291–2, 313
Ginzburg 354
Goethe, J. W. von 310, 375
graphite 68, 71, 136, 138, 208–9, 372
ground state 47, 54, 60, 62
group velocity 28–9, 108–9, 118
Gunn 130
    effect 130, 196–8

H

Hall
    coefficient 6, 20, 139, 142
    effect 5–6, 19–20, 113, 142, 199–200,
        262–4, 266, 369
Hamiltonian 73
hard magnetic materials 246, 248–52, 364
hardness 69–70
hard superconductors 364
Haynes–Shockley experiment 139–40
heat valve 367
heavy holes 192, 337
Heisenberg 35, 259
helimagnetism 260
helium 56–7, 62, 92, 142–3, 253, 280,
    301, 343, 365, 369, 371
    atom 32, 58, 92, 280
helium–neon laser 280–1, 323
High Electron Mobility Transistor
    (HEMT) 190, 192, 209
Hertz 34, 387
heterostructures 188–92, 291
high temperature superconductors
    367–71
Hilsum 196
Hockham 230
holes 2, 6–7, 20, 22–3, 107, 113–14,
    116–17, 120–1, 123, 125–8,
    130–2, 137–9, 141–2, 147, 163–4,
    177, 186–7, 190, 212, 284–8, 312,
    314, 336–7, 339, 341, 373, 380,
    382
    heavy 137, 142, 192, 337
    light 137, 142, 192, 337
holography 306–7, 318–21, 383–4
homopolar bond 68
Hooke's law 65, 224
Hund's rule 254
hybridization 378

hydride vapour phase epitaxy (HVPE)
    155
hydrofluoric acid 202
hydrogen
    atom 30, 50–62, 68, 77, 125, 154,
        253–4, 269, 334, 337, 377
    cyanide laser 281
    molecular ion 75, 77, 105
    molecule 41, 63, 68, 72, 75, 77–8
hydrodynamic model 4–5
hysteresis 229, 242–3, 245–7, 250–1, 334

I

impurities 119, 123–6, 129, 146, 149,
        151, 154, 160, 190, 198, 211, 246,
        250, 380
indirect gap 137, 145–6, 313
indium 124–6, 135, 152, 154, 359–60,
        380
    antimonide 20, 199, 266
    gallium nitride (InGaN) 152, 291,
        315–16
infrared detectors 313
inhomogeneous broadening 276
injection 140, 146, 163–5, 171, 189, 193,
        283, 285
insulators 113, 116, 210, 223
integrated circuits 149, 200, 227–8, 325
integrated optics 325–9
interdigital transducer 227–8
interference 24, 119, 139, 307–9, 318–19
intrinsic breakdown 223
intrinsic semiconductor 119–24, 126,
        128, 130–1, 154, 198, 212
inversion 177–8, 186–7, 289, 295, 301,
        309
inverted population 274–4, 280, 295, 309
ionic bond 66–7, 71, 133, 135, 215, 218
ion implantation 202, 266, 314
ionization energy 124
ionized acceptors 127, 154
ionized donors 126, 177, 319–20
ionosphere 13, 218
iron 18, 71, 190, 235, 237, 240–5, 247–8,
        252, 255, 259–60, 264, 270
    ferrite 248
isolators 267–8, 330
isotope
    effect 347, 372
    separation 305

J

Joking asides 1, 18, 33, 34, 41, 46, 68,
        70, 77, 88, 94, 110, 117, 119, 149,
        158, 163, 237, 258, 260, 264, 271,

295, 299, 301, 303, 335, 339,
  344, 367, 369, 371, 373, 380, 388
Josephson 362
  junction 363, 365–6, 369, 374
  tunnelling 362–3, 369–70
junctions
  capacity 165, 168
  Josephson 362–3, 365–6, 369, 374
  laser 283–92
  metal–insulator–semiconductor 175–7,
    192–3
  metal–metal 157, 172
  metal–semiconductor 171–3, 175–6,
    311
  p–n 156–66, 168, 171–2, 175–6,
    181–4, 188–9, 193, 196, 211–12,
    235, 284–5, 309, 311–12
  tunnel, superconducting 362

K

Kamerlingh Onnes 343
Kao 230
Kogelnik 294
Kompfner, R. 375
Kronig–Penney model 98–101, 107, 112,
  118

L

Landau–Ginzburg theory 354–60
Landau levels 262–4
Langevin function 216, 240–1
Larmor frequency 239
laser 82, 95, 152–3, 178, 191, 217, 224,
  271–309
  applications 301–8
  chemical 278, 283
  double heterojunction 286–7
  dye 281–2, 298, 381
  excimer 203, 283, 369
  fusion 305
  gas-dynamic 282–3
  gaseous discharge 280–1
  glass 224, 279
  machining 304
  modes 294–7
  quantum dot 290
  ruby 279–80
  semiconductor 117, 135, 146, 150,
    283–94, 313
  solid-state 279–81
  welding 304
Last Mile 305
lateral resonant tunnelling transistor 208

lattice
  ion 1–2, 67, 97–8, 101, 215, 223–4,
    258–9, 347
  spacing 2, 32, 128, 134–5, 139, 154,
    219, 290–2
  vibrations 70, 146, 344
lead 59, 61, 346, 360, 367
LED see light emitting diodes
Lenz's law 239, 346
Li 294
lifetime 136, 139, 146–7, 155, 163, 212,
    274–6, 309, 312, 337, 341
lighting 94, 313–16
light
  detectors 311–13
  holes 192, 337
  modulators 329–31, 339
light emitting diodes (LED) 152, 212,
    313–16
limericks 46, 344
liquid crystal 232–4, 330
Liquid Phase Epitaxy (LPE) 150–1
liquidus 147, 148
lithium 58–60, 64, 113, 118, 254, 305
  niobate (LiNbO$_3$) 317, 323, 326, 329
Little 344
lodestone 237
logic function 170–1, 187, 210, 365
London, F. 373
London, H. 373
Lorentzian lineshape 276
loss tangent 216, 230, 234
lossy material 8
Lucretius 237

M

Madelung constant 67, 136
magnesium boride 372
magnetic amorphous material 247–8
magnetic anisotropy 244, 246–7, 250–1
magnetic bubbles 264–6
magnetic domains 242–6, 270, 353,
  364
magnetic materials 237–70
magnetic resonance 260–4
Magnetic Resonance Imaging (MRI) 268,
  386
magnetic tunnel junction 267
magnetostriction 246–7
majority carriers 146, 160, 177
maser 253, 299–301, 309
mask 200–3, 266, 326, 381–2
mass
  action, law of 132
  effective 16, 20, 32, 108–10, 114, 118,
    120–1, 123, 125, 127, 129–30,

  138–9, 141–3, 153–5, 166, 190,
    196, 288, 337
matched filter 228
Maxwell 6–7, 34
Maxwell–Boltzmann distribution 2, 83,
    276
Maxwell's equations of 6–7, 21, 34, 216,
    235, 359, 384
MBE see molecular beam epitaxy
medical imaging 268–9, 341
Meissner effect 346–7
Memories
  ferroelectric 229–30
  holographic 307–8
  magnetic 264–7
  semiconductor 205
  superconducting 365
memory device 205, 207, 229, 264–6,
    334, 365,
MEMS see microelectro-mechanical
    systems
meson 77
metal–insulator–semiconductor junctions
    175–7, 192–3
metallic bond 67–9, 71, 258
metal–metal junctions 157, 172
Metal Organic Chemical Vapour
    Deposition (MOCVD) 150–2,
    192, 290, 292, 313–15
Metal–Oxide–Silicon Field Effect
    Transistor (MOSFET) 186–7,
    202–3
Metal–Oxide–Silicon Transistor (MOST)
    186–7, 202–3
metal–semiconductor junctions 171–3,
    175–6, 311
metamaterials 383–6
metrology 303–8, 366
microelectro-mechanical systems
    (MEMS) 205–6, 334–5
microelectronics 200–7, 210
microwaves 17, 142–3, 155, 182, 218,
    224, 267–8, 281, 299, 301, 303,
    309, 366, 370, 374–5, 385
minority carrier 146, 154–5, 163–6, 177,
    195, 311, 316
mobility 3, 20, 118, 130–1, 133, 139–42,
    144, 151–2, 154, 190–1, 196, 199,
    207, 313, 381
  measurement of 20, 139–42
MOCVD see Metal Organic Chemical
    Vapour Deposition
mode
  competition 295
  locking 296–7
MODFET see Modulation-Doped Field
    Effect Transistor

Modulation-Doped Field Effect Transistor (MODFET) 190–1
Molecular Beam Epitaxy (MBE) 150–2, 192, 290, 292, 314
molecular transistors 210
molten zone 148, 149
momentum operator 37, 47, 146
Moore's law 204
MOSFET see Metal–Oxide–Silicon Field Effect Transistor 188–9
MOST see Metal–Oxide–Silicon Transistor
MRI see Magnetic Resonance Imaging
Müller 367
multiple quantum well (MQW) structure 288–291, 292, 314, 339–41

N

Nakamura, Shuji 314
nano crystalline alloys 248
nanoelectronics 206–10
nanotubes 71–2, 208–9
Néel temperature 259
negative effective mass 109, 114
negative resistance 180–2, 196–8, 362
negative temperature 271, 275, 300–1
neodymium 224, 251, 279
neon 58, 254, 280
Newton 371
Newton's equations 34, 108
neutron 57, 72, 77, 80, 261, 305, 369
nickel 24, 244, 250, 255, 259, 270
niobium alloys 364, 371
nitrogen 58, 136, 142, 151–2, 255, 291, 367, 369
noise temperature 300–1
nonlinear optical materials 297, 302
n-type semiconductors 126–8, 140, 154–5, 171, 173–4, 176–7, 186, 193, 211–12, 284
nuclear forces 77
nuclear magnetic resonance 261–2, 268, 386

O

O'Donnell, K. P. 316
Ogg, R. A. 344
ohmic contact 175, 184, 195, 285
Ohm's law 1–4, 96, 142, 185, 196
OLED see organic light emitting device
operators
  average value 45
  Hamiltonian 73
  momentum 37, 47, 146
optical Darwinism 295
optical fibre amplifier 298–9

optical fibres 230–2, 251, 284, 298–9, 304–5, 326, 335
optical switching 334–6
optoelectronics 310–42
orbitals 377, 380
organic light emitting device (OLED) 380–2
organic semiconductors 377–82
  superconductors 373
orthonormal wave function 74
oxide superconductors 347, 367–8, 371–2

P

paramagnetic resonance 260–1
paramagnetism 240, 256–9, 299
parametric amplifier 183–4, 297–8
parametric oscillator 184, 297–8
Pauli 35, 56
Pauli's principle 56–8, 78, 84, 166, 254–5, 259, 345
penetration depth 20, 357–8, 373
periodic table 50, 56–60
Permalloy 247, 249, 264–6
permanent magnets 241, 248–52, 267–9
permeability 7, 238, 246, 248, 270, 385, 388
phase
  conjugate mirror 321–22
  conjugation 310, 318–22, 383
  diagram 147, 368–9
  shifter 203, 229, 326–7
  shifting mask 203
  transition 348–51, 365
  velocity 8, 26, 28, 342
philosophical implications of quantum mechanics 45–7
phonon 82, 146, 276, 344
phosphorus 59, 124, 126, 150, 202
photoconduction 136, 139, 147, 311
photodetector 304, 310–12, 325, 341
photodiode 311, 381
photoelectric effect 92–4
photoemission 136
photoengraving 264, 266
photon 28, 32, 45, 47, 49, 82, 93, 95, 120, 144–6, 213, 260, 272–5, 277–8, 281, 284, 286, 293, 308, 310–11, 336–9, 341, 380, 388
photonic band gaps 383–4
photonics 383–4
photorefractive materials 310, 317, 319–20, 323
photoresist 200–2, 205, 210
phototransistor 311–12
piezoelectric constant 224–5, 315
piezoelectricity 224–9, 246
piezomagnetism 246

pinch-off voltage 186
pin junction 311, 340–1
Pippard 354
planar technique 149, 200–5
Planck 24, 34, 37, 49, 272
Planck's constant 24, 29, 366, 387
plasma
  diagnostics 302
  frequency 17, 384–5
  heating 302
  physics 17
  waves 16–17, 32, 235
platinum 59, 61, 87, 343, 388
plutonium 59, 61, 372
p–n junction 156–66, 168, 171–2, 175–6, 181–4, 188–9, 193, 196, 211–12, 235, 284–5, 309, 311–12
  capacity 165, 168
Poisson's equation 17, 158, 191, 319
polarizability 215–16, 218, 220–1, 223, 234–5
polarization 214–17, 219–22, 224, 229, 233–4, 236, 316–17
polymer 378–80
population inversion 274, 277–8, 283, 296, 299–300
potassium 11, 59–61, 87
  chloride 79, 236
  chromium alum 257–8
potential
  barrier 38–41, 44, 47, 75–6, 80, 85, 89, 91, 95, 98, 100, 129, 161, 172–3, 175, 178–9, 182, 209, 336
  well 42–5, 47–8, 74, 80, 85, 98, 100, 166, 191–3, 208, 262, 286, 288–9, 336
Poynting vector 21, 49
proton 25, 30, 50–1, 54, 56–7, 60, 68, 72, 75–8, 80, 238, 253, 261, 268, 337, 387
  spin resonance 262, 270
p-type semiconductor 126, 128, 186, 198, 212, 284
pump 184, 274–5, 278–9, 281, 296–8, 300, 309, 321, 384

Q

Q switching 295–6
quantum
  cascade laser 293
  confined Stark effect 336, 338–9
  dots 290
  of energy 24
  Hall effect 262–4, 369
  numbers 55–6, 72, 253–6, 280
  well structures 310, 315, 336–41
  wires 290

quark 369
quartz 69, 94, 150–1, 200, 219, 225–6, 323
quartz–halogen lamps 94–7, 388

R

radar 31, 184, 198, 228, 301–3
radio waves 10, 13, 301
Rapid Single Flux Quantum devices 365
Rapid Solidification Technology (RST) 247
rare earth magnets 251
Rayleigh 227
recombination 131–2, 140, 147, 152, 155, 163, 166, 177, 193, 212, 284–6, 314, 316, 319, 336, 382
rectification 161–3, 235
rectifier equation 162, 178–9, 212
refractive index 70, 216–17, 219, 231, 285, 291, 296, 307, 317, 319, 326, 341, 360, 385
relative permeability 238
relaxation time 3, 221, 234–5, 283, 373
remanent flux 245
resonance
    cyclotron 13–16, 19, 108, 142–3, 151, 155, 262
    electron spin 261, 270
    ferrimagnetic 261
    ferromagnetic 261
    nuclear magnetic 261–2, 268, 386
resonator 197, 228, 277–9, 294–8, 308, 342, 370, 385
Ridley 196
ring laser 304
Rochelle salt 225
Rotoxane 210
RSFQ see Rapid Single Flux Quantum devices
RST see Rapid Solidification Technology
ruby 219, 274, 279, 299–300, 309

S

saturable absorber 297
scattering 128–30, 155, 190, 208, 218, 263, 266, 345
Schottky
    barrier diodes 173
    effect 88–90, 173, 339
Schrieffer 344
Schrödinger 34–5, 53
Schrödinger's equation 33–9, 41–2, 45, 48–50, 52, 56–7, 60, 62, 73–4, 76, 78, 80, 84, 95, 98, 166, 179, 191, 262

time-independent 48–50, 62, 73, 98
Scientific American 260
Seitz 85
semiconductors 1–4, 6, 14, 68, 97, 116, 119–55, 166, 171, 177–8, 190–2, 196, 198–200, 203, 209, 211, 213, 218, 252, 263–4, 267, 271, 283, 291, 313–14, 361–2, 377–2, 384
    degenerate 178, 284
    devices 136, 138, 146, 156–212, 233, 294, 325
    extrinsic 120, 124–8, 130, 144, 155, 166
    Fermi level 126, 128
    intrinsic 119–24, 126, 128, 130–1, 154, 198, 212
    lasers 117, 135, 146, 150, 283–94, 313
    measurements 139–47
    mobility 139–42
    table of properties 124, 134, 135, 141, 291
Shockley 2, 139–40, 166, 235
silicon 59, 63, 67–9, 97, 116, 119–20, 124–7, 132, 134, 137–8, 145–7, 149–50, 154, 173, 192, 199–200, 202, 204–6, 211–12, 235, 247, 286, 312
    covalent bond 68, 97, 124, 134
    controlled rectifier 195
    crystal structure 120, 137
    dielectric constant 125, 173
    effective mass 138, 152
    E–k curve 137
    energy gap 116, 138
    epitaxial deposition 149–51, 200–4
    holes in 147
    impurities in 119, 124, 126, 146–7, 152
    indirect gap 145
    in iron 247, 249
    melting point 135, 149
    metallurgical phase diagram 147–9
    microelectronic circuits 200–5
    mobility 130, 133, 141, 154, 381
    MOST 186–7, 202–3
    strain gauge 199
    tables of properties 124, 135, 141, 211
silicon dioxide 69, 151, 187, 193, 200–5, 206, 219, 225, 231–2, 285–7
silver 18, 59, 87, 254, 256, 381
silver halide 319
single crystals 24, 32, 138–9, 147–53
Single Electron Transistor 209
skin depth 9
sodium 18, 20, 58, 60, 64, 95, 133, 219, 258, 308
sodium chloride 64, 66–7, 133, 215, 219, 258

Sod's law 94
soft magnetic materials 246–50
Sommerfeld 80
solar cell 139, 152, 292, 312–13
solidus 147–8
source 184–92, 204, 207–9, 229, 267
spatial light modulator 329–30
specific heat 18–20, 22, 36, 84–5, 95, 351–3, 360
spin 35, 56, 68, 74, 78, 82, 86, 110, 113, 166, 254–5, 258, 260–1, 266–7, 270, 289, 299–300, 347, 369, 372–3
spintronics 266–7
split-off band 137, 142
spontaneous emission 272–3, 275–6, 287, 299, 301, 309
SQUID see Superconducting Quantum Interference Device
Steam engine 98, 210, 316
Stark effect 315, 336, 338–9
Stern–Gerlach experiment 256–8
stimulated emission 273–4, 278, 299, 308
Stoner 250
strain gauges 198–9
superconductors 269, 343–8, 350, 352–4, 356–73
    applications 364–7
    energy gap 360–3
    hard 364
    Josephson tunnelling 362–4, 369–70
    magnetometer 365–6
    memory elements 365
    oxide 347, 367–8, 371–2
    surface energy 352–4
    switches 365
    type I 358–9
    type II 358–9, 364
    vortex 359–60
Superconducting Quantum Interference Device (SQUID) 365
Supermalloy 247–9
surface acoustic wave (SAW) device 227–8
surface energy 353
surface states 173–5, 194
susceptibility
    dielectric 214
    magnetic 238–9, 253, 256, 269–70

T

Tamm states 175
tantalum 381
TBCCO (Tl2Ba2Ca2Cu3O10) 368, 371
TEGFET see Two-dimensional Electron Gas Field Effect Transistor
ternary compounds 151, 291

tetrahedral bonds 68–70, 134
thermionic emission 85–8, 90
thermal breakdown 223–4
thermal velocity 2, 4, 256
thoria 234–5
thought experiment 22
threshold wavelength 11
thyristor 195
tin 57, 200, 352, 360, 374, 380
transistor 165–71
transition
    elements 59, 259
    region *see* depletion region
transverse modes 294–5
tritium 305
tungsten 70, 87, 91–5, 250, 252, 279,
        388
tunnel diode 41, 178–82, 196, 208, 284,
        362
tunnelling 41, 47, 76, 91, 95–6, 178–82,
        194, 204, 207–9, 211, 267, 336,
        339, 360, 365, 374
    Josephson 362–3, 369–70
    between superconductors 361–2
Two-dimensional Electron Gas Field
        Effect Transistor (TEGFET) 190
Type I superconductors 358–9
Type II superconductors 358–9, 364

U

ultra violet 95, 135, 200–2, 218, 278,
        283, 290, 310–11
uncertainty relationship 29–32, 36, 44–7,
        275–6

V

valence band 116–17, 120–1, 123, 125–6,
        131, 136–8, 145, 154, 157, 166,
        171–2, 174–8, 180–2, 190, 192,
        199, 207, 212, 284, 286, 288, 293,
        313, 336
vanadium 59
van der Waals
    bond 70–1, 208, 266
    forces 70–1, 208
varactor diode 182–3, 229, 297
Variational calculus 389
velocity space 4
Vertical Cavity Surface Emitting Laser
        292
viscous force 4, 9
volume holography 307, 318–22, 383–4

W

water 217, 219, 223, 234, 262, 323, 348
Watkins 196
wave function 35, 39, 42–5, 48, 50, 53,
        55, 57, 62, 73, 74, 80, 103–4, 118,
        390
waveguide discontinuity 41, 328
wave number 20, 26–7, 29, 103–4, 331,
        342
    complex 8
wave packet 27–9, 32, 38–9, 44
Weiss 240, 242, 260
Weiss constant 240, 270
Wiedemann–Franz constant 18–19
Wohlfarth 250
work function 85, 87, 90, 93–5, 171, 173,
        175, 380

world wide web 299
wurtzite 136, 314

X

xenon 279
xerox process 139, 232
X-rays 25, 101–2, 118, 268, 383–4

Y

YAG *see* yttrium aluminium garnet
YBCO *see* yttrium–barium–copper oxide
yttrium 260, 264, 279
yttrium aluminium garnet (YAG) 279
    laser 279, 298, 303
yttrium barium copper oxide (YBCO)
        367–8
yttrium iron garnet (YIG) 260
Yukawa 77

Z

Zener breakdown 182, 184
Zener diode 181–2
Ziman model 101–4, 112, 118,
        128, 383
zinc 59, 61, 134–5, 147, 248,
        346
zincblende 134, 136, 291
    selenide 135, 141
    sulphide 135, 141
    telluride 135, 141
zone refining 119, 148–9